Nondifferentiable Optimization and Polynomial Problems

Nonconvex Optimization and Its Applications

Volume 24

The titles published in this series are listed at the end of this volume.

Nondifferentiable Optimization and Polynomial Problems

by

Naum Z. Shor

V.M. Glushkov Institute of Cybernetics,
Ukrainian National Academy of Sciences, Kiev

KLUWER ACADEMIC PUBLISHERS
BOSTON / DORDRECHT / LONDON

A C.I.P. Catalogue record for this book is available from the Library of Congress.

ISBN 0-7923-4997-0

Published by Kluwer Academic Publishers,
P.O. Box 17, 3300 AA Dordrecht, The Netherlands.

Sold and distributed in the U.S.A. and Canada
by Kluwer Academic Publishers,
101 Philip Drive, Norwell, MA 02061, U.S.A.

In all other countries, sold and distributed
by Kluwer Academic Publishers,
P.O. Box 322, 3300 AH Dordrecht, The Netherlands.

Printed on acid-free paper

Printed in the Netherlands.

CONTENTS

PREFACE ix

1 ELEMENTS OF CONVEX ANALYSIS, LINEAR ALGEBRA, AND GRAPH THEORY 1

1.1 Convex sets and convex functions 1

1.2 Properties of subgradient and ε-subgradient sets 8

1.3 Kuhn-Tucker theorem and dual bounds 19

1.4 Nonsmooth penalty functions 22

1.5 Polyhedral sets, matrices and quadratic functions 25

1.6 Elements of graph theory 30

2 SUBGRADIENT AND ε-SUBGRADIENT METHODS 35

2.1 The subgradient method 37

2.2 Fejer-type subgradient methods 51

2.3 ε-subgradient and bundle methods 57

2.4 The stochastic subgradient method 68

3 SUBGRADIENT-TYPE METHODS WITH SPACE DILATION 71

3.1 Heuristics of methods with space dilation 71

3.2 The Subgradient Method with Space Dilation in the Direction of Subgradient 74

3.3 The ellipsoid method and its generalizations 88

3.4 Methods with space dilation in direction of the difference of two successive subgradients in transformed space 100

4 ELEMENTS OF INFORMATION AND NUMERICAL COMPLEXITY OF POLYNOMIAL EXTREMAL PROBLEMS 113
4.1 Introduction to the complexity theory of optimization problems 113
4.2 Polynomial-time algorithms for linear programming 118
4.3 Interior point algorithms for linear programming and special convex programming problems 121
4.4 Review of main results on complexity theory of polynomial programs 133

5 DECOMPOSITION METHODS BASED ON NONSMOOTH OPTIMIZATION 141
5.1 Coordinate problems arising in decomposition on constraints 145
5.2 Nonsmooth problems linking with decomposition on variables 147
5.3 Examples of solving large-scale problems by using decomposition schemes in combination with r-algorithm on high-speed parallel computers 152

6 ALGORITHMS FOR CONSTRUCTING OPTIMAL ON VOLUME ELLIPSOIDS AND SEMIDEFINITE PROGRAMMING 169
6.1 Introduction 169
6.2 Simple algorithms for constructing optimal on volume circumscribed ellipsoids 178
6.3 Algorithms for constructing inscribed ellipsoids of maximum volume 187
6.4 Interior point methods for determination of optimal circumscribed and inscribed ellipsoids 193
6.5 Method of inscribed ellipsoids for minimization of convex functions on polytopes 196
6.6 Semidefinite programming problems 200
6.7 Semidefinite programming models in automatic control 210
6.8 Nonconvex quadratic programs and semidefinite programming 220

7 THE ROLE OF ELLIPSOID METHOD FOR COMPLEXITY ANALYSIS OF COMBINATORIAL PROBLEMS 227

7.1 Polynomial equivalence of the optimization and separation problems 227

7.2 Justification of the polynomial-time solvability of some combinatorial problems 232

7.3 Dual bounds for maximum weighted stable sets in graphs 239

7.4 Polynomial-time algorithm for perfect graphs 254

7.5 Using r-algorithm for obtaining upper bounds for the maximum weighted stable set problems 257

8 SEMIDEFINITE PROGRAMMING BOUNDS FOR EXTREMAL GRAPH PROBLEMS 265

8.1 The maximum cut problem 265

8.2 Graph partitioning problems 279

8.3 Upper bounds for the size of k-partite induced subgraphs of given graph and their applications to colouring problems 293

9 GLOBAL MINIMIZATION OF POLYNOMIAL FUNCTIONS AND 17-TH HILBERT PROBLEM 299

9.1 The problem of representation of nonnegative polynomials as sum of squares of polynomials 299

9.2 Constructing of quadratic problem with superfluous constraints for global minimization of one-dimensional polynomials 308

9.3 Necessary and sufficient conditions for dual quadratic bound to be exact. Multidimensional case 320

9.4 Instances of finding global minimum of polynomial functions using r-algorithm 330

REFERENCES 335

PREFACE

Polynomial extremal problems (PEP) constitute one of the most important subclasses of nonlinear programming models. Their distinctive feature is that an objective function and constraints can be expressed by polynomial functions in one or several variables.

Let $x = \{x_1, \ldots, x_n\}$ be the vector in n-dimensional real linear space R^n; $P_0(x), P_1(x), \ldots, P_m(x)$ are polynomial functions in R^n with real coefficients. In general, a PEP can be formulated in the following form:

$$\text{find } f^* = \inf P_0(x) \tag{0.1}$$

subject to constraints

$$P_i(x) = 0, \; i = 1, \ldots, m \tag{0.2}$$

(a constraint in the form of inequality can be written in the form of equality by introducing a new variable: for example, $P(x) \leq 0$ is equivalent to $P(x) + y^2 = 0$).

Boolean and mixed polynomial problems can be written in usual form by adding for each boolean variable z the equality:

$$z^2 - z = 0.$$

Let $\alpha = \{\alpha_1, \ldots, \alpha_n\}$ be integer vector with nonnegative entries $\{\alpha_i\}_{i=1}^n$.

Denote by $R[\alpha](x)$ monomial in n variables of the form:

$$R[\alpha](x) = \prod_{i=1}^{n} x_i^{\alpha_i};$$

$d(\alpha) = \sum_{i=1}^n \alpha_i$ is the total degree of monomial $R[\alpha]$. Each polynomial in n variables can be written as sum of monomials with nonzero coefficients:

$$P(x) = \sum_{\alpha \in A(P)} c_\alpha R[\alpha](x),$$

where $A(P)$ is the set of monomials contained in polynomial P.

The total degree $d(P)$ of polynomial P equal to $\max_{\alpha \in A(P)} d(\alpha)$, the total degree of the PEP (0.1),(0.2) equal to

$$\max_{\nu} d(P_\nu), \ \nu = 0, 1, \ldots, m.$$

Introducing new variables and using quadratic-type relations, for example, $z_{ij} = x_i x_j$; $y_i = x_i^2$, and so on, one can decrease the total degree of PEP,i.e. without loss of generality we may consider the quadratic-type PEP with quadratic or linear objective functions and quadratic (linear) constraints in the form of equalities or inequalities. Such problems were considered in the monography, N.Z.Shor, S.I.Stetsenko "Quadratic extremal problems and nondifferentiable optimization"- Kiev, Naukova dumka, 1989 (in Russian)[SSt 89].

There were many events in the areas of polynomial programmimng and non-smooth optimization from the date of edition of this book. The variety of applications of nondifferentiable optimization have expanded substantially.

The interest has arisen to using of quadratic-type models for solving the discrete optimization problems. After intensive development of interior point Dikin-Karmarkar's methods for solving linear programming problems ([Di 67],[Kar 84]) the far going generalizations of these methods were obtained for solving some important subclasses of convex programming problems. These subclasses include such problems as a problem of minimization of a convex quadratic function subject to the convex quadratic constraints, the problems of finding optimal parameters of circumscribed and inscribed ellipsoids, matrix parametric convex optimization problems and so on (see [NN 94]).

In many cases when we consider the extremal combinatorial problems in the form of quadratic-type non-linear programming model (boolean property of variable x is expressed by equation $x^2 - x = 0$) we can obtain the lower (or upper) bounds for optimal value of objective function by solution of certain matrix parametric convex optimization problems with constraints in the form of semidefiniteness of some symmetric matrices which entries are linearly dependent of optimized parameters. (Such problems are named as semidefinite programming problems) (see [Al 91], [Al 92b]). As a rule these bounds are much more exact in comparison with the similar bounds obtained from Lagrange relaxation of linear boolean models for the same problem.

For solving of semidefinite programming problems the interior point methods were developed which in many cases have polynomial-time effectiveness relatively to the length of the problem input and $\log \frac{1}{\varepsilon}$, where ε is permitted error of solution. Many mathematicians have hoped that by using interior point methods, good bounds for combinatorial problems could be found quickly, and in combination with "branch-and-bounds" methods would permit to expand essentially the class of practically solvable problems of discrete optimization.

It should be mentioned that similar problems of semidefinite programming arise also in investigation of complex systems stability and optimal control (see [BGFB 94]).

The typical functions that appear in semidefinite programming and other convex problems of matrix optimization are linked with maximal (minimal) eigenvalues or weighted sums of k largest eigenvalues of symmetric matrices. Since the eigenvalues of real symmetric matrices are nonsmooth functions of their entries, the problems of matrix optimization are an interesting subject for use of nondifferentiable optimization methods.

In this book we consider mainly the subgradient and subgradient-type methods with transformation of the space of variables, which were developed for nondifferentiable minimization problems in Kiev's Institute of Cybernetics of Ukrainian Academy of Sciences. Most these methods were described in English in the book: N.Z.Shor, Minimization Methods for Non-Differentiable Functions (Translated from the Russian by K.C.Kiwiel and A.Ruszczynski), Springer-Verlag, 1985 [Shor 85] (the Russian version of this book was published in 1979 [Shor 79]).

In the article [HWC 74] the subgradient-type methods were characterized as a key for solution of large dimensional optimization problems. In monographies [Shor 79] and [MTS 86] the applications of subgradient-type methods were discribed for realization of decomposition schemes when one meets with large-scale problems of special structure. Western specialists in the area of optimization are well acquainted with so-called ellipsoid method for solving of convex programming problems ([YN 76], [Shor 77]), because of L.Khachian, which in 1979 obtained the first polynomial time algorithm for LP problems, using this method. Note that the ellipsoid method is a very special case of the methods with space dilations in the direction of subgradients which was proposed by the author of this book in 1970 [Shor 70c].

The ellipsoid method is known as a good theoretical tool for investigation of numerical complexity of many mathematical programming problems including the discrete optimization problems (see [GLS 81]). But its speed of convergence is very slow as a rule.

In 1970 the class of the subgradient-type methods with space dilation in the direction of the difference of two successive subgradients was proposed [Shor 70b]. Such methods were named as "r-algorithms". These methods were tested on many hundreds of artificial and real-life complex problems of nondifferentiable optimization, and always demonstrated stable and relatively fast convergence. Therefore in this book we recommend using some checked program versions of r-algorithm for solvihg various classes of optimization problems.

One of the main sources of nondifferentiable optimization problems is the usage of nonsmooth penalty functions. Certain types of nonsmooth penalty functions are superior to the commonly used smooth penalty functions because usually they do not require penalty coefficients to tend to infinity, yielding exact solutions for sufficiently large, but finite values of these coefficients ([Er 67]), see also [Shor 85].

In many cases it is convenient to use Lagrange relaxation for solving complex problems of nonlinear programming. This technique consists of relocating some constraint functions to the subject function with unknown Lagrange multipliers. For fixed values of the multipliers we obtain a problem with smaller number of constraints. By this way we can obtain marginal function $\psi(u)$, that is value of Lagrange relaxation problem for a fixed vector of multipliers u. If one deals with the initial problem of minimization then function $\psi(u)$ (if it is proper) is concave and has convex domain.

The problem of finding $\psi^* = \sup_{u \in dom\psi} \psi(u)$ often is a problem of nonsmooth convex optimization. As a rule, $\psi(u)$ is an ill-behaved "gully-shaped" function, so the simple subgradient processes for minimization $\psi(u)$ possess slow convergence. In this case usage of r-algorithms is much more effective if the dimension of u is not very large.

In this book the mentioned technique for maximization of marginal functions $\psi(u)$ is frequently used both for realization of decomposition schemes for structured problems of linear and nonlinear programming, and also for obtaining dual bounds in discrete and multiextremal polynomial-type problems. In the last case the optimal value ψ^* may be often achieved at the boundary points of domain ψ. This property demands to develop special versions of r-algorithms with stable behavior near the boundary of $dom\psi$ (see [SSt 89]).

In many cases when we dealt with nonconvex polynomial problems the dual bound ψ^* can be essentialy improved by adding to the model of a problem new superfluous constraints which are consequences of initial constraints. In this book the reader will see many instances of such problems. The more interesting example is linked with the problem of representation of nonnegative polynomial function in the form of a sum of squares of polynomials. This question was initiated by D.Hilbert (see [Hi 88]) and generalized as 17-Hilbert's problem about the representation of a nonnegative rational form as sum of squares of rational forms [Art 27]. We found necessary and sufficient conditions for a real nonnegative polynomial to be decomposed into a sum of squares of real polynomials. These conditions correlate with the exactness of dual bound of the quadratic-type problem with superfluous constraints that arise when we try to find global minima of a given polynomial.

The main aim of the proposed book is to describe the modern state of nondifferentiable optimization and aburdance of its applications to various mathematical programming problems, partially, to polynomial-type extremal problems and linked with them matrix optimization problems. Great attention is given for information and numerical complexity of algorithms for solving polynomial extremal problems.

The reader shall find some known and new bounds for "classic" extremal graph problems such as determination of maximal weighted independent set, optimal colouring, max-cut, optimal partioning of graphs and so on.

The new methods of constructing optimal inscribed and circumscribed ellipsoids also were described in the book together with methods of solving of semidefinite problems arising in the theory of stability and optimal control of complex dynamic systems.

The content of the book is divided on 9 Chapters.

In the first Chapter we give necessary definitions and facts from the convex analysis, algebra, matrix theory and the theory of graphs. Further we consider Lagrange relaxation for nonlinear programming, generalized Kuhn-Tukker theorem, nonsmooth penalty function methods.

The second Chapter is devoted to the subgradient-type methods. We describe some modifications of the simple subgradient algorithm, ε-subgradient and aggregate methods, stochastic subgradient method, which belongs to the broad class of random search methods, i.e. iterative algorithms, that proceed along directions, resulting from random(pseudorandom) events.

In Chapter 3 we consider two families of subgradient-type methods with transformation of variable's space:

(I) methods with space dilation in the direction of a subgradient in transformed space (SDG methods);

(II) methods with space dilation in the direction of the difference of two successive subgradients in transformed space (r-algorithms).

The theory of convergence for SDG methods was done in detail for the case when one know the minimum value of minimized function. Mentioned methods may be used for solving systems of nonlinear nonsmooth equations and inequalities.

Further we consider the ellipsoid method for solving convex programming problems as a specific case of SDG-method (from the other point of view – as cutting-plane procedure with approximation of regions, locating minima, by ellipsoids). Some generalizations of central ellipsoid method are given and the theorem about the speed of convergence of those methods is proven.

Next we consider various forms of r-algorithms. The theory of convergence for r-algorithms is developed only for rather restricted classes of r-algorithms. But we developed some modifications of r-algorithms which demonstrated relatively fast and stable practical convergence for many hundreds of real-life and test problems of convex optimization. Note, that limit modifications of r-algorithm behave as certain implementations of conjugate gradient methods when we minimize the smooth functions.

In Chapter 4 we give elementary introduction to the theory of information and numerical complexity of polynomial extremal problems. We describe the first polynomial-time Khachiyan's algorithm for linear programming (LP) problems, based on special modification of ellipsoid methods. Next we give a short description of two interior point algorithms (of I.Dikin and Karmarkar) for solving LP problems, and also the review of interior point methods applications for solving special classes of convex programming problems, including quadratic-type problems (so-called problems of semidefinite programming). Then we give a review of general complexity results for polynomial extremal problems with continuous and(or) integer variables. This review is based mainly on the doctoral thesis of L.G.Khachiyan and his (or common with colleagues) articles (see [Kh 82]).

Chapter 5 is devoted to the realization of decomposition schemes for solving large-scale structured extremal problems by using algorithms of non-differentiable optimization. After presentation of the main decomposition schemes (decomposition on the set of constraints and decomposition on the set of variables) we give some instances of decomposition of certain LP and convex programming structured problems which was solved by using of subgradient-type algorithms.

In particular, we describe numerical experiments realizing a decomposition scheme of variables by r-algorithm for large-scale quadratic-type problems provided on a high performance computer [Mai 91] and the results of using a decomposition scheme for solving large-scale mixed boolean-continuous problem of net constructing on a high speed multiprocessor computer. In this case the parallel implementation of r-algorithm was used for solving coordinating problems successfully [DMi 95].

In Chapter 6 we consider the problems of constructing optimal on n-dimensional volume circumscribed ellipsoids containing a given finite set of points in E^n, and the problems of constructing maximal on volume ellipsoid, inscribed in a convex body, given by a system of linear inequalities. For these problems we consider dual problems and the methods of nonsmooth penalty functions.

Then we describe simple original algorithms for solving mentioned problems in which operations of space dilation play the main role. We also review polynomial-time algorithms used for solving the problems of constructing optimal inscribed and circumscribed ellipsoids. Next we explain the so-called "inscribed ellipsoid method" for solving the problem of minimization of convex functions determined on convex sets given by linear inequalities [KTE 88]. This method demands $O(n \log \frac{1}{\varepsilon})$ iterations for receiving optimal value of subject function with relative error $\varepsilon > 0$. Besides, we consider some semidefinite programming problems that arise in the theory of stability and optimal control of dynamic systems, and describe applications of nonsmooth optimization algorithms for solving mentioned problems. Next we describe possible ways of reducing nonconvex quadratic programs to semidefinite programming.

In Chapter 7 we present the general theory (see [GLS 88]) of the role of the ellipsoid method for analysis of complexity for various classes of combinatorial problems. In particular, the weighted maximal independent vertex set problem for graphs is considered and polynomial-time algorithm for the class of perfect graphs is described (see [GLS 88]).

Chapter 8 is devoted to the algorithms for obtaining dual bounds for quadratic-type problems that arise in the models, describing multiextremal combinatorial problems such as the maximal cut problem, optimal graph partitioning, maximal size k-partite induced subgraph of given graph, and so on. In this Chapter the role of superfluous constraints for improving of dual bounds is considered.

The results of numerical tests for obtaining dual bounds in various models of combinatorial optimization are demonstrated.

In Chapter 9 the problem of finding global minima of a polynomial function $P(x), x \in R^n$, is investigated. This problem can be reduced to the special type of quadratic optimization problems with superfluous constraints. Using Lagrange relaxation we obtain dual lower bound ψ^* for objective function of the quadratic problem. We prove that the dual bound $\psi^* = \inf_{x \in R^n} P(x) = P^*$ iff the nonnegative polynomial $P(x) - P^* = \overline{P}(x)$ can be represented as sum of squares of polynomials. This result is tightly linked with 17-th Hilbert's problem of decomposition nonnegative rational functions into a sum of squares of rational functions.

The possibilities of r-algorithm obtaining dual quadratic bounds for global minimum value of polynomials is considered.

Thus, the main aim of this book is to show that nevertheless the fast development of various directions of optimization theory and its computational aspects in last years the nondifferentiable optimization remains till now as a "nuclear" part of mathematical programming which connected by many linrs with linear, nonlinear and discrete programming, global optimization, it is a basis for solving various real-life large-scale optimization problems by using of the decomposition schemes for stochastic programming, image recognition, estimation of regression parametres, design of experiments, for matrix optimization in analysis and control of dynamic systems, for theory of games and decision making in conflict situations.

The works of many Western mathematicians have influenced the contents of this book. Among them George Dantzig, Jack Edmonds, Ralph Gomory, Alan Hofman, Harold Kuhn, Olvi Mangasarian, George Nemhauser, Richard Bellman, Terry Rockafellar, M.J.D.Powell, Martin Grötschel, Laslo Lovasz, Bernard Korte, Jan Lenstra, Philip Wolfe, Michael Held, Martin Beal, Ray Fulkerson, Al Tucker, M.Todd, David Goldfarb, Claude Lemarechal, Victor Klee, Ellis Johnson, Narendra Karmarkar, Krzysztof Kiewil, Panos Pardalos, F. Al-Khayyal, J.L.Goffin, S.Polyak, E.Pollak, Richard Karp, S.Cook, Jean Renegar, Roger Fletcher, Stephen Boyd, Peter Kall, Athanasios Migdalos, Hoang Tuy.

I am grateful to many colleagues from the former Soviet Union for discussions on the topic of this book, in particular to Boris Pshenichnii, Boris Polijak, Yuri Ermol'ev, Yuri Danilin, Vladimir Mihalevic, Vladimir Trubin, Arkadii Chikrii, Roman Poljak, Matvei Primak, Yuri Nesterov, Arkadii Nemirovskii, Leonid Khachiyan, Evgenii Goldschtein, Ilia Dikin, Ivan Sergienko, Vladimir Demjanov, Alexandr Zykov, Valeriy Bulatov, Leonid Chernoys'ko, to my students Nicolai Gurbenko, Lidiya Shabashova, Sergei Stetsenko, Vladimir Gershovich, Sergei Rjevskii, Peter Stetsjuk, Aleksei Likhovid, Aleksei Kuntsevich, Elena Kiseleva, Dmitrii Solomon for their participation on the theoretical background and practical applications of nondifferentiable optimization.

My special thanks are due to Aleksei Likhovid, Inna Naydenova, Tanya Trushko, Peter Stetsjuk for the careful typing and patient retyping of the text in TEX. Sergei Stetsenko gave aid in improving the language of the book. I am very grateful to Scientific Editor of this book prof.Panos Pardalos whose advices play an essential role in improving of the text of the book. I also thank John Martindale for his collaboration and help in preparing the manuscript. I hope that my collaboration with Kluwer Academic Publishers will continue.

Kiev, 1997 N.Z.Shor

1

ELEMENTS OF CONVEX ANALYSIS, LINEAR ALGEBRA, AND GRAPH THEORY

1.1 CONVEX SETS AND CONVEX FUNCTIONS

We shall review a number of fundamental properties of convex sets and functions which will be usefull in the following chapters. This review is based on the latest monographies in convex analysis and optimization, mainly, [Psh 69], [HUL 93], [Roc 70], [Roc 82a], [IT 79], [DV 85].

1.1.1 Convex sets

Definition 1 *Banach space is a complete metric linear space with distinct function $d(x, y)$ given by norm $\|\cdot\|_B$:*

$$d(x, y) = \|x - y\|_B,$$

for all pairs $x, y \in B$.

The norm possesses the following properties:

(1) $\|x\| = 0$ if and only if $x = 0$;

(2) $\|\alpha x\| = |\alpha| \|x\|$ for any $\alpha \in R$;

(3) $\|x + y\| \leq \|x\| + \|y\|$.

Definition 2 *The space B^* of continuous linear functionals $x^*(x)$ determined on B is called the conjugate to B space.*

In B^* one can define the (strong) norm:

$$\|x^*\| = \sup_{\|x\| \leq 1} x^*(x).$$

The dual formula is true:

$$\|x\| = \max_{x \in S^*} x^*(x), \text{ where } S^* = \{x^* : \|x^*\| \leq 1.$$

For the finite dimensional linear spaces the topological properties generated by different norms are equivalent. We shall consider mainly n-dimensional Eucledean space E^n, the n-dimensional space in which the bilinear symmetric function $f(x, y) = (x, y)$ (scalar product) is determined. Two vectors $x, y \in E^n$ are called orthogonal, if $(x, y) = 0$. The norm in E^n is introduced by formula:

$$\|x\| = \sqrt{(x, x)}$$

If $x = (x^{(1)}, x^{(2)}, \ldots, x^{(n)})$, $y = (y^{(1)}, y^{(2)}, \ldots, y^{(n)})$ are coordinates x and y in some orthonormal basis, then

$$f(x, y) = (x, y) = \sum_{i=1}^{n} x^{(i)} y^{(i)}.$$

The space E^n is self-conjugate, i.e. $(E^n)^* = E^n$.

Let X be a Banach space. A set $C \subset X$ is convex if for any two points $x, y \in C$ it contains the whole segment $[x, y]$ linking these points, i.e.

$$(1 - \alpha)x + \alpha y \in C \text{ for any } \alpha \in [0, 1]. \tag{1.1}$$

Let $x_1, \ldots, x_m$ be any finite subset of X. We call x a convex combination of $\{x_i\}_{i=1}^m$, if $x = \sum_{i=1}^m \alpha_i x_i$, $\sum_{i=1}^m \alpha_i = 1$, $\alpha_i \geq 0$, $i = 1, \ldots, m$. If set $C \subset X$ is convex, then any convex combination of the points of C belongs to C.

If C is convex, then $\lambda C = \{\lambda x \mid x \in C\}$ is a convex set for any $\lambda \in R$.

When C_1 and C_2 are convex, their algebraic sum (i.e. the set of all sums $x = x_1 + x_2$, $x_1 \in C_1$, $x_2 \in C_2$) is also a convex set.

We consider that empty set $\emptyset$ convex by definition. Then the intersection of arbitrary family of convex sets is a convex set.

The closure ($cl\, C$ or $\overline{C}$) and the interior ($int\, C$) of a convex set C are always convex.

In the partial case, where X has a finite dimension n, convex set C either has nonempty interior, or is contained in some affine manifold of dimension less than n. In the last case the intersection of all linear subspaces of the form $C - x_0$, where x_0 is any point from C, is called **carrier subspace** of C and denoted $Lin\, C$. The dimension of C $dim\, C = dim\,(Lin\, C)$.

A set of points $x \in C$ such, that $x + Lin\, C \cap (\varepsilon B) \subset C$, where B is a unit ball in X, $\varepsilon > 0$, is called **relative interior** of C and denoted by $ri\, C$.

Let $M \subset X$. Intersection of all convex (closed convex) sets in X, containing M, is called convex hull (convex closure) of M, and denoted by $conv\, M$ ($\overline{conv}\, M$). M and $\overline{conv}\, M$ are convex sets, $conv\, M$ may be expressed in the following form:

$$conv\, M = \{\sum_{k=1}^{m} \lambda_k x_k \mid x_k \in M;\ \lambda_k \geq 0;\ \sum_{k=1}^{m} \lambda_k = 1\}$$

where m is an arbitrary positive integer.

Definition 3 *The point $c \in C$ (C is a convex set) is called* **extreme** *if it cannot be expressed as an interior point of a segment linking any two points $x, y \in C$.*

Theorem 1 (Theorem of Caratheodory) *Any point of a compact convex set $C \subset E^n$ can be expressed as* **a convex combination** *of its no more than $(m+1)$ extreme points, where m is dimension of C.*

A set $K \subset X$ is called **cone**, if $x \in K$ implies $\lambda x \in K$ for all $\lambda > 0$. Cone K is convex if $x_1, x_2 \in K$ implies $x_1 + x_2 \in K$.

Let X^* be the space of linear continuous functionals on X. If $x^* \in X^*$, we denote $x^*(x)$ as $< x^*, x >$. One can associate for each convex cone K in X the corresponding **conjugate cone** $K^* \subset X^*$ defined by the following way:

$$K^* = \{x^* \in X^* \mid < x^*, x > \geq 0\ \forall x \in K\},$$

which is weakly closed convex cone in X^*.

If $X \equiv E^n$, then $X^* \equiv E^n$, so for $K \subset E^n$,

$$K^* = \{y \in E^n \mid (y, x) \geq 0, \ \forall x \in K\}. \tag{1.2}$$

Theorem 2 *If $K \subset E^n$ is a convex cone, then $(K^*)^* = \overline{K}$, where $\overline{K}$ is a closure of cone K.*

The intersection of all convex cones in X containing a set $M \subset X$ and origin 0 is a convex cone denoted as *con* M:

$$con\, M = \cup_{\lambda \geq 0} \lambda conv\, M.$$

Let C be a nonempty closed convex set in X. The asymptotic, or **recession cone** $R(C)$ of the set C is given by the following formula:

$$R(C) = \{d \in X \mid \exists x \in C, d \in X \text{ such that } x + \lambda d \in X \text{ for all } \lambda \geq 0\}.$$

$R(C)$ is always a closed convex cone. A closed convex set is compact if and only if $C_\infty = 0$.

Now let us formulate one of the main theorems of convex analysis.

Theorem 3 (Separability theorem) *Let C^* be a convex weakly closed set in X^*. Then for any $x_0^* \in C^*$ there exists $x_0 \in X$ such that*

$$< x^*, x_0 > \leq < x_0^*, x_0 > -\varepsilon$$

for all $x^* \in C^*$ and some $\varepsilon > 0$.

In n- dimensional Euclidean space E^n the separability theorem can be expressed in the following form:

Theorem 4 *Let $M \subset E^n$ be a closed convex set and $x_0 \notin M$ be a given point. Then there exist $g \in E^n$, $a \in E^n$ such that:*

$$(g, x - a) < 0 \text{ for all } x \in M; \text{ and } (g, x_0 - a) > 0. \tag{1.3}$$

The hyperplane $P = \{x : (g, x - a) = 0\}$ is called stricly separating M from x_0.

The following theorm is a generalization of Theorem 4.

Theorem 5 *Let M_1 and M_2 be closed convex nonintersecting sets in E^n and at least one of them is bounded. Then there exist $g \in E^n$ and $a \in E^n$ such that*

$$(g, x - a) > 0 \text{ for all } x \in M_1;$$

$$(g, x - a) < 0 \text{ for all } x \in M_2.$$

Theorem 6 *Let $W_1, W_2 \subset E^n$ be convex and closed sets with disjoint interiors and assume that the interior of W_1 is nonempty. Then there exist $g \in E^n$, $g \neq 0$, and $a \in E^n$ such that*

$$(g, x - a) \leq 0 \text{ for all } x \in W_1;$$

$$(g, x - a) \geq 0 \text{ for all } x \in W_2.$$

A hyperplane P passing through a boundary point x of a convex set $W \subset E^n$ is called a **supporting hyperplane** of W at x, if W lies in one of two closed halfspaces with common boundary P.

Theorem 7 *At any boundary point of a convex set $W \in E^n$ there exists at least one supporting hyperplane.*

1.1.2 Convex functions

Let X be a Banach space. Consider the function $f : X \to R \cup \{\pm\infty\}$ and two sets associated with it:

$$dom\, f = \{x \in X \mid f(x) < +\infty\}$$

(the effective domain or domain of f),

$$epi\, f = \{(x, \lambda) \in X \times R \mid \lambda \geq f(x)\}$$

(epigraph of the function f).

One can see that

$$f(x) = \inf\{\lambda | (x, \lambda) \in \text{epi} f\}.$$

Definition 4 *Function f is called convex, if the set epif is convex in the space $X \times R$. If domf $\neq \emptyset$ and $f(x) > -\infty$ for all x, then function f is called* **proper.**

Function φ is called **concave** *(proper concave) if function $f = -\varphi$ is convex (proper convex).*

The proper function f is convex if and only if the following inequality is true:

$$f(\lambda x_1 + (1-\lambda)x_2) \leq \lambda f(x_1) + (1-\lambda)f(x_2) \tag{1.4}$$

for all $x_1, x_2 \in X, \lambda \in [0,1]$.

We say that f is **strictly convex** on C when (1.4) holds as a strict inequality for $x_1 \neq x_2$, $\lambda \in int[0,1]$.

Definition 5 *f is* **strongly convex** *on C (with modulus of strong convexity c) if there exists $c > 0$ such that $f(\alpha x + (1+\alpha)x') \leq \alpha f(x) + (1-\alpha)f(x') - \frac{1}{2}c\alpha(1-\alpha)||x - x'||^2$ for all $(x, x') \in C \times C$ and $\alpha \in [0,1]$.*

Proposition 1 *The sum $f_1 + f_2$ of two proper convex functions f_1 and f_2 is a convex function.*

Definition 6 *Function f is called* **closed** *if epif is closed in $X \times R$.*

The **closeness** of the function f is equavalent to the semicontinuety of this function that is the same as the closeness of level sets $\{x : f(x) \leq \alpha\}$ for all $\alpha \in R$.

The function

$$\text{cl}f = \inf\{\lambda | (x, \lambda) \in \text{epi}f\}.$$

cl$f$ is convex for convex f.

Function $f(x) = \sup\{f_i(x) | i \in I\}$ is called upper enveloping (or max-function) of family functions $\{f_i(x)\}, i \in I$.

Proposition 2 *If all $f_i(x), i \in I$, are convex, then $f(x)$ is convex, because epif $= \bigcap_{i \in I}$ epif_i is a convex set.*

Definition 7 *Let f and g be convex functions.* **Infimal convolution** $f \bigtriangledown g$ *of functions f and g is the function defined by the condition:*

$$(f \bigtriangledown g)(x) = \inf_{\lambda}\{(x, \lambda) \in \text{epi} f + \text{epi} g\}$$

$f \bigtriangledown g$ is a convex function, and for proper f and g

$$f \bigtriangledown g(x) = \inf_{y}\{f(x-y) + g(y)\}. \tag{1.5}$$

Each function $f : X \to R \cup \{\pm\infty\}$ can be associated with **conjugate** to it function

$$f^*(x^*) = \sup_{x}\{< x^*, x > -f(x)\}, x^* \in X^*.$$

Function f^* is always convex and closed in weak topology of space X^*.

Theorem 8 (Fenchel-Moreau Theorem) *If f is a proper convex function, then*

$$f(x) = f^{**}(x),$$

where

$$f^{**}(x) = \sup_{x^*}\{< x^*, x > -f^*(x^*)\} \tag{1.6}$$

When $X \equiv E^n$, f and f^* are defined on the same space E^n and:

$$f(x) = f^{**}(x) = \sup_{y}\{(y, x) - f^*(y)\}$$

By $\delta(x/C)$ we denote an indicator function

$$\delta(x/C) = \begin{cases} 0 & , \quad \text{if } x \in C \\ +\infty & , \quad \text{if } x \ni C. \end{cases}$$

Proposition 3 *The indicator function of a convex set is convex.*

By **support function** *of the set C we denote the function*

$$S_C(x^*) = \sup\{< x^*, x > \mid x \in C\}.$$

For convex set C, $S_C(x^*)$ is convex, and functions $\delta(x/C)$ and $S_C(x^*)$ are mutually conjugate.

Let C_i^*, $i \in I$ be sets in X^*, where I is an arbitrary set of indeces, and

$$C^* = \cup_{i \in I} C_i^* .$$

Then

$$S_{C^*}(\cdot) = \sup_{i \in I} S_{C_i^*}(\cdot)$$

1.2 PROPERTIES OF SUBGRADIENT AND ε-SUBGRADIENT SETS

Proposition 4 *The proper closed convex function f is continuous at each point of* int dom f. *If $X = R^n$ we have a stronger assertion: convex function f is continuous on the set* ri dom f.

Proposition 5 *If proper function f is convex, then the following conditions are equivalent:*

1) *f is bounded from above in the neighborhood of x;*

2) *f is continuous in x;*

3) *f is Lipschitz's function in neighborhood of x, i.e. there exists a number $l > 0$ such that*

$$|f(x_1) - f(x_2)| \leq l \|x_1 - x_2\|$$

for all x_1, x_2 from some neighborhood of x.

Definition 8 *The directional derivative in direction $\overline{x}$ at the point $x \in X$ is defined as*

$$f'(x, \overline{x}) = \lim_{\varepsilon \to +0} \frac{1}{\varepsilon} [f(x + \varepsilon \overline{x}) - f(x)]$$

if the limit (finite or infinite) exists.

Proposition 6 *For proper convex function f in any point $x \in$ dom f there exists the derivative $f'(x, \overline{x})$ in any direction $\overline{x} \in X$. This derivative is a convex, positively homogeneous function in $\overline{x}$.*

Definition 9 *The set*

$$\partial f(x_0) = \{x^* \in X^* \mid f(x) - f(x_0) \geq (x^*, x - x_0),\ \forall x \in X\}$$

is called **subdifferential** *of convex function f in the point $\overline{x}$.*

Proposition 7 *The subdifferential $\partial f(x)$ is a convex, weakly* closed set in X^*. If f is continuous in the point x, then $\partial f(x)$ is nonempty convex weakly* compact set and*

$$f'(x, \overline{x}) = \max\{< x^*, \overline{x} > \mid x^* \in \partial f(x)\}$$

Theorem 9 (Theorem of Rockafellar-Moreau) *Let f_1, f_2 be proper convex functions and in a point $x_0 \in dom\ f_1 \cap dom\ f_2$ function f_1 is continuous. Then*

$$\partial(f_1 + f_2)(x) \subset \partial f_1(x) + \partial f_2(x).$$

For n-dimensional Eucleadean space $X \equiv E^n$ we can formulate more exact results. Below in this section we shall consider $X \equiv E^n$.

Definition 10 *Let f be a convex function with domain M and let x_0 be an interior point of M. A vector $g_f(x_0)$ satisfying*

$$f(x) - f(x_0) \geq (g_f(x_0), x - x_0) \text{ for all } x \in M \tag{1.7}$$

is called a **subgradient** *or a* **generalized gradient** *of f at x_0.*

The set of subgradients of function f in the point x_0 is denoted as $G_f(x_0)$.

Theorem 10 *The set of generalized gradients $G_f(x_0)$ (or subgradient set, or subdifferential) of a convex function f, defined on E^n at any interior point x_0 of the domainM is nonempty, bounded, convex and closed. If $G_f(x_0)$ consists of a single point $g_f(x_0)$, then $g_f(x_0)$ is the gradient of f in x_0.*

Theorem 11 *A convex function f has a directional derivative in any direction at any interior point x_0 of its domainM. This derivative*

$$f'_\eta(x_0) = f'(x_0, \eta) = \max_{g \in G_f(x_0)} (g, \eta).$$

Theorem 12 *A function f, defined finitely on E^n or on some open subset $M \subset E^n$ is convex if and only if its directional derivative exists at all points of M and $f'_\eta(x+t\eta)$ is a nondecreasing function of t for any x and η, $x+t\eta \in M$. If a function f is twice differentiable then it is convex if and only if its Hessian $H(x)$ is a positive semidefinite matrix at any point $x \in M$.*

Definition 11 *A direction η^*, $||\eta^*|| = 1$ will be called a* **direction** *of* **steepest descent** *of a convex function f at an interior point x_0 of the domain M, if*

$$\min_{||\eta^*||=1} f'_\eta(x_0) = f'_{\eta^*}(x_0).$$

Theorem 13 *Suppose that $0 \notin G_f(x_0)$ and let $\overline{\eta}$ be the unique element of $G_f(x_0)$ that is nearest to the origin. Then $\eta^* = -\frac{\overline{\eta}}{||\overline{\eta}||}$ is a direction of steepest descent at x_0, and $f'_{\eta^*}(x_0) = -||\overline{\eta}||$.*

Theorem 14 *For an interior point x_0 of the domain M of f to be the point of minimum, it is necessary and sufficient that $0 \in G_f(x_0)$.*

Let $f(x)$ be a convex function determined on E^n, and $\varepsilon \geq 0$ is given.

Definition 12 *Vector $g \in E^n$ is called ε-subgradient of the function f in the point $\overline{x} \in domf$ if the inequality*

$$f(x) - f(\overline{x}) \geq (g, x - \overline{x}) - \varepsilon \tag{1.8}$$

holds for all $x \in domf$.

Note that for $\varepsilon = 0$ this definition coincides with the definition of usual subgradient. We denote ε-subgradient of the function f in the point $\overline{x}$ as $g^\varepsilon_f(\overline{x})$. The set of ε-subgradients in the point $\overline{x}$ is denoted as $G^\varepsilon_f(\overline{x})$.

Proposition 8 *If $\overline{x} \in intdomf$, then for any $\varepsilon \geq 0$ the set $G^\varepsilon_f(\overline{x})$ is nonempty, convex, closed and bounded.*

It is clear that if $\varepsilon_1 > \varepsilon_2$, $G^{\varepsilon_1}_f(\overline{x}) \supseteq G^{\varepsilon_2}_f(\overline{x})$.

Proposition 9 $0 \in G_f^\varepsilon(\overline{x})$ *iff* $f(\overline{x}) - \inf_{x \in E^n} f(x) \le \varepsilon$.

Definition 13 *If* $0 \notin G_f^\varepsilon(\overline{x})$, *then* ε*-steepest descent direction* $d_\varepsilon(\overline{x})$ *is determined by the formula:*

$$d_\varepsilon(\overline{x}) = -\frac{g^*}{\|g^*\|},$$

where

$$g^* = \arg \min_{g \in G_f^\varepsilon(\overline{x})} \|g\|.$$

Proposition 10 *Unlike point-to-set mapping* $x \to G_f(x)$ *that is upper semi-continuous for nonsmooth convex functions* f, *the mapping* $x \to G_f^\varepsilon(x)$ *is continuous for* $\varepsilon > 0$ *at any point* $\overline{x} \in intdomf$.

So, for $\varepsilon > 0$ the multifunction $x \to d_\varepsilon(x)$ is continuous at all points $\overline{x} \in domf$, where $d_\varepsilon(\overline{x})$ is defined by Definition 13.

Taking into account that for $\varepsilon > 0$ $G_f^\varepsilon(\overline{x})$ contains a large amount of information about the behavior of f in neighborhood of $\overline{x}$, its structure is much more complex than $G_f(\overline{x})$ even for the simple functions f. So in implemented versions of ε-descent algorithms the rough approximations of $G_f^\varepsilon(\overline{x})$ are used (see [Le 75], [Wol 75]).

1.2.1 Some methods for computing subgradients

We shall consider basic operations with respect to which the class of convex functions is closed: 1) the operation of forming linear combinations of functions with nonnegative coefficients; 2) the operation of taking the pointwise maximum over a family of functions. In applications, complicated convex functions are usually obtained from simpler ones (e.g. linear or quadratic) by the superposition of the two indicated operations. The operation of taking the maximum is the main factor in the origination of nonsmooth convex functions from smooth ones. We shall formulate theorems describing the subdifferentials of functions which are linear combinations of convex functions with nonnegative coefficients, and the subdifferentials of functions which are pointwise maxima (max functions).

Theorem 15 *Let $f(x) = \sum_{i=1}^{k} a_i f_i(x)$, where $a_i \geq 0$ and the functions f_i are proper and convex on E^n, $i = 1, \ldots, k$. Then f is convex and its subdifferential at any point x_0 belonging to intdomf consists of all the vectors of the form*

$$g(x_0) = \sum_{i=1}^{k} a_i g_i(x_0), \; g_i(x_0) \in G_{f_i}(x).$$

Let J be an arbitrary index set and $\{f_j\}$ be a collection of convex proper functions from E^n to R, and we assume that

$$f(x) := \sup\{f_j(x), \; j \in J\} < +\infty \text{ for all } x \in E^n \tag{1.9}$$

We already know that f is convex. Let $J(x) := \{j \in J; \; f_j(x) = f(x)\}$. We are interested in computing its subdifferential.

Theorem 16 *Assume that J is a compact set (in some metric space), on which the functions $j \to f_j(x)$ are upper semicontinuous for each $x \in E^n$. Then $G_f(x) = conv\,\{\cup G_{f_j}(x), \; j \in J(x)\}$. If we assume also that each f_i is* **differentiable**, *then*

$$G_f(x) = conv\,\{g_{f_j}(x), \; j \in J(x)\}.$$

The following theorem can be applied in some decomposition schemes.

Theorem 17 *For some compact set $Y \subset R^p$, let $F : \; E^n \times Y \to R$ be a function satisfying the following properties:*

- *for each $x \in E^n$, $F(x, \cdot)$ is upper semicontinuous;*
- *for each $y \in Y$, $F(\cdot, y)$ is convex and differentiable;*
- *the function $f := \sup_{y \in Y} g(\cdot, y)$ is finite-valued on E^n;*
- *at some $\overline{x} \in E^n$, $F(\overline{x}, \cdot)$ is maximized at a unique $y(\overline{x}) \in Y$.*

Then f is differentiable at this $\overline{x}$ and its gradient is

$$g_f(\overline{x}) = \frac{\partial F}{\partial x}(\overline{x}, y(\overline{x})).$$

(where $\frac{\partial F}{\partial x}(x, y)$ denotes the partial gradient of the function $F(\cdot, y)$ at x).

The most general formula for subgradient set of a finite-valued sup-function is given in the following theorem :

Theorem 18 *Let*

$$S_\delta := \cup_x \{G_{f_j}(x) : j \in J_\delta(x_0);\ x \in B(x_0, \delta)\},$$

where $B(x_0, \delta)$ is the ball in E^n of radius δ. Then if $f(x) = \sup_{j \in J} f_j(x)$, where $f_j(x)$ are proper convex functions on E^n for all $j \in J$ and $f(x)$ is finite-valued, then for given $x_0 \in E^n$

$$G_f(x_0) = \cap_{\delta > 0} \overline{conv} S_\delta .$$

Although in some cases the subgradient set of nonsmooth function f in the point x_0 of nondifferentiability has a complex structure, in many important practical cases the computation of a subgradient does not essenteally differ in complexity from the computation of a gradient. It is also worth mentioning the following result of Rademeister (see [Bus 58]).

Theorem 19 *Let f be a proper convex function defined on E^n. The subset of $dom f$, where f fails to be differentiable is of zero (Lebesque) n-dimensional measure. Moreover, the gradient of a convex function is continuous on the set of points at which it exists.*

Theorem 20 *For proper convex function f, let S be a convex compact subset of ri dom f. Then there exists $L = L(S) > 0$ such that*

$$|f(x) - f(x')| \leq L\|x - x'\| \text{ for all } x \text{ and } x',$$

i.e. the proper convex functions are locally Lipschitzian.

1.2.2 Almost differentiable functions

Most works devoted to gradient-type methods or generalized gradient methods assume respectively either continuity of the gradient or convexity of the objective function. But in many practical problems one can hardly postulate the smoothness (or convexity) of the objective function. For example, in economic planning some components of the objective function are usually defined as piecewise-smooth and not necessarily convex functions of a parameter which characterizes the productivity (the output capacity) of a certain unit.

Nondifferentiability and nonconvexity are typical of a broad class of minimax problems.

Therefore a need arises to consider minimization problems with a class of functions which is broad enough to contain piecewise-smooth functions and functions encountered in minimax problems, and narrow enough to allow a natural generalization of the gradient and an application of gradient-type methods for finding local minimum.

These requirements are met by the class of almost differentiable functions introduced in [Shor 72].

Definition 14 *A function f defined on n- dimensional Euclidean space E^n is called almost differentiable if :*

(a) *in any bounded area it satisfies the Lipschitz condition;*

(b) *it is differentiable almost everywhere;*

(c) *its gradient is continuous on its domain M.*

Note that property **(b)** is not independent. We have the following theorem [Sl 66]:

Let a real-valued function f defined on an open set $D \subset E^n$ have finite Dini derivatives in all directions, i.e.

$$\limsup_{t \to 0} \left| \frac{f(x+tv) - f(x)}{t} \right| < +\infty$$

for any $x \in D$ and any $v \in E^n$. Then f is differentiable almost everywhere on D.

It is clear that if a function f satisfies the Lipschitz condition in any open bounded area $S \subset E^n$, i.e.

$$|f(x) - f(y)| \leq L\|x - y\|, \;\; L > 0; \;\; x, y \in S,$$

then

$$\limsup_{t \to 0} \left| \frac{f(x+tv) - f(x)}{t} \right| \leq L\|v\| < \infty$$

for all $x \in S$. Therefore the local Lipschitz property implies the differentiability almost everywhere.

Definition 15 *An almost-gradient of an almost differentiable function f at a point x_0 is an accumulation point of a sequence of gradients $g(x_1), g(x_2), \ldots$, such that $\{x_k\}_{k=1}^{\infty}$ converges to x_0 and f is differentiable at each x_k, $k = 1, 2, \ldots$.*

Theorem 21 *The set $AG(x)$ of almost-gradients of an almost differentiable function f at any point $x \in E^n$ is nonempty, bounded and closed.*

Proof. Property **(a)** implies the boundedness of the sequence $\{g(x_k)\}$, which in turn implies that $G(x)$ is nonempty and bounded. The closedness follows from the definition of almost-gradients as limit points. Indeed, let $\{\widehat{g}^{(k)}(x)\}$ be a sequence of almost-gradients converging to g^*. Then for each k one can find $x^{(k)}$ such that f is differentiable at $x^{(k)}$, $\|x - x^{(k)}\| \leq 1/k$ and $\|\widehat{g}^{(k)}(x) - g(x^{(k)})\| \leq 1/k$. Then

$$\begin{aligned}\|g^* - g(x^{(k)})\| &\leq \|g^* - \widehat{g}^{(k)}(x)\| + \|\widehat{g}^{(k)}(x) - g(x^{(k)})\| \\ &\leq \|g^* - \widehat{g}^{(k)}(x)\| + \frac{1}{k}.\end{aligned}$$

Since $\lim_{k\to\infty} \|g^* - \widehat{g}^{(k)}(x)\| = 0$, it follows that g^* is an almost-gradient of f at x.□

Theorem 22 *A convex function f defined on E^n is almost differentiable and at any point its almost-gradients coincide with subgradients.*

Proof. By Rademeister's theorem (see [Bus 58]) any convex function defined on E^n is almost everywhere continuously differentiable. It follows from well-known properties of convex functions that directional derivatives of f are uniformly bounded in any bounded area. Therefore f is almost differentiable.

Let $\widehat{g}_f(x_0) = \lim_{k\to\infty} g_f(x_k)$ be an almost-gradient, where each $g_f(x_k)$ is the gradient of f at x_k and $\lim_{k\to\infty} x_k = x_o$. It follows from the convexity of f that for all $x \in E^n$

$$f(x) - f(x_k) \geq (g_f(x_k), x - x_k).$$

Letting k approach infinity we obtain

$$f(x) - f(x_0) \geq (\widehat{g}_f(x_0), x - x_0),$$

i.e.$\widehat{g}_f(x_0)$ is a subgradient of f at x_0. This completes the proof. □

The class of almost differentiable functions is broad enough.

Theorem 23 *If functions f_1 and f_2, defined on E^n are almost differentiable, then the functions $f_3(x) = f_1(x) + f_2(x)$, $f_4(x) = f_1(x)f_2(x)$ and $f_5(x) = \max\{f_1(x), f_2(x)\}$ are also almost differentiable.*

Definition 16 *Any vector in the closure of the convex hull of the set of almost-gradients of f at x is called a generalized gradient of f at x. The set of generalized gradients of function f at the point x is denoted $\bar{G}_f(x)$, so $\bar{G}_f(x) =$ conv $AG(x)$.*

Theorem 24 *If f is a proper convex function, then at each $x \in$ intdom f the subgradient set $G_f(x)$ is equal to $\bar{G}_f(x)$.*

In general, the $\bar{G}_f(x)$ is convex, bounded and closed.

In modern theory of nonconvex analysis the slightly another concept of generalized gradient is used. This concept was introduced by F.H. Clarke [Cl 75]. In this paper the class of locally Lipschitz functions is considered and the following definition is formulated:

The subdifferential of f at x, denoted by $\partial f(x)$, is the convex closure of the set of limit points of all sequences of the form $\{g_f(x+h_i)\}_{i=1}^{\infty}$ with $\lim_{i\to\infty} h_i = 0$.

Our definition of the class of almost differentiable functions differs from the class of locally Lipschitzian function with the additional requirement that the gradient be continuous on its domain.

In [Cl 73], [Cl 75] (see [Cl 83], F.H. Clarke considered the class of locally Lipschitz functions and the following analogue of the directional derivative

$$f^0(x,d) = \lim_{\substack{h\to 0 \\ t\to +0}} \sup \frac{1}{t}[f(x+h+td) - f(x+h)].$$

The **generalized gradient set** (called **generalized gradient** by Clarke) of f at x is defined by

$$\partial f(x) = \{g \in E^n : (g,d) \leq f^0(x,d) \text{ for all } d \in E^n\}$$

and has the following properties:

(a) $\partial f(x)$ is a nonempty convex compact set;

(b) $f^0(x,d) = \max_{g \in \partial f(x)}(g,d)$;

(c) f is differentiable almost everywhere. Each point of the set $\partial f(\overline{x})$ can be obtained as a convex combination of the limit points of gradients $\{g_f(x_k)\}_{k=1}^{\infty}$ calculated in the points x_k of differentiability of f converging to $\overline{x}$.

(d) if $\{x_k\}$ converges to x and $g_k \in \partial f(x_k)$ for all k, then the sequence $\{g_k\}$ is bounded and all its accumulation points belong to $\partial f(x)$, i.e. the point-to-set mapping $x \rightarrow \partial f(x)$ is bounded and upper semicontinuous;

(e) an analogue of the mean value theorem holds: for any two points y and z one can find $\lambda \in (0;1)$ and $g \in \partial f(y + \lambda(z-y))$ such that $f(z) - f(y) = (g, z-y)$.

The class of Lipschitzian functions is too broad for constructing meaningful general theory of necessary and sufficient conditions for the points of local minimum, or local maximum. In the work [DR 80] the more restricted, but wide enough subclass of functions was proposed, which was named as subclass of **quasidifferentiable functions** (QUD-functions).

Let on the open set $S \subset E^n$ the finitely valued function $f(x)$ be given. We say that function $f(x)$ is quasidifferentiable at the point $x_0 \in S$ if it differentiable on any direction g in this point and there exist convex compact sets $\underline{\partial f(x_0)} \subset E^n$ (subdifferential) and $\overline{\partial f(x_0)}$ (superdifferential) such that

$$\begin{aligned} \frac{\partial f(x_0)}{\partial g} &= \lim_{\alpha \rightarrow +0} \frac{f(x_0 + \alpha g) - f(x_0)}{\alpha} = \\ &= \max_{v \in \underline{\partial f(x_0)}} (v,g) + \min_{w \in \overline{\partial f(x_0)}} (w,g). \end{aligned}$$

The pair of sets $Df(x_0) = [\underline{\partial f(x_0)}, \overline{\partial f(x_0)}]$ is called as quasidifferential of function $f(x)$ in the point x_0.

$Df(x_0)$ is determined nonuniquely because for arbitrary compact B

$$A(x_0) = [\underline{f(x_0)} + B, \overline{\partial f(x_0)} - B]$$

is also quasidifferential of function $f(x)$ in the point x_0. Yet for specific subclasses of QUD-functions "natural" representations of $Df(x_0)$ are prefered. For example, if $f(x)$ is convex (concave) function then it is natural to choose

$$\begin{aligned} Df(x_0) &= [G_f(x_0), \emptyset], \\ Df(x_0) &= [\emptyset, \overline{G_f(x_0)}], \end{aligned}$$

where $G_f(x_0)$ (correspondingly $\overline{G_f(x_0)}$) are the set of subgradients (the set of supergradients) in the point x_0. For convex-concave functions $f(z) = f(x, y)$ it is natural to choose $Df(z_0)$ as

$$Df(x_0, y_0) = \left[\frac{\partial f(z_0)}{\partial x}, \frac{\partial f(z_0)}{\partial y}\right],$$

where $\frac{\partial f(z_0)}{\partial x}$(correspondingly, $\frac{\partial f(z_0)}{\partial y}$) are partial subdifferential on x (partial superdifferential on y) in the point z_0.

The notion of quasidifferential is very useful for the theory of multiextremal nonsmooth problems of mathematical programming, especially for the problems of the type

$$\min_{x_1}(\max_{x_2}) \dots \min_{x_{k-1}} \max_{x_k} f(x_1, \dots, x_k)$$

with multiple sequential using of operations of "maxima" and "minima" on the parts of variables. In the book [DV 85] many interesting properties of QUD-functions and quasidifferentials are given. Note, that in general the problem of finding "good" algorithms for minimization (maximization) of QUD-functions is very hard.

Definition 17 *A function f is semiconvex at a point $x \in E^n$ if:*

(a) *f is Lipschitz in a neighborhood of x;*

(b) *f is quasidifferentiable at x;*

(c) *$f'(x, d) \geq 0$ implies $f(x + d) \geq f(x)$,*

(see [Mi 77a], [Mi 77b]).

A function which is both semiconvex and differentiable is called **pseudoconvex** [DV 85].

E.A. Nurminski [Nu 73] introduced the class of **weakly convex functions.**

Definition 18 *A continuous function f defined on E^n is called weakly convex if for any $x \in E^n$ one can find a nonempty set $M(x)$ of vectors g such that for all $y \in E^n$*

$$f(y) - f(x) \geq (g, y - x) - r(x, y),$$

where $r(x, y)/||x - y|| \to 0$ *as* $y - x \to 0$, *uniformly in all compact subsets of* E^n.

Weakly convex functions form a subclass of the family of quasidifferentiable functions. Both classes are closed with respect to the operation of taking the pointwise maximum (see [Nu 79]), which motivates their application in the analysis of minimax problems.

1.3 KUHN-TUCKER THEOREM AND DUAL BOUNDS

The general form of convex programming problems(CPP) that we shall study is the following: to find

$$f_0^* = \inf f_0(x), \; x \in E^n, \tag{1.10}$$

$$x \in X \subseteq E^n; \tag{1.11}$$

$$f_i(x) \le 0, \; i \in I = \{1, \dots, m\}. \tag{1.12}$$

where X is a closed convex subset of E^n, or$X \equiv E^n$, $f_\nu(x)$ are convex functions defined on open convex set $X' \supset X$ (it may be $X' \equiv E^n$) for all $\nu \in I \cup \{0\}$.

Let us costruct the Lagrange function $L(x, u)$ defined on $X' \times U$(where $u = \{u_i\}_{i \in I}$ is the variable vector of Lagrange multipliers, $U = \{u : u_i \ge 0 \; \forall i \in I\}$) in the following form:

$$L(x, u) = f_0(x) + \sum_{i \in I} u_i f_i(x)$$

The major necessary and sufficient conditions for the point $x \in X$ to be optimal in CPP (1.10)–(1.12) are given by the Kuhn–Tucker theorem (on saddle point):

Theorem 25 *Let* $x^* \in X$ *and exists a vector of Lagrange multipliers* $u^* \ge 0$ *such that the pair* (x^*, u^*) *is the saddle point of the Lagrange function in the following sense:*

$$L(x^*, u) \le L(x^*, u^*) \le L(x, u^*) \; for \; all \; x \in X, u \in U \tag{1.13}$$

Then x^* *is an optimal point of CPP problem(1.10)–(1.12).*

If additionally the Slater condition is satisfied, i.e. exists $\overline{x} \in X$ *such that*

$$f_i(\overline{x}) < 0 \; for \forall i \in I, \tag{1.14}$$

then if the point x^* *is optimal for the problem (1.10)–(1.12) there exists Lagrange multipliers vector* u^* *forming with* x^* *a saddle point* (x^*, u^*) *of the Lagrange function in the sense (1.13). Moreover, at this point*

$$x_i^* u_i^* = 0 \; for \; all \; i \in I. \tag{1.15}$$

Proof. The proof is divided in two parts.

(I) Let (x^*, u^*) be a saddle point of the Lagrange function in the sense (1.13). Since $L(x^*, u^*) \geq L(x^*, u)$ for all $u \in U$, $f_i(x^*) \leq 0$ for all $i \in I$ and if $f_{\overline{i}}(x^*) < 0$ for some $\overline{i} \in I$, then $u_{\overline{i}}^* = 0$. Hence, $x^* \in X$ is a feasible point for the problem (1.10)–(1.12) and $u_i^* f_i(x^*) = 0$ for all $i \in I$. So, $L(x^*, u^*) = f_0(x^*)$.

Consider an arbitrary feasible point $\overline{x}$. Since $\overline{x} \in X$ and $f_i(\overline{x}) \leq 0$ for all $i \in I$, using (1.13) we have:

$$f_0(x^*) = L(x^*, u^*) \leq L(\overline{x}, u^*) = f_0(\overline{x}) + \sum_{i \in I} u_i^* f_i(\overline{x}) \leq f_0(\overline{x}).$$

So $f_0(x^*) \leq f_0(\overline{x})$, and x^* is an optimal point of the problem (1.10)–(1.12).

(II) Let x^* be an optimal point for the problem (1.10)–(1.12). Consider the family of the ε-perturbed problems: find

$$F^*(\varepsilon) = \inf f_0(x), x \in E^n, \varepsilon = \{\varepsilon_i\}_{i \in I} \tag{1.16}$$

$$x \in X \subseteq E^n; \tag{1.17}$$

$$f_i(x) + \varepsilon_i \leq 0, i \in I = \{1, \ldots, m\}. \tag{1.18}$$

We shall use the following.

Lemma 1 *If the Slater condition is satisfied, then* $F^*(\varepsilon)$ *is a proper convex function in some neighborhood of the point* $0 \in E^m$.

Proof. From (1.14) one can deduce that there exists $\overline{x} \in X$ which is feasible for all ε-perturbed problems with $\varepsilon_i \leq f_i(\overline{x}) - f_i(x^*)$. Let $x^{(1)} \in X$

be a feasible point for ε-perturbed problem with $\varepsilon = \varepsilon^{(1)}$, $x^{(2)}$ be a feasible point for a similar problem with $\varepsilon = \varepsilon^{(2)}$.

Then for arbitrary α_1, α_2 such that $\alpha_1 + \alpha_2 = 1$, $\alpha_1 \geq 0; \alpha_2 \geq 0$, and $i \in I$

$$0 \geq \alpha_1 \left[f_i\left(x^{(1)}\right) + \varepsilon_i^{(1)}\right] + \alpha_2 \left[f_i\left(x^{(2)}\right) + \varepsilon_i^{(2)}\right]$$
$$\geq f_i\left(\alpha_1\left(x^{(1)}\right) + \alpha_2\left(x^{(2)}\right)\right) + \left[\alpha_1\varepsilon_i^{(1)} + \alpha_2\varepsilon_i^{(2)}\right].$$

So the domain of the function $F^*(\varepsilon)$ is convex. Further,

$$f_0\left(\alpha_1\left(x^{(1)}\right) + \alpha_2\left(x^{(2)}\right)\right) \leq \alpha_1 f_0\left(x^{(1)}\right) + \alpha_2 f_0\left(x^{(2)}\right) \tag{1.19}$$

The inequality (1.19) is true for arbitrary feasible points $x^{(1)}, x^{(2)}$ of the corresponding ε-perturbed problems.

So, $F^*\left(\alpha_1\varepsilon^{(1)} + \alpha_2\varepsilon^{(2)}\right) \leq \alpha_1 f_0\left(x^{(1)}\right) + \alpha_2 f_0\left(x^{(2)}\right)$. Since $x^{(1)}$ and $x^{(2)}$ are arbitrary feasible points of the corresponding ε-perturbed problems,

$$F^*\left(\alpha_1\varepsilon^{(1)} + \alpha_2\varepsilon^{(2)}\right) \leq \alpha_1 F^*\left(\varepsilon^{(1)}\right) + \alpha_2 F^*\left(\varepsilon^{(2)}\right),$$

i.e. F^* is a convex function, besides, the point $0 \in E^m$ is an interior point of $domF^*$. **Lemma is proved.**

In the point $\varepsilon = 0$ the problem (1.16)–(1.18) equals the initial problem (1.10)–(1.12).

Let $g_{F^*}(0)$ be a subgradient of the function F^* at the point 0. Let $g_{F^*}(0) = \{u_i^*\}_{i=1}^m = u^*$. It is easy to verify that F^* is a nondecreasing function for each $\varepsilon_i, i \in I$, so $u_i^* \geq 0$ for $\forall i \in I$. If $f_{\bar{i}}(x^*) < 0$ for some $\bar{i} \in I$, then the constraint $f_{\bar{i}}(x) \leq 0$ is not essential for the problem (1.10)–(1.12), and for the problem (1.16)–(1.18) with small enough $\varepsilon_{\bar{i}}$, because the function $f_{\bar{i}}(x)$ is continuous in the point x^*. Therefore for such $\bar{i}$, $u_{\bar{i}}^* = 0$, i.e.

$$u_i^* f_i(x^*) = 0 \text{ for all } i \in I.$$

Consider the inequality for $g_{F^*}(0)$:

$$F^*(\varepsilon) - F^*(0) \geq (g_{F^*}(0), \varepsilon) = \sum_{i \in I} u_i^* \varepsilon_i$$

for arbitrary ε.

Let $\bar{x} \in X$ and for all $i \in I$, $\varepsilon_i = -f_i(\bar{x})$. Then $\bar{x}$ is a feasible point for ε-perturbed problem, and $f_0(\bar{x}) \geq F^*(\varepsilon)$. So,

$$f_0(\bar{x}) - f_0(x^*) \geq -\sum_{i \in I} u_i^* f_i(\bar{x}),$$

and

$$f_0(\overline{x}) + \sum u_i^* f_i(\overline{x}) \geq f_0(x^*) + \sum u_i^* f_i(x^*)$$

For arbitrary $u \geq 0$, $\sum_{i \in I} u_i f_i(x^*) \leq 0$, because $f_i(x^*) \leq 0$ for all $i \in I$. So

$$f_0(x^*) + \sum_{i \in I} u_i f_i(x^*) \leq f_0(x^*) \leq f_0(x^*) + \sum_{i \in I} u_i^* f_i(x^*)$$

Thus, if the problem (1.10)–(1.12) has an optimal solution x^* and Slater's condition is true, one can find a vector of multipliers $u^* = \{u_i^*\}_{i \in I}$ such that the Lagrange function

$$L(x, u) = f_0(x) + \sum_{i \in I} u_i f_i(x)$$

has the following properties:

$$f_0(x^*) + \sum_{i \in I} u_i f_i(x^*) \leq f_0(x^*) + \sum_{i \in I} u_i^* f_i(x^*) \leq f_i(x) + \sum_{i \in I} u_i^* f_i(x)$$

for all $u \geq 0$ and $x \in X$, i.e. the (x^*, u^*) is the saddle-point of the Lagrange function in the sense (1.14).

The Kuhn-Tucker theorem is proved.

Remark 1. $\psi(u)$ is a result of minimization of the Lagrange function $L(x, u)$ which is linear in u with respect to parameter vector $x \in X$. So $\psi(u)$ is a concave function. Let $X(u) \subseteq X$ be a minimum set of the inner problem:

$$\inf_{x \in X} L(x, u).$$

Since for $u \geq 0$ $L(x, u)$ is convex in x, $X(u)$ is a convex subset of X. The vector $\{f_i(x(u))\}_{i=1}^m$ is the supergradient ψ in the point u for any $x \in X(u)$. If the solution of inner problem $x(u)$ is unique, then ψ is differentiable in the point u, and $g_\psi(u) = \{f_i(x(u))\}_{i=1}^m$, otherwise the supergradient set $G_\psi(u) = \cup_{x(u) \in X(u)} \{f_i(x(u))\}_{i \in I}$.

1.4 NONSMOOTH PENALTY FUNCTIONS

Let $s = \{s_i\}_{i \in I} \geq 0$ be a vector of penalty multipliers. One can consider the penalty function

$$S_1(x, s) = f_0(x) + \sum_{i \in I} s_i \max\{0, f_i(x)\},$$

associated with the CPP (1.10)–(1.12).

Suppose that this problem has a set X^* of optimal solutions and optimal Lagrange multipliers $u^* = \{u_i^*\}_{i\in I}$. Let

$$\rho^* = \min_{x\in X} S_1(x,s) \tag{1.20}$$

Theorem 26 *If $s_i \geq u_i^*$ for all $i \in I$, then $f_0^* = s^*$ and $X^* \subseteq X_1^*$ where X_1^* is the set of the minimum points of $S_1(x,s)$. If $s_i > u_i^*$ for all $i \in I$,then $X^* = X_1^*$, i.e. the CPP (1.10)–(1.12) is completely equivalent to the problem (1.20).*

Proof. For all feasible x and all $i \in I, \max\{0, f_i(x)\} = 0$, so if we denote by ρ_1^* the minimum of $S_1(x,s)$ at feasible x, $\rho_1^* \leq f_0^*$. But for all nonfeasible $x \in X$

$$s_i \max\{0, f_i(x)\} - u_i^* f_i(x) \geq (s_i - u_i^*)\max\{0, f_i(x)\} \geq 0 \tag{1.21}$$

Thus, $\rho^* = f_0^*$. If $s_i > u_i^*$, then for nonfeasible x $S_1(x,s) > L(x,u^*) \geq f_0^*$, and X_1^* has empty intersection with the set $X_2 = \{x \in X,\ x \text{ infeasible}\}$. That is why $X^* \equiv X_1^*$ in this case.

Let us consider another form of penalty function, namely, the maximum of the discrepancies in the constraints (1.12).

Choose a number $p > 0$ and set

$$S_{\max}(x,p) = f_0(x) + p\max[0, \max_{i\in I} f_i(x)] \tag{1.22}$$

Let $F(x) = \max_{x\in X}[0, \max_{i\in I} f_i(x)]$.

Theorem 27 *The $\min_{x\in X} S_{\max}(x,\overline{p})$ equals to f_0^* if $\overline{p} \geq \sum_{i\in I} u_i^*$ where $u = \{u_i^*\}_{i\in I}$ is an optimal vector of Lagrange multipliers for associated CPP (1.10)–(1.12). If $\overline{p} > \sum_{i\in I} u_i^*$, then all minimum points of $S_{\max}(x,\overline{p})$ are the optimal solutions of (1.10)–(1.12).*

Proof. For the set X_1 of feasible x $F(p) = 0$, so

$$\inf_{x\in X_1} f(x,\overline{p}) = \inf_{x\in X_1} f_0(x) = f_0^*.$$

For $x \in X \setminus X_1$

$$f_0(x) + \overline{p}F(x) \geq f_0(x) + \left(\sum_{i \in I} u_i^*\right) F(x) \geq f_0(x) + \sum_{i \in I} u_i^* f_i(x) \geq f_0^*. \quad (1.23)$$

So $\inf_{x \in X \setminus X_1} S_{\max}(x, \overline{p}) \geq f_0^*$, and $\inf_{x \in X} S_{\max}(x, p) = f_0^*$. If $p > \sum_{i \in I} u_i^*$ the inequality (1.23) is strong, and for all $x \in X \setminus X_1$, $f_0(x) + \overline{p}F(x) > f_0^*$, i.e. nonfeasible points are not the points of minimum of $S_{\max}(x, \overline{p})$. This proves Theorem 27.

One can use for CPP (1.10)–(1.12) the more general outer convex penalty functions.

Following [Bert 75a], let us define the function for CPP (1.10)–(1.12)

$$S(x) = f_0(x) + \sum_{i \in I} p_i[f_i(x)] \quad (1.24)$$

where $p_i(t), i \in I$ are convex functions satisfying $p_i(t) = 0$ for $t \leq 0$ and $p_i(t) > 0$ for $t > 0$. Suppose that there exists a point x^* which minimizes $S(x)$ for all $x \in I$.

Theorem 28 *For x^* to be a solution of original CPP (1.10)–(1.12) it is necessary that*

$$\lim_{t \to +0} \frac{p_t}{t} \geq \overline{u}_i^*, \ \forall i \in I,$$

where $\overline{u} = \{\overline{u}_i^*\}_{i \in I}$ is optimal Lagrange multiplier vector of the CPP (1.10)–(1.12). Moreover, if $\lim_{t \to +0} \frac{p_t}{t} \geq \overline{u}_i^*$, $\forall i \in I$, then the sets of minimum points of $S(x)$ are equal.

Theorem 28 can be generalized for a class of nonconvex problems of mathematical programming. Let us consider a family of mathematical programming problems MPP(b) (where $b = \{b_i\}_{i \in I}$ is parameter vector) of the following type: to find

$$\varphi^*(b) = \inf f_0(x), \ x \in X \subseteq E^n \quad (1.25)$$

subject to the constraints

$$f_i(x) + b_i \leq 0, \ i \in I \quad (1.26)$$

where $f_\nu(x)$ are continuous functions, $\nu \in \{0\} \cup I$.

Let $F(t) = \varphi^*(\bar{b}_1, \ldots, \bar{b}_j)$, where all $\bar{b}_i = t, i \in I$. It is clear that $F(t)$ is a monotone nonincreasing function in t.

Set

$$\mu = \sup_{t>0} \frac{F(0) - F(t)}{t} \geq 0. \tag{1.27}$$

Suppose, that $\mu < \infty$, and consider the penalty function

$$S_{\max}(x, p) = f_0(x) + p \max\{0, \max_{i \in I} f_i(x)\}.$$

Theorem 29 *[Psh 93]. If $\bar{p} \geq \mu$, then $S^*(\bar{p}) = \inf_{x \in X} S(x, \bar{p}) = \varphi^*(0)$. If $\bar{p} > \mu$, then all optimal solutions of the problem of minimization of $S(x, \bar{p})$ are the optimal solutions of original MPP (1.25)–(1.26) for $b = 0$.*

Proof. Let $X = X_1 \cup X_2$, where X_1 is the set $\{x \in X : f_i(x) \leq 0, \forall i \in I\}$ and $X_2 = X \backslash X_1$. It is obvious that $S(x, \bar{p}) = f_0(x)$ for $\forall x \in X_1$, so $\inf_{x \in X_1} S(x, \bar{p}) = \varphi^*(0)$. If $\bar{x} \in X \backslash X_1$, then $S(\bar{x}, p) = f_0(\bar{x}) + \bar{p} f_{i(\bar{x})}(\bar{x})$, where $i(\bar{x}) \in I$, $f_{i(\bar{x})}(\bar{x}) = \max_{i \in I} f_i(\bar{x})$. Since $\bar{x} \in X \backslash X_1$, $f_{i(\bar{x})}(\bar{x}) > 0$. Further, $f_i(\bar{x}) \leq f_{i(\bar{x})}(\bar{x})$. Set $\bar{b} = \{\bar{b}_i\}_{i \in I}; \bar{b}_i = 0$ when $f_i(\bar{x}) \leq 0$ and $\bar{b}_i = f_i(\bar{x})$, when $f_i(\bar{x}) > 0$. We can obtain, using(1.27):

$$f_0(\bar{x}) - \varphi^*(0) \geq \varphi^*(\bar{b}) - \varphi^*(0) \geq F[\max_{i \in I} f_i(\bar{x})] - \varphi^*(0) \geq -\mu \max_{i \in I} f_i(\bar{x})$$

So $S(\bar{x}, p) = f_0(\bar{x}) + p \max_{i \in i} f_i(\bar{x}) \geq \varphi^*(0) + (\bar{p} - \mu) \max_{i \in I} f_i(\bar{x}) \geq \varphi^*(0)$, if $\bar{p} = \mu$, and $> \varphi^*(0)$ if $p > \mu$. Therefore

$$\inf_{x \in X} S(x, \bar{p}) = \varphi^*(0) \text{ if } \bar{p} \geq \mu.$$

If $\bar{p} > \mu$, $S(x, \bar{p}) > \varphi^*(0)$ for $x \in X \backslash X_1$, so all minimal points of $S(x, \bar{p})$ are feasible. Theorem 29 is proved.

1.5 POLYHEDRAL SETS, MATRICES AND QUADRATIC FUNCTIONS

We need some definitions linking algebraic and geometrical descriptions of combinatorial problems.

Let A be a real $m \times n$-matrix and $b \in R^m$. The solution set of the system of linear inequalities $\{x \in R^n | Ax \leq b\}$ is called a **polyhedron**. A **bounded** polyhedron P (i.e. $P \subseteq S(0, R)$ for some $R > 0$) is called a **polytope**. Every polyhedron is the intersection of a finite number of halfspaces of the form: $\{x \in R^n | (a, x) \leq a_0; a \in R^n \backslash 0\}$. The hyperplane $(a, x) = a_0$ is equivalent to intersection of two halfspaces: $(a_0, x) \leq a_0$ and $(-a_0, x) \leq -a_0$. So linear equalities can be a part of the polyhedron description.

An inequality $(a, x) \leq a_0$ is called **valid** with respect to P, if polyhedron $P \subseteq \{x | (a, x) \leq a_0\}$. A set $F = \{x \in P | (a, x) = a_0\}$ if there exists a valid inequality $(a, x) \leq a_0$ for P such that $F = \{x \in P | (a, x) = 0\}$. We say that F is **induced** by $(a, x) \leq a_0$. The point $v \in P$ is called a vertex of P, if v is a face of P. A polyhedron is called **pointed** if it has a vertex. A nonempty polyhedron is pointed iff the corresponding matrix A has full column rank. Nonempty face of P of dimension $dim(P) - 1$ is called **facet**.

If the polyhedron P is full dimensional, then P has a representation $P = \{x \in R^n | Ax \leq b\}$ such that each of the inequalities $A_i x \leq b_i$ defines a facet of P and each facet of P is defined by exactly one of these inequalities. If we normalize the inequalities so that $||A_i||_\infty = 1$ for all rows of A, then this representation is unique and called **standard representation**.

If the polyhedron is not full-dimensional, then one can find a system $Cx = d$ of linear independent equations whose solution set L is the affine hull of P. We can consider P as a full dimensional polyhedron due to the corresponding affine set L and accordingly may introduce the **standard representation** for P.

Let $A \in R^{m \times n}$ be a matrix of rank r. A submatrix B of A is called a **basis** of A if it is a nonsingular $r \times r$ matrix. Let $A \in R^{m \times n}$ have full column rank and let $b \in R^n$. For any basis $B = A_I$, let b_I be the subvector of b, corresponding to B. Then the vector $B^{-1}b_I$ is called a bf basic solution of $Ax \leq b$. If $B^{-1}b_I$ satisfies $Ax \leq b$ it's called a bf basic feasible solution of this system. It is easy to see that a vector is a vertex of polyhedron $P = \{x | Ax \leq b\}$, A with full column rank, iff it is a basic feasible solution of $Ax \leq b$ for some basis B.

Every polytope is a convex hull of its vertices. Moreover, every point in a polytope $P \subseteq R^n$ is a convex combination of at most $dim(P) + 1$ affinely independent vertices of P (Caratheodory's theorem).

For nonempty set $S \subseteq R^n$

$$rec(S) := \{y \in R^n | x + \lambda y \in S, \forall x \in S, \forall \lambda \geq 0\}.$$

denotes **recession cone** of S, $rec(\emptyset) = \{0\}$. Every vector in $rec(S)$ represents a "direction to infinity" in S. A generalization of the Caratheodory theorem states that every point in a d-dimensional polyhedron P can be represented as a convex combination of affinely independent points $v_1, \ldots, v_s, v_1 + e_1, \ldots v_1 + e_t$, where $s + t \leq d + 1$, the points $v_1, \ldots, v_s$ are elements of minimal nonempty faces of P, and the points $e_1, \ldots, e_t$ are elements of minimal nonzero faces of the recession cone of P.

If polyhedron $P = \{x | Ax \leq b\}$ is nonempty

$$\begin{aligned} rec(P) &= \{x | Ax \leq 0\} \\ lineal(P) &= \{x | Ax = 0\} \end{aligned}$$

For any set $S \subseteq R^n$,

$$S^0 := \{y \in R^n | (y, x) \leq 0, \ \forall x \in S\}$$

is called the **polar cone** of S, and

$$S^* := \{y \in R^n | (y, x) \leq 1, \ \forall x \in S\}$$

is called the polar of S. S is a closed convex iff $(S^0)^0 = S$. S is a closed convex set containing zero iff $(S^*)^* = S$.

If $P = conv(V) + cone(E)$ is a nonempty polyhedron, then

$$\begin{aligned} P^0 &= \{x | (y, x) \leq 0, \ \forall y \in V \cup E\}; \\ P^* &= \{x | (v, x) \leq 1, \ \forall v \in V, \ (e, x) \leq 0, \ \forall e \in E\} \end{aligned}$$

For any set $S \subseteq R^n_+$, its **blocker** is

$$bl(S) := \{y \in R^n_+ | (y, x) \geq 1 \forall x \in S\}$$

and its **antiblocker** is

$$abl(S) := \{y \in R^n_+ | (y, x) \leq 1, \forall x \in S\}.$$

$bl(bl(S)) = S$ if and only if $S \subseteq R^n_+$ is closed, convex and up-monotone (i.e. $\forall y \in S$ all $x \geq y$ are in S). $abl(ablS)) = S$ iff $S \subseteq R^n_+$ is nonempty, closed, convex, and down-monotone in R^n_+ (i.e. for each $i \in S$ all vectors $x \in R^n_+$, $x \leq y$ are in S).

Now we give necessary notations and fragments of matrix and quadratic form theory (see [Ga 53]). Our notations are similar to ones that are appkied in the book [GLS 88].

For any set R, $R^{m \times n}$ denotes the set of $m \times n$-**matrices** with entries in R. For a matrix $A \in R^{m \times n}$, we usually assume that the row index set of A is $\{1, \ldots, m\}$ and that the column index set is $\{1, \ldots, n\}$. Unless specified otherwise, the elements or entries of $A \in R^{m \times n}$ are denoted by a_{ij}, $1 \leq i \leq m$, $1 \leq j \leq n$; we write $A = (a_{ij})$. Vectors with n components are also considered as $n \times 1$-matrices.

If I is a subset of the row index set M of a matrix A and J a subset of the column index set N of A, then A_{IJ} denotes the submatrix of A induced by those rows and columns of A whose indices belong to I and J, respectively. Instead of A_{MJ} (A_{IN} resp.) we frequently write $A_{.J}$ ($A_{I.}$ resp.). A submatrix A of the form A_{II} is called a **principal submatrix** of A. If $K = \{1, \ldots, k\}$, then A_{KK} is called the k-th **leading principal submatrix** of A. $A_{i.}$ is the i-th row of A (so it is a row vector), and $A_{.j}$ is the j-th column of A.

The **identity matrix** is denoted by I or, if we want to stress its dimension, by I_n. The symbol 0 stands for any appropriately sized matrix, which has all entries equal to zero, and similarly for any zero vector. The j-**th unit vector** in R^n, whose j-th component is one while all other components are zero, is denoted by e_j. If $x = (x_1, \ldots, x_n)^T$ is a vector, then the $n \times n$-matrix with the entries $x_1, \ldots, x_n$ on the main diagonal and zeros outside the main diagonal is denoted by $diag(x)$.

The **determinant** of an $n \times n$-matrix A is denoted by $\det(A)$. The **trace** of an $n \times n$-matrix A, denoted by $tr(A)$, is the sum of the elements of the main diagonal of the matrix A.

When using functions like det, tr, or $diag$ we often omit the brackets if there is no danger of confusion, i.e., we frequently write $\det A$ instead of $\det(A)$ etc.

The inverse matrix of an $n \times n$-matrix A is denoted by A^{-1}. If a matrix has an inverse matrix, then it is called **nonsingular**, and otherwise **singular**. An $n \times n$-matrix A is nonsingular if and only if $\det A \neq 0$.

The $n \times n$ real matrix O is called **orthogonal** if $O \cdot O^T = O^T \cdot O = I$. Each $n \times n$ real matrx A can be represented as $O_1 \cdot S_1$ or $S_2 \cdot O_2$, where S_1, S_2 are symmetric matrices, O_1, O_2 are orthogonal matrices.

If A is an $n \times n$-matrix, then every complex number λ with the property that there is a nonzero vector $u \in C^n$ such that $Au = \lambda u$ is called an **eigenvalue** of A. The vector u is called an **eigenvector** of A associated with λ. The function $f(\lambda) := \det(\lambda I_n - A)$ is a polynomial of degree n, called the **characteristic polynomial** of A. Thus the equation

$$\det(\lambda I_n - A) = 0$$

has n (complex) roots (multiple roots counted with their multiplicity). These roots are the (not necessarily distinct) n eigenvalues of A.

We will often consider **symmetric matrices** (i.e., $n \times n$-matrices $A = (a_{ij})$ with $a_{ij} = a_{ji}$, $1 \leq i \leq j \leq n$). It is easy to see that all eigenvalues of real symmetric matrices are real numbers.

There are useful relations between the eigenvalues $\lambda_1, \ldots, \lambda_n$ of a matrix A, its determinant and its trace, namely

$$\det A = \prod_{i=1}^{n} \lambda_i, \tag{1.28}$$

$$trA = \sum_{i=1}^{n} \lambda_i. \tag{1.29}$$

For any symmetric matrix S with eigenvalues $\lambda_1, \ldots, \lambda_n$ there is an orthogonal matrix O such that

$$OSO^T = diag\{\lambda, \ldots, \lambda_n\}$$

The $n \times n$ matrix U is called fully **unimodular** if all its quadratic induced submatrices have determinant with values 0, 1, or -1.

An $n \times n$-matrix A is called **positive definite (positive semidefinite** if A is symmetric and if $x^T Ax > 0$ for all $x \in R^n \backslash \{0\}$ ($x^T Ax \geq 0$ for all $x \in R^n$). If A is positive definite, then A is nonsingular and its inverse is also positive definite.

A is positive definite (semidefinite) iff

(i) All eigenvalues of A are positive (nonnegative) real numbers;

(ii) $A = B^T B$ for some nonsingular (any) matrix B;

(iii) $\det A_k > 0$ ($\det A_k \geq 0$) for all principal $k \times k$ submatrices of A ($k = 1, \ldots, n$).

If A is positive definite, then A^{-1} is also positive definite.

If A is positive semidefinite there is a positive definite $k \times k$ principal submatrix of A, where $k = rank A$. The set of all $n \times n$ positive semidefinite matrices is a convex closed cone in the linked space of all symmetric matrices.

With any $n \times n$ symmetric matrix A one can associate a quadratic form

$$Q_A(x) = \sum_{i,j=1}^{n} a_{ij}x_ix_j = (Ax, x).$$

The quadratic function $K(x) = (Ax, x) + (c, x) + d$ is convex (strongly convex) iff A is positive semidefinite (definite).

Each convex quadratic form $Q_A(x)$ can be represented as sum of k squares of linear forms where $k = rank(A)$.

1.6 ELEMENTS OF GRAPH THEORY

We describe main notions of graph theory similarly as it is made in [GLS 88].

A **graph** $G = (V, E)$ consists of a finite nonempty set V of **nodes(vertices)** and a finite set E of **edges**. With every edge, an unordered pair of nodes, called its **endnodes**, is assocoated and we say that an edge is incident to its endnodes. Note that we usually assume that the two endnodes of an edge are distinct, i.e., we do not allow loops, unless specified otherwise. If there is no danger of confusion we denote an edge e with endnodes i and j by ij. Two edges are called **parallel** if they have the same endnodes. A graph without parallel edges is called **simple**. The number of nodes of G is called the **order** of G.

A node that is not incident to any edge is called **isolated**. Two nodes that are joined by an edge are called **adjacent** or **neighbors**. For a node set W, $\Gamma(W)$ denotes the set of neighbors of nodes in W. We write $\Gamma(v)$ for $\Gamma(\{v\})$. The set of edges having a node $v \in V$ as one of their endnodes is denoted by $\delta(v)$. The number $|\delta(v)|$ is the **degree** of node $v \in V$. More generally, if $W \subseteq V$, then $\delta(W)$ denotes the set of edges with one endnote in W and the other endnote in $V \backslash W$. Any edge set of the form $\delta(W)$, where $\emptyset \neq W \neq V$, is called a **cut**.

A **matching** (or **1-matching**) M in a graph $G = (V, E)$ is a set of edges such that no two edges of M have a common endnode. A matching M is called **perfect** if every node is contained in one edge of M.

A simple graph is called **complete** if every two of its nodes are joined by an edge. The (up to isomorphism unique) complete graph of order n is denoted by K_n. A graph G whose node set V can be partitioned into two nonempty disjoint sets V_1, V_2 with $V_1 \cup V_2 = V$ such that no two nodes in V_1 and no two nodes in V_2 are adjacent is called **bipartite**. The node sets V_1, V_2 are called **color classes**, a bf 2-coloring, or a **bipartition** of V. If G is simple and bipartite, $|V_1| = m, |V_2| = n$, and every node in V_1 is adjacent to every node in V_2, then G is called **complete bipartite** and is denoted by $K_{m,n}$. The complete bipartite graph $K_{1,n}$ is called a **star**, and the star $K_{1,3}$ a **claw**.

If G is a graph, then the **complement** of G, denoted by $\overline{G}$, is the simple graph which has the same node set as G and in which two nodes are adjacent if and only if they are nonadjacent in G.

The **line graph** $L(G)$ of a graph G is the simple graph whose node set is the edge set of G and in which two nodes are adjacent if and only if the corresponding edges of G have a common endnode.

A **stable set** (**clique**) in a graph $G = (V, E)$ is a set of nodes any two of which are nonadjacent (adjacent). A **coloring** (**clique covering**) of a graph $G = (V, E)$ is a partition of V into disjoint stable sets (cliques).

Clearly, every graph $G = (V, E)$ can be drawn in the plane by representing nodes as points and edges as lines linking the two points which represent their endnodes. A graph is called **planar**, if it can be drawn in the plane in such a way that no two edges (i.e., the lines representing the edges) intersect, except possibly in their endpoints.

A **directed graph** (or **digraph**) $D = (V, A)$ consists of a finite nonempty set V of **nodes** and a set A of **arcs**. With every arc a, an ordered pair (u, v) of nodes, called its **endnodes**, is associated; u is the **initial endnode** (or **tail**) and v the **terminal endnode** (or **head**) of a. As in the undirected case, loops (u, u) will only be allowed if explicitly stated. If there is no danger of confusion we denote an arc a with tail u and head v by (u, v); we also say that a **goes from** u **to** v, that a is **incident from** u and **incident to** v, and that a **leaves** u and **enters** v. If there is an arc going from u to v, we say that u is a **predecessor** of v and that v is a **successor** of u.

If $D = (V, A)$ is a digraph and $W \subseteq V$, $B \subseteq A$, then $V(B)$ is the set of nodes occuring at least once as an endnode of an arc in B, and $A(W)$ is the set arcs with head and tail in W. Deletion and contraction of node or arc sets is defined in the same way as for undirected graphs.

If $D = (V, A)$ is a digraph, then the graph $G = (V, E)$ having an edge ij whenever $(i, j) \in A$ or $(j, i) \in A$ is called the **underlying graph** of D. A digraph has an "undirected property" whenever its underlying graph has this property. For example, a digraph is planar if its underlying graph is planar.

If $v \in V$, then the set of arcs having v as initial (terminal) node is denoted by $\delta^+(v)$ ($\delta^-(v)$); we set $\delta(v) := \delta^+(v) \cup \delta^-(v)$. The numbers $|\delta^+(v)|$, $|\delta^-(v)|$, and $|\delta(v)|$ are called the **outdegree, indegree**, and **degree** of v, respectively.

For any set $W \subseteq V$, $\emptyset \neq W \neq V$, we set $\delta^+(W) := \{(i, j) \in A | i \in W, j \notin W\}, \delta^-(W) := \delta^+(V \setminus W)$, and $\delta(W) := \delta^+(W) \cup \delta^-(W)$. If s, t are two different nodes of a digraph $D = (V, A)$, then an arc set $F \subseteq A$ is called an (s, t)-**cut** ([s,t]-**cut**) in D if there is a node set W with $s \in W$ and $t \notin W$ such that $F = \delta^+(W) (F = \delta(W))$. An arc set of the form $\delta^+(W), \emptyset \neq W \neq V$, is called a **directed cut** or **dicut** if $\delta^-(W) = \emptyset$, i.e., $\delta(W) = \delta^+(W) = \delta^-(V \setminus W)$. If $r \in V$, then every arc set of the form $\delta^-(W)$, where $\emptyset \neq W \subseteq V \setminus \{r\}$ is called an r-**rooted cut** or just r-**cut**.

In a graph or digraph, a **walk** is a finite sequence $W = v_0, e_1, v_1, e_2, v_2, \ldots, e_k, v_k$ ($k \geq 0$), beginning and ending with a node, in which nodes v_j and edges (arcs) e_j appear alternately, such that for $i = 1, 2, \ldots, k$ the endnodes of every edge (arc) e_i are the nodes v_{i-1}, v_i. The nodes v_0 and v_k are called the **origin** and the **terminus**, respectively, or the **endnodes** of W. The nodes $v_1, \ldots, v_{k-1}$ are called the **internal nodes** of W. The number k is the **length** of the walk. If (in a digraph) all arcs e_i are of the form (v_{i-1}, v_i), then W is called a **directed walk** or **diwalk**. An edge (arc) connecting two nodes of a walk but not contained in the walk is called a **chord** of the walk.

A walk in which all nodes (edges or arcs) are distinct is called a **path (trail)**. A path in a digraph that is a diwalk is called a **directed path** or **dipath**. If a node s is the origin of a walk (diwalk) W and t the terminus of W, then W is called an $[s, t]$-**walk** ((s, t)-**diwalk**).

Two nodes s, t of a graph G are said to be **connected** if G contains an $[s, t]$-path. G is called **connected** if every two nodes of G are connected. A digraph D is called **strongly connected** (or **diconnected**) if for every two nodes s, t of D there are an (s, t)-dipath and a (t, s)-dipath in D.

A graph G (digraph D) is called **k-connected** (**k-diconnected**) if every pair s, t of nodes is connected by at least k $[s,t]$-paths ((s,t)-dipaths) whose sets of internal nodes are mutually disjoint. The **components** of a graph are the maximal connected subgraphs of the graph. An edge e of G is called a **bridge** (or **isthmus**) if G–e has more components than G. A **block** of a graph is a node induced subgraph (W, F) such that either $F = \{f\}$ and f is a bridge, or (W, F) is 2-connected and maximal with respect to this property.

A walk is called **closed** if it has nonzero length an its origin and terminus are identical. A closed walk (diwalk) in which the origin and all internal nodes are different and all edges (arcs) are different is called a **circuit** (**dicycle** or **directed cycle**). A circuit (dicycle) of odd (even) length is called **odd** (**even**). A circuit of length three (five) is called a **triangle** (**pentagon**).

A walk (diwalk) that traverses every edge (arc) of a graph (digraph) exactly once is called an **Eulerian trail** (**Eulerian ditrail**). We refer to a closed Eulerian trail (ditrail) as an **Eulerian tour**. An **Eulerian graph** (**Eulerian digraph**) is a graph (digraph) containing an Eulerian tour.

A circuit of length n in a graph of order n is called a **Hamiltonian circuit**. A graph G that contains a Hamiltonian circuit is called **Hamiltonian**. Similarly, a digraph D is called **Hamiltonian** if it contains a Hamiltonian dicycle. Hamiltonian circuits or dicycles are often called **(Hamiltonian) tours**.

We shall use also the words "path, circuit, dipath, dicycle, Eulerian tour" to denote the edge or arc set of a path, circuit, dipath, dicycle, Eulerian tour. Thus, whenever we speak of the incidence vector of a circuit etc., we mean the incidence vector of the edge (arc) set of the circuit etc.

A **forest** is an edge set in a graph which does not contain a circuit. A connected forest is called a **tree**. A **spanning tree** of a graph is a tree containing all nodes of the graph. A digraph or arc set which does not contain a dicycle is called **acyclic**.

A **branching** B is an arc set in a digraph D that is a forest such that every node of D is the head of at most one arc of B. A branching that is a tree is called an **arborescence**. A branching that is a spanning tree of D is called a **spanning arborescence** of D. Clearly, in a spanning arborescence B of D every node of D is the head of one arc of B except for one node. This node is called the **root** of B. If r is the root of arborescence B we also say that B is **rooted at r** or **r-rooted**.

2

SUBGRADIENT AND ε-SUBGRADIENT METHODS

Let us consider a convex programming problem (CPP):

$$\text{find } f^* = \inf f_0(x), \; x = (x^{(1)}, \dots, x^{(n)}) \in E^n, \tag{2.1}$$

subject to constraints:

$$f_i(x) \quad \leq 0, \quad i \in \{1, 2, \dots, m\} = I; \tag{2.2}$$

$$x \in X \quad \subseteq E^n, \tag{2.3}$$

where $F_\nu(x), \nu \in \{\{0\} \cup I\}$, are proper convex fuctions determined on convex open neigborhood X^1 of a given convex set X. We introduce the notion of "information portion" for this problem.

Definition. *The portion of information $\Pi(\overline{x})$, $\overline{x} \in X^1$, for the problem (2.1)–(2.3) is the list of function and subgradient values, and answer for a question $\overline{x} \in X$ or not.*

$$\Pi(\overline{x}) = \Big\{ f_\nu(\overline{x}), \; \nu \in I \cup \{0\}; \; g_{f_\nu}(\overline{x}) \in G_{f_\nu}(\overline{x}), \; \nu \in I \cup \{0\}$$

$$\text{and } \chi(\overline{x}) \in \{0, 1\}, \; \chi(\overline{x}) = \left\{ \begin{matrix} 0 & , \textit{if } \overline{x} \notin X \\ 1 & , \textit{if } \overline{x} \in X \end{matrix} \right\} \Big\}$$

We suppose that all above mentioned values can be given with any necessary accuracy. This condition is very often used in the so-called information theory of algorithmic complexity, where we are not interested in time for calculating $\Pi(\overline{x})$, we only want to know how many information portions must be obtained to guarantee the necessary accuracy of results. Such a point of view can also be acceptable for description of principal schemes of algorithms (see [YN 76]).

Otherwise, for more detailed study of executable algorithms one can use the approximate information portion $\Pi_{\varepsilon,\delta}(\overline{x})$ that can be interpreted, for example, in the following way for given $\varepsilon, \delta > 0$:

$$\Pi_{\varepsilon,\delta}(\overline{x}) = \{f_\nu^\varepsilon(\overline{x}), g_{f_\nu}^\varepsilon(\overline{x}), \nu \in I \cup \{0\}, \chi_\delta(\overline{x})\}, \tag{2.4}$$

where

$$|f_\nu^\varepsilon(\overline{x}) - f_\nu(\overline{x})| \leq \varepsilon, \quad \min_{g \in G_{f_\nu}(\overline{x})} \|g_{f_\nu}^\varepsilon - g\| \leq \varepsilon;$$

$$\chi_\delta(\overline{x}) = \begin{cases} 1, & \text{if } \min_{x \in X} \|x - \overline{x}\| \leq \delta \\ 0, & \text{otherwise} \end{cases} \tag{2.5}$$

Now we intend to describe in general terms the structure of subgradient-type methods for solving CPP. From the beginning we must define the subroutines for obtaining the portion of information $\Pi_{\varepsilon,\delta}(\overline{x})$ for $\overline{x} \in X$ and for choosing the starting point $\overline{x}_0 \in X$ (or several starting points $\overline{x}_0^{(i)}, i \in I_0$). Then we need: a) calculation of the starting points corresponding to information portions; b) processing phase of the information portions. As a result of this phase we decide what portions of information or their parts we must record in memory for the next calculations and choose several points (or a point) in which we calculate new portions of information; c) checking the stopping criterion and output of the results.

The phases a), b), and c) are cyclically repeated untill after the phase c) we obtain the signal to stop for one of the following reasons:

1) the CPP is infeasible;

2) it is shown that point $\overline{x}$ obtained by the algorithm satisfies conditions of approximate optimal solution with given accuracy;

3) we cannot continue the processing because the time given for calculations is exhausted. Then we try to output an approximate solution $\overline{x}$ and the constant which characterizes the accuracy of the objective function of the outputed solution $\overline{x}$ with respect to the optimal value, and the level of satisfying the constraints.

In the framework of the described general scheme it is possible to construct an enormous number of subgradient-type algorithms. Below we shall describe **the most known of them**.

2.1 THE SUBGRADIENT METHOD

For simplicity let us consider the unconstrained problem of minimization of the convex function $f(x)$ defined on n-dimensional Eucledian space E^n.

The **subgradient method** is an algorithm which generates a sequence $\{x_k\}_{k=0}^{\infty}$ according to the formula:

$$x_{k+1} = x_k - h_k(x_k)g_f(x_k), \tag{2.6}$$

where x_0 is a given starting point, $g_f(x_k)$ is an arbitrary subgradient from the subgradient set $G_f(x_k), k = 0,1,2,\ldots$. At first sight formula (2.6) does not differ from the formulae defining the usual gradient methods for smooth optimization, especially if we take into account that the subgradient $g_f(x)$ is almost everywhere identical with the gradient, because a convex function is continuosly differentiable almost everywhere. But the rules for determining stepsize $h_k(x_k)$ must be entirely different here. Indeed, the stepsize rules used for continuosly differentiable functions, e.g. $h_k(x_k) = h$ for some sufficiently small constant h, as in the simple gradient method, or the minimization of $f(x_k - hg_f(x_k))$ with respect to h, as in the method of steepest descent, do not apply to all convex functions. Constant stepsizes may be unsuitable because the function f may be nondifferentiable at the optimal point and then $\{g_f(x_k)\}_{k=0}^{\infty}$ does not necessarily tend to zero, even if $\{x_k\}_{k=0}^{\infty}$ converges to the optimal point. On the other hand the exact directional minimization may cause the process to converge to a nonstationary point, as shown in example below.

Example [Wol 75]. Consider the function of two variables

$$f(x_1,x_2) = \begin{cases} 5(9x_1^2+16x_2^2)^{1/2} & \text{, for } x_1 > |x_2|, \\ 9x_1 + 16|x_2| & \text{, for } x_1 \leq |x_2|. \end{cases}$$

If $x_1 > 0$ then $f(x_1,x_2) = \max\{5(9x_1^2+16x_2^2)^{1/2}, 9x_1+16|x_2|\}$. Thus f is convex as the pointwise maximum of two convex functions. We shall prove that f is continuously differentiable in this area. The set of possible discontinuities of the gradient satisfies the condition $x_1 = |x_2|$. Let us compute the gradients

$$\partial[5(9x_1^2+16x_2^2)^{1/2}] = \left\{\frac{5\cdot 9x_1}{(9x_1^2+16x_2^2)^{1/2}}; \frac{5\cdot 16x_2}{(9x_1^2+16x_2^2)^{1/2}}\right\};$$

$$\partial[5(9x_1^2+16x_2^2)^{1/2}] = \{9; 16\mathrm{sign}x_2\} \text{ for } x_1 = |x_2|;$$

$$\partial[9x_1 + 16|x_2|] = \{9; 16\mathrm{sign}x_2\}.$$

Thus there is no discontinuity of the gradient on the line $x_1 = |x_2|$, so f is continuously differentiable on the halfplane $x_1 > 0$. If we choose a starting point in the area $x_1 > |x_2| > (\frac{9}{16})^2 x_1$ then, as one can easily verify, the method of steepest descent will follow a polygonal path having successive orthogonal segments whose vertices converge to the point $\{0,0\}$. We know that for continuously differentiable functions the set of accumulation points of a sequence generated by the method of steepest descent consists of stationary points. But in our example the method converges to a nonstationary point $\{0,0\}$. This results from the discontinuity of the gradient at that point (the gradient is discontinuous on the ray $x_2 = 0, x_1 \leq 0$, while $\lim_{x_1 \to -\infty} f(x_1, 0) = -\infty$). This example shows that the method of steepest descent may be unsuitable for minimizing nondifferentiable convex functions.

Therefore, when the subgradient method is applied to nonsmooth functions, specific difficulties in stepsize selection arise. The most useful property that gives us a possibility to construct a good sequence of stepsizes is the following:

if x^* is a minimum point of $f(\cdot)$, $\overline{x}$ is not a minimum point, then the angle made by vectors $g_f(\overline{x})$ and $\overline{x} - x^*$ is acute. That is why if we move from the point $\overline{x}$ in direction opposite to subgradient with small enough step, then distance to x^* is cut down. To get an insight into this subject let us consider the subgradient method with constant stepsizes, i.e. with

$$h_k(x_k) = \frac{h}{||g_f(x_k)||}, h > 0, k = 0, 1, 2, \ldots; \; g_f(x_k) \neq 0 \tag{2.7}$$

Theorem 30 *[Shor 64]. Let f be a convex function with a nonempty set of minimum points X^*. Suppose that a sequence $\{x_k\}$ is calculated by the subgradient method (2.6) with stepsizes (2.7). Then for any $\varepsilon > 0$ and any $x^* \in X^*$ one can find $k = \overline{k}$ and $\overline{x}$ such that $f(\overline{x}) = f(x_{k^*})$ and $||\overline{x} - x^*|| < h(1+\varepsilon)/2$.*

Proof. If $g_f(x_{k^*}) = 0$ for some k^*, then $f(x_{k^*}) = f(x^*)$, for all $x^* \in X^*$, and one may take $\overline{x} = x^*$ for any $x^* \in X^*$. If $g_f(x_k) \neq 0$ for all $k = 0, 1, \ldots$, then

$$x_{k+1} = x_k - \frac{h g_f(x_k)}{||g_f(x_k)||}, k = 0, 1, 2, \ldots$$

For any $k, k = 0, 1, 2, \ldots$ let $z_k(x^*) = x_k - x^*$. Then

$$||z_{k+1}(x^*)||^2 = ||z_k(x^*) - \frac{h g_f(x_k)}{||g_f(x_k)||}||^2 = \tag{2.8}$$

$$= ||z_k(x^*)||^2 + h^2 - 2h \left(z_k(x^*), \frac{g_f(x_k)}{||g_f(x_k)||} \right).$$

We shall estimate $(z_k(x^*), \frac{g_f(x_k)}{\|g_f(x_k)\|})$, denoted by $a_k(x^*)$ which is equal to the distance from x^* to the supporting hyperplane $L_k = \{x : (g_f(x_k), x - x_k) = 0\}$. Define $U_k = \{x : f(x) = f(x_k)\}$ and $b_k(x^*) = \min_{x \in U_k} \|x - x^*\|$. Since both the set U_k and the point x^* lie on the same side from L_k and any segment joining x^* with a point of L_k passes through U_k, we have

$$a_k(x^*) \geq b_k(x^*).$$

From (2.8) we obtain:

$$\|z_{k+1}(x^*)\|^2 \leq \|z_k(x^*)\|^2 - 2h\left(b_k(x^*) - \frac{h}{2}\right)$$

Let us introduce some $\varepsilon > 0$. If $b_k(x^*) \geq \frac{h}{2}(1+\varepsilon)$ for all $k = 0, 1, 2, \ldots$, then

$$\|z_{k+1}(x^*)\|^2 \leq \|z_k(x^*)\|^2 - \varepsilon h^2 \leq \|z_0 - x^*\|^2 - \varepsilon(k+1)h^2 \qquad (2.9)$$

for all k. But $\|z_{k+1}(x^*)\|^2 \geq 0$ for $k = 0, 1, 2, \ldots$ that contradicts to (2.9). Hence there exists k^* such that

$$b_k^* = \min_{x \in U_k^*} \|x - x^*\| < \frac{h(1+\varepsilon)}{2}$$

as required.

From the above theorem we can easily deduce the following results.

Corollary 1 *Under the assumptions of Theorem 30 for any $\delta > 0$ there exists $h_\delta > 0$ such that if the subgradient method is used with stepsizes $h_{k+1}(x_k) = \frac{h}{\|g_f(x_k)\|}$, $0 < h < h_\delta$, then either $x_{k^*} \in X^*$ for some k^* or there is a subsequence $\{k_i\}_{i=1}^{\infty}$ such that $f(x_{k_i}) - \min_{x \in E^n} f(x) < \delta$.*

Corollary 2 *If the set X^* of minimum points of the function f contains a sphere with a radius $r > \frac{h}{2} > 0$ and the subgradient method (2.7) is applied with $h_{k+1}(x_k) = \frac{h}{\|g_f(x_k)\|}$ then there exists k^* such that $x_k^* \in X^*$.*

It can be seen from the proof of Theorem 30 that if we use subgradient method and move with the constant stepsizes small enough we can get to any level set $L(c) = \{x : f(x) \leq \min_{x \in X} f(x) + c, c > 0\}$. To obtain convergence theorems in standard form it is necessary to require that in formula $h(x_k) = \frac{h_k}{\|g_f(x_k)\|}$, $\{h_k\}_{k=0}^{\infty}$ tends to zero. The reduction of steplengths, however, should not be too rapid. In particular, if the sum S of the series $\sum_{k=0}^{\infty} h_k$ is finite the sequence $\{x_k\}_{k=0}^{\infty}$ has a limit, but this limit lies outside X^* if $\min_{x^* \in X} \|x_k - x^*\| > S$. So we have arrived at the "classical" conditions: $h_k > 0, h_k \to 0$ as $k \to \infty$ and $\sum_{k=1}^{\infty} h_k = +\infty$.

Theorem 31 *[Shor 69] Let f be a convex function defined on E^n, which has a bounded set of minimum points X^* and let a sequence of positive numbers $\{h_k\}_{k=0}^{\infty}$ $k = 0, 1, 2, \ldots$ satisfy the conditions:*

$$\lim_{k \to \infty} h_k = 0, \sum_{k=0}^{\infty} h_k = +\infty.$$

Then for any $x_0 \in E^n$ the sequence $\{x_k\}_{k=0}^{\infty}$ generated according to the formula

$$x_{k+1} = x_k - h_k \frac{g_f(x_k)}{\|g_f(x_k)\|}$$

satisfies one of the following conditions: either an index $\bar{k}$ exists such that $x_{\bar{k}} \in X^*$, or $\lim_{k \to \infty} \min_{y \in X^*} \|x_k - y\| = 0$ and $\lim_{k \to \infty} f(x_k) = \min_{x \in E^n} f(x) = f^*$.

There are many alternative proofs of this theorem or its modifications (see [Erm 66], [Po 67], [Shor 69]). All the proofs are based on the analysis of the behavior of the sequence $\min_{x \in X^*} \|x_k - x\|^2$ and are similar to the proof of Theorem 30. One can find the proof of Theorem 31 in the book [Shor 85].

A similar result holds for another stepsize rule which does not use the scaling factors $\|g_f(x_k)\|^{-1}$ for normalizing the subgradient.

Theorem 32 *[Shor 69] Let the assumptions of the theorem (32) be satisfied and suppose that a sequence $\{x_k\}_{k=1}^{\infty}$ is generated according to the formula:*

$$x_{k+1} = x_k - h_k g_f(x_k), \; k = 0, 1, 2, \ldots, \tag{2.10}$$

where $x_0 \in E^n$ is an arbitrary starting point, $h_k > 0, h_k \to 0$, and $\sum_{k=1}^{\infty} h_k = +\infty$.

Then either

(a) *the sequence $\{g_f(x_k)\}_{k=0}^{\infty}$ is bounded and the algorithm converges in the sense that*

$$\lim_{k \to \infty} \min_{x \in X^*} \|x_k - x\| = 0, \; \lim_{k \to \infty} f(x_k) = f^*,$$

or

(b) *the sequence* $\{g_g(x_k)\}_{k=0}^{\infty}$ *is unbounded and there is no convergence.*

Let us observe that if the function f is piecewise linear(polyhedral) with a finite number of pieces, then $\{g_f(x_k)\}_{k=0}^{\infty}$ is always bounded, i.e. case (a) holds.

Note also that in general case for any x_0 one can always find $\delta > 0$ such that if $\max h_k \le \delta$ then case (a) must occur. Indeed, consider for some $a > 0$ the sets $U(a) = \{x : f(x) = f(x_0) + a\}$ and $S(a) = \{x : f(x) \le f(x_0) + a\}$. Since f is defined on the whole of E^n and X^* is bounded, the sets $U(a)$ and $S(a)$ are compact. Consider the minimal closed sphere S^* centered at $x^* \in X^*$ and containing $S(a)$, and define $\bar{c} = \max_{x \in S^*}\{\|g_f(x)\|\}$ and distances from set to set $\rho_1 = \rho(U(a), U(\frac{a}{2}))$, $\rho_2 = \rho(x^*, U(\frac{a}{2}))$. Take $\delta = \min(\frac{\rho_1}{c}, \frac{\rho_2}{c})$. If $h_k \le \delta$ then $\bar{h}_k = h_k\|g_f(x_k)\| \le \min(\rho_1, \rho_2)$ for all $x_k \in S^*$. We shall prove that $x_k \in S^*$ for all k. Indeed, $x_0 \in S^*$. If $x_k \in S(a) \backslash S(a/2)$ then (2.8) yields $\rho(x_{k+1}, x^*) \le \rho(x_k, x^*)$ and $x_{k+1} \in S^*$. If $x_k \in S(a/2)$ then $x_{k+1} \in S(a) \subseteq S^*$, since $\|x_{k+1} - x_k\| \le \rho_1$. Thus we always have x_k in S^*. But then the sequence $\{\|g_f(x_k)\|\}_{k=1}^{\infty}$ is bounded and case (a) occurs.

The above considerations make it possible to modify Theorem 32 as follows.

Theorem 33 *Under the assumptions of Theorem 32, for any starting point* x_0 *the sequence* $\{x_k\}_{k=0}^{\infty}$ *generated according to the formula:*

$$x_{k+1} = \begin{cases} x_k - h_k g_f(x_k) & , \quad \text{if } h_{k+1}\|g_f(x_k)\| \le c \\ x_0 & , \quad \text{otherwise} \end{cases}$$

where c *is a positive number,* $h_k > 0, h_k \to 0$ *and* $\sum_{k=1}^{\infty} h_k = +\infty$*, satisfies relations*

$$\lim_{k\to\infty} \min_{x \in X^*} \|x_k - x\| = 0; \ \lim_{k\to\infty} f(x_k) = f^*.$$

We can modify the Theorem 33 for variable step multipliers.

Theorem 34 *Suppose that the set* X^* *of minimum points of* f *contains a sphere* S_r *of radius* $r > 0$ *and the subgradient method in the form (2.6) uses step multipliers* $h_k > 0$ *satisfying* $\sum_{k=0}^{\infty} h_k = +\infty$ *and* $\lim_{k\to\infty} \sup h_k < 2r$*. Then for any* $x_0 \in E^n$ *there exists a finite index* $k(x_0)$ *such that* $x_{k(x_0)} \in X^*$*.*

The previous theorems have a relation to the problems of unconstrained minimization of convex functions. Using the methods of nonsmooth penalty functions we can often reduce the CPP to an unconstrained problem. But there is another approach to solution of CPP based on so-called projection subgradient method proposed by B.T. Polyak [Po 67]. Originally this method was designed for minimization of convex functionals defined on a Hilbert space, subject to rather general convex constraints. We shall interpret these results for the case of finite dimensional CPP and obtain the generalization of Theorem 31 for constrained case.

Let f be a convex function defined on a convex subset Q of n-dimensional Eucleadean space E^n. Assume that $Q = Q_1 \cap Q_2$, where $Q_1 = \{x : \varphi(x) \leq 0\}$, φ is a convex function defined on E^n, Q_1 has a nonempty interior $Q_1^{(0)}$, Q_2 is convex and closed, and $Q_1^{(0)} \cap Q_2 \neq \emptyset$.

For closed set $M \subset E^n$ we denote by $P_M(x)$ the projection of a point x on M, i.e.

$$P_M(x) \in M, \|x - P_M(x)\| = \inf_{y \in M} \|x - y\|$$

The subgradient projection method consists in generating $\{x_k\}_{k=1}^{\infty}$ by the formula

$$x_{k+1} = P_{Q_2}(x_k - h_k \frac{g(x_k)}{\|g(x_k)\|}) \tag{2.11}$$

where

$$g(x_k) = \begin{cases} g_f(x_k), & \text{if } x_k \in Q_1, \\ g_\varphi(x_k), & \text{if } x_k \notin Q_1, \end{cases}$$

x_0 belongs to Q_2.

The following theorem is a partial case of convergence results for projective subgradient method [Po 67].

Theorem 35 *Suppose that $h_k > 0, k = 0, 1, 2, \ldots, \lim_{k\to\infty} h_k = 0$ and $\sum_{k=0}^{\infty} h_k = +\infty$. Then for any x_0 one can find a subsequence $\{x_{n_k}\}$ such that $x_{n_k} \in Q$ and $\lim_{k\to\infty} f(x_{n_k}) = f^* = \inf_{x\in Q} f(x)$.*

Proof. One can see that operation of projection of the point $\bar{x}$ on closed convex set M does not increase the distance to any point $y \in M$. Indeed, if $\bar{x} \in M$, then $P_M(\bar{x}) = \bar{x}$, and so $\|y - \bar{x}\| = \|y - P_M(\bar{x})\|$. If $\bar{x} \notin M$, then $P_M(\bar{x}) - \bar{x}$ is the gradient of a supporting plane in the point $P_M(\bar{x})$, which separates M from $\bar{x}$. So the angle in the point $P_M(\bar{x})$ of the triangle made by the points $\overline{x}, P_M(\overline{x})$ and $y \in \overline{M}$ is more than $\frac{\pi}{2}$, (except for the trivial case when $y = P_M(\overline{x})$). Hence $\|\overline{x} - y\| > \|P_M(\overline{x}) - y\|$, as required. Now we can continue the proof of the theorem.

Take any $\alpha > f^*$ and define the set $S_\alpha = \{x : x \in Q_1, f(x) \leq \alpha\}$. Denote by $S_\alpha^{(0)}$ the interior of S_α. One can always find a point $\tilde{x} \in Q_2 \cap S_\alpha^{(0)}$. Take $\rho > 0$ such that $x \in S_\alpha$ for all $\|x - \tilde{x}\| \leq \rho$.

Suppose that $x_k \notin S_\alpha$ for all k. If $x_{\overline{k}} \notin Q_1$ then $S_\alpha \subset \{x : f(x) \leq f(x_{\overline{k}})\}$. In both cases $(-g(x_{\overline{k}}), x_k) \leq (-g(x_{\overline{k}}), \tilde{x} + \frac{\rho g(x_{\overline{k}})}{\|g(x_{\overline{k}})\|})$. Hence, $\|x_{k+1} - \tilde{x}\|^2 = \|P_{Q_2}(x_k - h_k \frac{g(x_k)}{\|g(x_k)\|} - \tilde{x}\|^2 \leq \|x_k - h \frac{g(x_k)}{\|g(x_k)\|} - \tilde{x}\|^2 = \|x_k - \tilde{x}\|^2 + h_k^2 + 2h_k(g(x_k), x_k - \tilde{x} + \rho \frac{g(x_k)}{\|g(x_k)\|}) - 2\rho h_k \|g(x_k)\| \leq \|x_k - \tilde{x}\|^2 + \|h_k^2\| - 2\rho h_k \frac{g(x_k)}{\|g(x_k)\|}$. Take N such that $h_k \leq \rho$ for all $k \geq N$ and sum up the above inequalities from $k = N$ to $k = N + m$.

We obtain

$$0 \leq \|x_{N+m+1} - \tilde{x}\|^2 \leq \|x_N - \tilde{x}\|^2 +$$

$$+ \sum_{k=N}^{N+m} h_k(h_k - r_\rho) \leq \|x_N - \tilde{x}\|^2 - \rho \sum_{k=N}^{N+m} h_k, m = 1, 2, \ldots,$$

which contradicts to the divergence of $\sum_{k=1}^{\infty} h_k$. Consequently, for every $\alpha > f^*$ one can find $k(\alpha)$ such that $x_{k(\alpha)} \in S_\alpha$, i.e. a subsequence $\{x_{n_i}\}_{i=1}^{\infty}$ exists such that $x_{n_i} \in Q$ and $\lim_{i \to \infty} f(x_{n_i}) = f^*$. **The proof is complete.**

Note that in general case of the set M the calculation of $P_M(\overline{x})$ is not an easy problem. Therefore the operation of projection is used in subgradient method as a rule when the set Q_2 is simple for projection: for example, when Q_2 is sphere, or the set $\Pi_{(a,b)} = \{x : a \leq x \leq b\}$, or $Q^+ = \{x : x \geq 0\}$.

For example, $P_{\Pi[a,b]}(\overline{x}) = \overline{y}$, where $\overline{y} = (\overline{y}_1, \ldots, \overline{y}_n)$, $\overline{y}_i = a_i$, if $\overline{x}_i \leq a_i$; $\overline{y}_i = b_i$, if $\overline{x}_i \geq b_i$; $\overline{y}_i = \overline{x}_i$, if $a_i \leq \overline{x}_i \leq b_i$. $P_{Q^+}(\overline{x}) = \{y_1, \ldots, y_n\}$, where $y_i = \overline{x}_i$, if $\overline{x}_i \geq 0$, and $y_i = 0$, if $\overline{x}_i \leq 0$.

Under certain additional assumptions it is possible to construct the versions of the subgradient method which converge at a linear rate. To begin with we introduce the notion of **generelized condition number** for the convex function f defined on E^n. Let us recall the classical definition of **condition number** for a positive definite symmetric $n \times n$ matrix $A = \{a_{ij}\}_{i,j=1}^n$. Let $\lambda_1(A)$ and $\lambda_n(A)$ be correspondingly maximal and minimal eigenvectors of matrix A.

Then **condition number** $\rho(A)$ of A is given by the following formula:

$$\rho(A) = \frac{\lambda_1(A)}{\lambda_n(A)}.$$

For a nonsymmetric $n \times n$ real matrix $B = \{b_{ij}\}_{i,j=1}^n$ **condition number** $\rho(B)$ is defined as

$$\rho(B) = \sqrt{\frac{\lambda_1(BB^T)}{\lambda_n(BB^T)}}$$

The notion of condition number plays an important role in analysis of systems of linear algebraic equations and in the theory of solution stability for studying convergence of iterative methods (see [FF 60]).

For example, L.V. Kantorovich proved the theorem about the rate of convergence of steepest descent method for finding the minimum of positive definite quadratic functions [Kan 47].

Theorem 36 *Let $f(x) = \frac{1}{2}(Ax, x), x \in E^n$, where A is positive definite symmetric matrix, $\{x_x\}_{k=0}^{\infty}$ is the sequence generated by the method of steepest descent with starting point x_0. Then*

$$\|x_k\| \leq \sqrt{\frac{M}{m}} \left(\frac{M-m}{M+m}\right)^k \|x_0\| \tag{2.12}$$

where $M = \lambda_1(A)$, $m = \lambda_n(A)$; or

$$\|x_k\| \leq \sqrt{\rho(A)} \left(\frac{\rho(A)-1}{1+\rho(A)}\right)^k \|x_0\| \tag{2.13}$$

When we use the method of steepest descent we move from the point $\overline{x}$ in the direction $-g_f(\overline{x}) = -A\overline{x}$ while the direction to the minimum point is $-\overline{x}$. Let $\varphi(\overline{x})$ be an angle made by vectors $-g_f(\overline{x})$ and $-\overline{x}$. Let us calculate the minimal value of $\cos\varphi(\overline{x})$ for all $\overline{x} \neq 0$

$$\cos\varphi(\overline{x}) = \min \frac{(Ax, x)}{\|Ax\| \|x\|}$$

Now choose a vector $\overline{y}$ different from all eigenvectors of A and consider the two dimensional subspace $E(\overline{y})$ spanned by $\overline{y}$ and $A\overline{y}$. Let us denote by $\tilde{A}$ the matrix of the two dimensional quadratic form defined on $E(\overline{y})$ such that

$(\tilde{A}x, x) = (Ax, x)$ for $x \in E(\overline{y})$. It easy to see that the operator $\tilde{A}$ defines the positive definite form on the subspace $E(\overline{y})$ and its eigenvalues u_1 and $u_2, (u_1 \geq u_2)$ satisfy the inequalities $\lambda_1(A) \geq u_1 \geq u_2 \geq \lambda_n(A)$. Let x_1 and x_2 be the orthonormal eigenvectors of $\tilde{A}$ in $E(\overline{y}_1)$ and assume that $||\overline{y}|| = 1$. Let us express $\overline{y}$ as $\overline{y} = \alpha_1 x_1 + \alpha_2 x_2$, where $\alpha_1^2 + \alpha_2^2 = 1$. Then

$$\frac{(A\overline{y}, \overline{y})}{||A\overline{y}||||\overline{y}||} = \frac{\alpha_1^2 u_1 + \alpha_2^2 u_2}{\sqrt{\alpha_1^2 u_1^2 + \alpha_2^2 u_2^2}} \tag{2.14}$$

Minimizing the right side of (2.14) subject to the constraint $\alpha_1^2 + \alpha_2^2 = 1$, we obtain the inequality

$$\frac{(A\overline{y}, \overline{y})}{||A\overline{y}||||\overline{y}||} \geq \frac{2\sqrt{u_1 u_2}}{u_1 + u_2}$$

with equality occuring for

$$\alpha_1^2 = \frac{u_2}{u_1 + u_2}, \quad \alpha_2^2 = \frac{u_1}{u_1 + u_2}$$

This immediately implies that

$$\min_{x, x \neq 0} \frac{(Ax, x)}{||Ax||||x||} = \frac{2\sqrt{\lambda_1(A)\lambda_n(A)}}{\lambda_1(A) + \lambda_n(A)}$$

with the minimum attained at

$$\overline{x} \in \overline{X} = \overset{+}{-}\sqrt{\frac{\lambda_n(A)}{\lambda_1(A) + \lambda_n(A)}} s_1 \overset{+}{-}\sqrt{\frac{\lambda_1(A)}{\lambda_1(A) + \lambda_n(A)}} s_n \tag{2.15}$$

where $\overset{+}{-}s_1$ and $\overset{+}{-}s_n$ are the orthonormal eigenvectors of A, which correspond to the largest and smallest eigenvalues, respectively.

Thus, $\overline{\varphi}(A) = \arccos \frac{2\sqrt{\lambda_1(A)\lambda_n(A)}}{\lambda_1(A)+\lambda_n(A)}$ is the maximal angle that is made by direction of antigradient and direction to the minimum of positive definite quadratic form with matrix A from any point $\overline{x} \neq 0$. It is easy to calculate that

$$\sin \overline{\varphi}(A) = \sqrt{1 - \cos^2 \overline{\varphi}(A)} = \frac{\lambda_1(A) - \lambda_n(A)}{\lambda_1(A) + \lambda_n(A)} = \frac{\rho(A) - 1}{\rho(A) + 1}$$

This implies that $\rho(A) = \frac{1+\sin \overline{\varphi}(A)}{1-\sin \overline{\varphi}(A)}$.

Remark. It is interesting to note that denominator $q(A) = \frac{\rho(A)-1}{\rho(A)+1}$ that characterizes the speed of convergence of the steepest descent method in formula (2.13) is equal to $\sin \overline{\varphi}(A)$.

For the starting point $x_0 = c\overline{x}$, where $c \neq 0$, $\overline{x}$ is one of the vectors given by formula (2.15), the following equality holds:

$$||x_{k+1}|| = \sin\overline{\varphi}(A) \cdot ||x_k||, k = 0, 1, \dots$$

It is known (see [Ak 59], [Em 77]) that if $n \geq 3$, then for almost all starting points x_0

$$\lim_{k\to\infty} \frac{||x_{k+1}||}{||x_k||} = q(A)$$

Thus, the estimate (2.13) **cannot be improved essentially**.

Now we can introduce the notion of **generalized condition number** for convex function $f(x)$, defined on E^n.

Definition. *Let x^* be the unique minimum point of f, $S(x^*, R) = \{x : ||x - x^*|| \leq R\}$, $\varphi(\overline{x}) = \arccos \frac{(g_f(\overline{x}), \overline{x} - x^*)}{||g_f(\overline{x})|| ||\overline{x} - x||}$, where $\overline{x}$ is any point different from x^*, $\overline{x} \in S(x^*, R)$.*

Let $\overline{\varphi}_R(f)$ be equal to $\sup_{\overline{x} \in M_f(R)} \varphi(\overline{x})$, where $M_f(R) = S(x^*, R) \cap D_f, D_f$ is the set of points in which f is differentiable. We shall say that the function f is essentially gully-shaped(EGS), if $\overline{\varphi}_R(f) = \frac{\pi}{2}$.

If the function is not EGS then the **generalized condition number** *$\rho_R(f)$ is defined as*

$$\rho_R(f) = \frac{1 + \sin\overline{\varphi}_R(f)}{1 - \sin\overline{\varphi}_R(f)}$$

(for EGS-function f we can set $\rho_R(f) = +\infty$).[Shor 70c]

Let us minimize non-EGS convex function f defined on E^n with unknown unique minimum point x^*, and suppose we know the starting point $x_0 \in E_n$ such, that $||x_0 - x^*|| \leq R$, ($R > 0$ is a given number), and some number q that satisfies to inequalities $\sin\varphi_R(f) \leq q < 1$. Then we can organize the following process of minimization that converges with the speed of geometrical progression:

$$x_{k+1} = x_k - h_k \frac{g_f(x_k)}{||g_f(x_k)||} \tag{2.16}$$

(if $g_f(x_k) = 0$, then $x_{k+1} = x_k = x^*$)

where

$$h_0 = \begin{cases} R\sqrt{1-q^2} & \text{, if } q \geq \frac{\sqrt{2}}{2} \\ \frac{R}{2\sqrt{1-q^2}} & \text{, if } q \leq \frac{\sqrt{2}}{2} \end{cases} \tag{2.17}$$

$$h_{k+1} = \begin{cases} h_k \cdot q & \text{, if } q \geq \frac{\sqrt{2}}{2} \\ \frac{h_k}{2\sqrt{1-q^2}} & \text{, if } q \leq \frac{\sqrt{2}}{2} \end{cases} \tag{2.18}$$

The following theorem can be proved [Shor 68].

Theorem 37 *If the convex function $f(x)$ satisfies the condition*

$$\sin \varphi_R(f) \leq q < 1 \tag{2.19}$$

and we use the subgradient process (2.16)–(2.18) then

$$\|x_k - x^*\| \leq \begin{cases} Rq^k & \text{, } for\ q \geq \frac{\sqrt{2}}{2} \\ R \cdot (\frac{1}{2\sqrt{1-q^2}})^k & \text{, } for\ q \leq \frac{\sqrt{2}}{2} \end{cases} \tag{2.20}$$

Proof. The proof seems to be very simple if we use a geometrical interpretation of the algorithm. At the first step we have the sphere $S(x_0, R)$. Calculate $g_f(x_0)$ and consider the cutting plane $P(x_0) = \{x : (g_f(x_0), x - x_0)\}$ and half-space $H(x_0) = \{x : (g_f(x_0), x - x_0) \leq 0\}$. The intersection $H(x_0) \cap S(x_0, R) = S_H(x_0, R)$ gives us the region of localization of the point x^* in the form of half-sphere. The condition (2.19) enables us to reduce the region by moving off the subset of points w such that angle made by $-g_f(x_0)$ and $x^* - x_0$ is more than $\arcsin q$. The remaining set $T(x_0, R, q)$ of localization of x^* can be represented as intersection of $S_H(x_0, R)$ and the spherical cone with the vertex in x_0, central axis of it is coinsided with a ray that has an origin in the point x_0 and direction equal to $-g_f(x_0)$, and has generatrixes which form the angle $\varphi = \arcsin q$ with $-g_f(x_0)$.

At the pictures (a) (b) one can see the two-dimensional cut of $T(x_0, R, q)$ passing through the central axis. The picture (a) corresponds to the case $q \geq \frac{\sqrt{2}}{2}$, or $\varphi \geq \frac{\pi}{4}$. In this case the circle with center in the point x_1 and radius $R_1 = \|x_1 - x_2\|$ contains the sector $\{x_0, x_R, x_b, x_1, x_0\}$, $R_1 = R \cdot \sin \varphi = Rq$; $\|x_0 - x_1\| = R\cos\varphi = R\sqrt{1-q^2}$. It is clear that $S(x_1, R_1)$ contains $T(x_0, R, q)$, and we can cyclically repeat our calculations and obtain (2.20).

The picture (b) corresponds to the case $q \leq \frac{\sqrt{2}}{2}$ $(\varphi \leq \frac{\pi}{4})$. In this case we must construct the circle $C(x_1, R_1)$ that circumscribes the triangle $\{x_L, x_0, x_R\}$ and contains sector $\{x_L, x_0, x_R, x_b, x_L\}$, $R_1 = \|x_0 - x_1\| = \frac{R}{2\cos\varphi} = \frac{R}{2\sqrt{1-q^2}}$. It is clear that in this case $S(x_1, R_1)$ contains $T(x_0, R, q)$. Cyclically repeating calculations we obtain (2.20) for this case.

For a quadratic positive definite function using the subgradient method (2.16)–(2.18) if we know $\sin\overline{\varphi}(f)$ and take $q = \sin\overline{\varphi}(f)$, we can receive the same speed of convergence as in the steepest descent method in the case $\overline{\varphi}(A) \geq \frac{\pi}{4}$.

Remark. Theorem 37 can be generalized for the case of nonunique set of optimal points X^*. We must change in all formulas expressions of the type $(x - x^*)$ by expressions $x - X^*(x)$ where $X^* = \arg\min_{y \in X^*} \|x - y\|$ (see [Shor 85], p.30–31).

We shall now formulate a theorem analogous to Theorem 37 directly in terms characterizing the shape of the level surfaces.

Theorem 38 *Let f be a convex function defined on E^n and let x^* be its unique minimum point. Assume that for a given starting point x_0 numbers σ and h_1 are chosen such that $\sigma \geq \sqrt{2}, h_1 \geq \|x_0 - x^*\|/\sigma$. Furthermore, let $Y = \{y : \|y - x^*\| \leq \sigma h_1\}$ and let for every pair of points $x, z \in Y$ satisfying $f(x) = f(z) \neq f(x^*)$ the following inequality holds:*

$$\|x - x^*\| \leq \sigma \|z - x^*\|. \tag{2.21}$$

Then the sequence $\{x_k\}_{k=0}^{\infty}$ generated recursively by

$$\begin{aligned} x_{k+1} &= x_k - h_{k+1} \frac{g_f(x_k)}{\|g_f(x_k)\|}, \\ h_{k+1} &= h_k \frac{\sqrt{\sigma^2 - 1}}{\sigma}, \quad k = 1, 2, \ldots, \end{aligned}$$

converges to x^* at a linear rate:

$$\|x_k - x^*\| \leq h_{k+1}\sigma = \left(\frac{\sqrt{\sigma^2 - 1}}{\sigma} \right)^k \cdot h_1 \sigma. \tag{2.22}$$

except for the case in which $g_f(x_{\overline{k}}) = 0$ for some $\overline{k}$, i.e. $x_{\overline{k}} = x^*$.

Proof. For $k = 0$ inequality (2.22) holds, since $x_0 \in Y$. If $x_k \in Y$ and $x_k \neq x^*$ then (2.21) holds:

$$(g_f(x_k), x_k - x^*) = a_k(x^*)\|g_f(x_k)\|, \tag{2.23}$$

where $a_k(x^*)$ is the projection of $x_k - x^*$ on $g_f(x_k)$. From the proof of Theorem 30 we know that $a_k(x^*) \geq b_k(x^*)$, where $b_k(x^*)$ is the distance from x^* to the level surface $f(x) = f(x_k)$. Thus $(g_f(x_k), x_k - x^*) \geq b_k(x^*)\|g_f(x_k)\|$, and (2.21) yields $b_k(x^*) \geq \|x_k - x^*\|/\sigma$. Hence

$$(g_f(x_k), x_k - x^*) \geq \frac{1}{\sigma}\|g_f(x_k)\|\|x_k - x^*\|. \tag{2.24}$$

The condition (2.24) corresponds to the condition (2.19) with $q = \frac{\sqrt{\sigma^2-1}}{\sigma}$ in Theorem 37. Taking into acount that for $\sigma \geq \sqrt{2}$, $q \geq \frac{\sqrt{2}}{2}$, one may obtain, using Theorem 37, that $\|x_{k+1} - x^*\| \leq \frac{\sqrt{\sigma^2-1}}{\sigma}\|x_k - x^*\|$ for all k, as required.

Let us note that for a convex quadratic function $f(x) = (Ax, x)$ the number σ involved in previous Theorem may be taken equal to $\sqrt{\rho}$ (if $\rho \geq 2$), ensuring linear convergence with the ratio

$$\tilde{q} = \frac{\sqrt{\sigma^2 - 1}}{\sigma} = \sqrt{\frac{\rho - 1}{\rho}} \approx 1 - \frac{1}{2\rho} \text{ (for } \rho >> 1)$$

Similar to our results on convergence of subgradient processes with the speed of geometrical progression were obtained later by J.L. Goffin (see [Go 77], [Go 78]).

We shall now consider another version of the subgradient method, in which the stepsize remains constant for a certain number of iterations and then is divided by two (see [Shor 85], [SSh 68]).

Theorem 39 *Let for a convex function f the assumptions of Theorem 38 be fulfilled with* $\sigma \geq 2$. *For a given* x_0 *consider the iterative algorithm:*

$$x_{k+1} = x_k - h_{k+1}\frac{g_f(x_k)}{\|g_f(x_k)\|},$$

where $h_{k+1} = h_0 2^{-[(k+1)/N]}$. *If* h_0 *is large enough and* $N \geq 3\sigma^2 + 1$ *then*

$$\|x_k - x^*\| \leq 2\sigma h_{k+1}, \quad k = 0, 1, 2, \ldots. \tag{2.25}$$

Proof. Let $h_0 \geq \|x_0 - x^*\|/2\sigma$. Just as in the proof of Theorem 30, we obtain

$$\|x_{k+1} - x^*\|^2 \leq \|x_k - x^*\|^2 + h_{k+1}^2 - 2h_{k+1}b_k(x^*).$$

If $b_k(x^*) \geq h_{k+1}/2$, then

$$\|x_{k+1} - x^*\| \leq \|x_k - x^*\|, \tag{2.26}$$

while for $b_k(x^*) \geq h_{k+1}$ we get

$$\|x_{k+1} - x^*\|^2 \leq \|x_k - x^*\|^2 - h_{k+1}^2. \tag{2.27}$$

Set $h_i = h_0, i = 1, \ldots, [3\sigma^2] + 1$. Suppose that (2.27) holds for all iterations. Since $\|x_0 - x^*\|^2 \leq 4\sigma^2 h_0^2$, we obtain

$$\|x_N - x^*\|^2 \leq 4\sigma^2 h_0^2 - ([3\sigma^2] + 1)h_0^2 \leq \sigma^2 h_0^2.$$

If inequality (2.27) is violated for a certain $\overline{k} < N$, then $b_{\overline{k}}(x^*) \leq h_0$ and thus $\|x_{\overline{k}} - x^*\| \leq \sigma h_0$. Then two cases are possible.

(a) The inequality $b_k(x^*) \geq h_{k+1}/2$ holds for $k = \overline{k} + 1, \ldots, N$. In this case (2.26) yields $\|x_N - X^*\| \leq \sigma h_0$.

(b) At a certain step $\overline{\overline{k}}, \overline{k} \leq \overline{\overline{k}} \leq N$, one has $b_{\overline{\overline{k}}}(x^*) < h_0/2$. Then $\|x_{\overline{\overline{k}}} - x^*\| \leq h_0\sigma/2$ and $\|x_k - x^*\| \leq (h_0\sigma/2) + h_0 \leq h_0\sigma$ for $\overline{\overline{k}} \leq k \leq N$ (recall that $\sigma \geq 2$).

Therefore, in all cases

$$\|x_N - x^*\| \leq \sigma h_0; \ \|x_i - x^*\| \leq 2\sigma h_0, \ i = 0, \ldots, N.$$

Setting $x_0^{(1)} = x_N$ and $h_0^{(1)} = h_0/2$, we obtain the inequality $h_0^{(1)} \geq \|x_0^{(1)} - x^*\|/2\sigma$, analogous to (2.25),which implies that $\|x_i - x^*\| \leq \sigma h_0,,, i = N + 1, \ldots, 2N - 1$;

$$\|x_{2N} - x^*\| \leq \frac{\sigma h_0}{2}.$$

By induction,

$$\|x_{rN+i} - x^*\| \leq 2\sigma h_0 2^{-r}, \quad i = 0, \ldots, N-1, \quad r = 1, 2, \ldots,$$

as required.

Remark. Subgradient method with the similar step regulation rule given in Theorem 39 was proposed in 1962 by the author for the solution of dual transportation problem in net form. It was the first known example of using generalized gradient methods for minimization of nonsmooth convex function [Shor 62].

Remark. If values $\overline{\varphi}(f)$ and $\sigma(f)$ are unknown one may build adaptive procedures for finding correct upper bounds for them in the process of calculations in acordance with algorithms given in Theorems 37 and 38. The ideas of such procedures are described in [Shor 85], p.35–36. Note that subgradient-type methods can be applied also for nonconvex functions (see [Baz 72]).

2.2 FEJER-TYPE SUBGRADIENT METHODS

We shall establish useful relations between the subgradient method and the method of Fejer-type approximations developed by I.I. Eremin (see [Er 62], [Er 65], [Er 66], [Er 68a], [Er 68b], [Er 70]).

Definition. *A mapping $\varphi : E^n \to E^n$ is called M-Fejer ($E^n \supset M \neq \emptyset$), if*

$$\varphi(y) = y \text{ and } ||\varphi(x) - y|| < ||x - y|| \text{ for all } y \in M \text{ and } x \notin M. \tag{2.28}$$

Using M-Fejer maps, one can construct Fejer-type algorithm of the form $x_{k+1} = \varphi(x_k)$, which converges to M.

Theorem 40 *If M-Fejer map φ is continuous then for any starting point x_0 the sequence $\{x_k\}$ generated by $x_{k+1} = \varphi(x_k),, k = 0, 1, \ldots,$ converges to an element of M (see [EA 76]).*

Proof. Choose $y \in M$. By (2.28), $||x_k - y|| \leq ||x_0 - y||$, i.e. the sequence $\{x_k\}_{k=0}^{\infty}$ is bounded. Let x^* be an accumulation point of this sequence. Suppose that $x^* \notin M$. Then $||\varphi(x^*) - y|| < ||x^* - y||$ and it follows from the continuity of φ that one can find a sufficiently large $\overline{k}$ for which $||\varphi(x_{\overline{k}-1}) - y|| = ||x_{\overline{k}} - y|| < ||x^* - y||$. On the other hand, $||x^* - y||$ is the limit of a decreasing sequence $\{||x_k - y||\}$, i.e. $||x^* - y|| \leq ||x_k - y||$ for all $k = 0, 1, \ldots$. We have got a contradiction. Thus $x^* \in M$ and the proof is complete.

B.T. Polyak [Po 69] suggested versions of the subgradient method which are essentially Fejer-type algorithms.

Let f be a convex function defined on E_n, and let $f^* = \inf f(x), c \geq f^*, M(c) = \{x \in E_n : f(x) \leq c\}$. Consider the following iterative algorithm:

$$x_{k+1} = x_k - \frac{\gamma[f(x_k) - c]}{||g_f(x_k)||^2} g_f(x_k) = \varphi_c(x_k). \tag{2.29}$$

This algorithm can be considered as generalization of relaxation method for linear inequalities (see [A 54], [MSh 54], [Te 82]).

Theorem 41 *If* $0 < \gamma < 2$ *then for any* $x_0 \in E_n$ *either* $x_{k^*} \in M(c)$ *for some* k^* *or* $\lim_{k\to\infty} x_k \in M(c)$.

Proof. Let $x_k \notin M(c)$ and $y \in M(c)$. Then

$$||x_{k+1} - y||^2 = ||x_k - y||^2 - 2\gamma \frac{f(x_k) - c}{||g_f(x_k)||^2}(x_k - y, g_f(x_k)) + \gamma^2 \frac{[f(x_k) - c]^2}{||g_f(x_k)||^2}.$$

Observe that

$$0 \leq f(x_k) - c \leq f(x_k) - f(y) \leq (g_f(x_k), x_k - y),$$

which yields

$$||x_{k+1} - y||^2 \leq ||x_k - y||^2 - \gamma(2 - \gamma)\frac{[f(x_k) - c]^2}{||g_f(x_k)||^2} < ||x_k - y||^2. \tag{2.30}$$

Setting $\varphi_c(y) = y$ for all $y \in M(c)$, we conclude that $\varphi_c(x)$ is an $M(c)$-Fejer map. Actually, Theorem 40 is not directly applicable here, because the map$\varphi_c(x)$ is not necessarily continuous (the subgradient mapping g_f may be discontinuous).

Let us observe, however, that the monotonicity of $\{||x_k - y||\}$ implies the boundedness of $\{x_k\}_{k=0}^{\infty}$ and $\{g_f(x_k)\}_{k=0}^{\infty}$. If an accumulation point x^* of $\{x_k\}_{k=0}^{\infty}$ satisfies $f(x^*) > c$, then there exist $\delta > 0$ and an infinite sequence of indeces $k_1 < k_2 < \ldots$ such that $f(x_{k_i}) - c > \delta$, i.e.

$$||x_{k_{i+1}} - y||^2 \leq ||x_{k_i} - y||^2 - \varepsilon, \; i = 1, 2, \ldots,$$

where $\varepsilon > 0$. Hence $||x_{k_{i+1}} - y||^2 \leq ||x_{k_0} - y||^2 - i\varepsilon$. We have arrived at a contradiction. Therefore, $x^* \in M(c)$. The uniqueness of x^* results from the monotone convergence of the distance $||x_k - x^*||$. The proof is complete.

Theorem 42 *Let ψ be a convex function defined on E^n, and let* $\min_{x \in E^n} \psi(x) = d > 0$). *Consider the sequence $\{x_k\}$ generated by the algorithm*

$$x_{k+1} = x_k - \frac{\gamma \psi(x_k)}{\|g_\psi(x_k)\|^2} g_\psi(x_k)$$

from a starting point x_0. Then

$$\lim_{k \to \infty} \min_{0 \le i \le k} \psi(x_i) = \frac{2d}{2-\gamma}.$$

If $f(x_k) - \overline{f} \ge d > 0$ for $k = 0, 1, 2, \ldots$, then $\overline{f} < f^*$ and the estimate f should be increased. But this test cannot be applied in practice, because there is no algorithm which could detect from a finite part of a sequence whether this sequence is convergent. One can suggest here only some heuristic rules having limited applicability (see also [AHKS 87]).

A group of Italian scientists [CFM 75] proposed a modification of the subgradient method, which uses both an estimate of the minimum value of the function and the information about the direction applied at the previous step. Although their paper deals with the minimization of a piecewise-linear convex function, the suggested algorithm may be applied with no essential changes to any convex function.

The main idea of the method consists in computing the current search direction as a linear combination of the current subgradient and the direction used at the previous step (as in the conjugate gradient method).

Let f be a convex function on E^n, and let $f^* = \min_{x \in E^n} f(x) = f(x^*)$. Consider the following procedure:

$$x_{k+1} = x_k - h_k s_k; \quad h_k = \frac{[f(x_k) - f^*]\gamma_k}{\|s_k\|^2}; \tag{2.31}$$

where

$$s_0 = g_f(x_0); \quad s_k = g_f(x_k) + \beta_k s_{k-1}; \quad k = 1, 2, \ldots.$$

Theorem 43 *If $0 < \gamma_k \le 1$ and $\beta_k \ge 0$ for all $k = 0, 1, \ldots$, then $(x_k - x^*, s_k) \ge (x_k - x^*, g_f(x_k))$ for all k.*

Proof. We shall prove the theorem by induction. Suppose that the assertion of the theorem is true for some $k = m$. To prove it for $k = m + 1$ observe that

$$(x_{m+1} - x^*, s_{m+1}) = (x_m - h_m s_m - x^*, g_f(x_{m+1} + \beta_m s_m) \ge$$

$$(x_{m+1} - x^*, g_f(x_{m+1})) + \beta_m[(x_m - x^*, s_m) - (f(x_m) - f^*)].$$

By the inductive assumption

$$(x_m - x^*, s_m) \geq (x_m - x^*, g_f(x_m)) \geq f(x_m) - f^*.$$

Since $\beta_m \geq 0$, the two preceding inequalities yield

$$(x_{m+1} - x^*, s_{m+1}) \geq (x_{m+1} - x^*, g_f(x_{m+1})),$$

which complete the proof.

With a particular choice of β_k one can obtain a stronger result.

Theorem 44 *[CFM 75]. Under the conditions of the preceding theorem, let*

$$\beta_k = \begin{cases} -\alpha_k \frac{(s_{k-1}, g_f(x_k))}{\|s_{k-1}\|^2} & , \quad \textit{if } (s_{k-1}, g_f(x_k)) < 0 \\ 0 & , \quad \textit{otherwise} \end{cases}$$

$$0 \leq \alpha_k \leq 2.$$

Then

$$\frac{(x_k - x^*, s_k)}{\|s_k\|} \geq \frac{(x_k - x^*, g_f(x_k))}{\|g_f(x_k)\|}. \tag{2.32}$$

Proof. If $\beta_k = 0$ then $s_k = g_f(x_k)$ and (2.32) holds. Consider the case

$$(s_{k-1}, g_f(x_k)) < 0,$$
$$\beta_k = \frac{-\alpha_k(s_{k-1}, g_f(x_k))}{\|s_{k-1}\|^2}.$$

Then

$$\|s_k\|^2 - \|g_f(x_k)\|^2 =$$
$$= \beta_k^2\|s_{k-1}\|^2 + 2\beta_k(s_{k-1}, g_f(x_k)) = \frac{(s_{k-1}, g_f(x_k))^2}{\|s_{k-1}\|^2}(\alpha_k^2 - 2\alpha_k) \leq 0.$$

Hence $\|s_k\| \leq \|g_f(x_k)\|$. Applying Theorem 43, we obtain

$$\frac{(s_k, x_k - x^*)}{\|s_k\|} \geq \frac{(g_f(x_k), x_k - x^*)}{\|g_f(x_k)\|},$$

as required.

We see from the proof of the theorem that, under certain conditions, the direction $-s_k$ computed by (2.31) forms a smaller angle with the direction towards the minimum than does the direction opposite to the subgradient, thus enhancing the speed of convergence for a given f^*. In [CFM 75] the use of $\alpha_k \sim 1.5$ is recommended and computational results are given, which indicate that the algorithm (2.31) is superior to the subgradient method(2.29).

M.B. Shchepakin [Shc 87] proposed the generalization in certain sense of the describe above version of subgradient-type method. Let $f(x)$ be a convex function defined on E^n with known number $f^* = \min_{x \in E^n} f(x)$, and $\overline{x} \in E^n$ is given as well as $\{f(\overline{x}), g_f(\overline{x})\}$. This information gives us a possibility to reduce the search region for the set X^* of minimum points by inequality:

$$X^* \subseteq H(\overline{x}) \equiv \{x : (g_f(\overline{x}), x - \overline{x}) + f(\overline{x}) - f^* \leq 0\} \tag{2.33}$$

Consider the point:

$$x_1 = \overline{x} - \frac{f(\overline{x}) - f^*}{\|g_f(\overline{x})\|^2} g_f(\overline{x})$$

The subgradient relaxation step gives us the point

$$x_2^s = x_1 - \frac{f(x_1) - f^*}{\|g_f(x_1)\|^2} g_f(x_1) \tag{2.34}$$

If $x_2^s \in H(\overline{x})$ (this corresponds to the case $(g_f(\overline{x}), g_f(x_1)) \geq 0$, which implies $f(x_1) < f(\overline{x})$ then set $x_2 = x_2^s$. Otherwise,set

$$x_2 = P_{H(\overline{x})}(x_2^s) \tag{2.35}$$

(P_M denotes the orthogonal projection on the set M). In the last case x_2 can be represented as:

$$x_2 = x_1 - t_1 [g_f(x_1) \frac{(g_f(x_1), g_f(\overline{x}))}{\|g_f(\overline{x})\|} g_f(\overline{x})], \tag{2.36}$$

where $t_1 = \frac{f(x_1) - f^*}{\|g_f(x_1)\|^2}$, i.e. from the point x_1 we move in the direction that is opposite to convex combination of subgradients $g_f(\overline{x})$ and $g_f(x_1)$ because $(g_f(x_1), g_f(x_2)) < 0)$.

One may continue the process of minimizing using the cutting plane of the type (2.33) by constructing $H(x_1), H(x_2), \ldots, H(x_k)$ and iterative formula:

$$x_{k+1} = P_{H_k}(x_{k+1}^s), \tag{2.37}$$

where

$$x_{k+1}^{(s)} = x_k - \frac{f(x_k) - f^*}{||g_f(x_k)||^2} g_f(x_k) \tag{2.38}$$

$$H_k = \cap_{i \leq k} H(x_i) \tag{2.39}$$

For a such algorithm the operation of projection P_H is reduced to a special problem of convex quadratic programming for which practically fast convergent methods exist (see, for example, [Wol 76]). But for large dimensional case it is not easy to solve this problem at each iteration. That is why M.B. Schepakin proposed to approximate the search region of the type $\cap_\alpha H(x_\alpha)$ by one that is cut out by several inequalities of the type

$$H_I = (g_i, x - x_i) \leq 0, \; i \in I, \tag{2.40}$$

where g_i are mutually orthogonal vectors. As we know the projection of a point on such region is a simple operation.

If we have the search region in the form (2.40) we can move from recurrent point x using formulas of the type (2.34)–(2.35) (H_k changes to H_I). If $x_{k+1} \notin H_I$, and $|I| < n$ we find the orthogonal complement $\overline{g}$ of $g_f(x_k)$ on the subspace generated by system $\{g_i\}_{i \in \overline{I}}$ and introduce a new constraint in the form:

$$(\overline{g}, x) + \overline{a} \leq 0,$$

that is added to system (2.40) (here $\overline{I}$ is a certain subset of I). Besides, at each iteration we may remove some of the inequalities (2.40). The removing is necessary if $|I| = n$ and we want to introduce a new constraint. The Fejer property of the algorithm implies its convergence to minimum point.

If f^* is unknown one can try to use some estimate $\overline{f}$ of f^* and change this estimate depending on the results of previous iterations. One of the possible technique of such change is considered in [Akg 84].

The details of the algorithms and the test experiments are published in [Er 68a], [Po 69], [Go 78], [Go 80], [Er 70].

2.3 ε-SUBGRADIENT AND BUNDLE METHODS

As we already know the steepest descent method cannot be used for minimization of nonsmooth convex functions. For choosing a stable direction of descent from nonminimum point $\bar{x}$ one has to calculate the portions of information not only in $\bar{x}$ but in some neighborhood of this point and to use a "bundle" of this information.

In the pioneer works on this subject (Lemaréchal [Le 74], [Le 75] and Wolfe [Wol 75]) the idea of approximation of the so-called ε-steepest descent direction was essentially exploited.

Let $f(x)$ be a convex function defined on E^n, and $\varepsilon \geq 0$ be given. By $G_f^\varepsilon(\bar{x})$ we denote the ε-subgradient set of f in the point $\bar{x}$ for $\varepsilon \geq 0$ (see [Roc 70]):

$$G_f^\varepsilon(\bar{x}) = \{g \in E^n : f(x) - f(\bar{x}) \geq (g, x - \bar{x}) - \varepsilon \text{ for all } x\}.$$

If $0 \in G_f^\varepsilon(\bar{x})$, then $f(\bar{x}) - f^* \leq \varepsilon$, where $f^* = \inf_{x \in E^n} f(x)$. For $\varepsilon = 0$, $G_f^\varepsilon(\bar{x}) = G_f(\bar{x})$.

Definition. *[Le 74] The ε-steepest descent direction $d_\varepsilon(\bar{x})$ is determined by the formula:*

$$\begin{aligned} d_\varepsilon(\bar{x}) &= -\frac{g^*}{\|g^*\|}, \text{ where} \\ g^* &= \arg \min_{g \in G_f^\varepsilon(\bar{x})} \|g\| \end{aligned}$$

For $\varepsilon = 0$, $d_0(\bar{x})$ is coinsided with the direction of the speedest descent in the point $\bar{x}$.

Taking into account that $G_f^\varepsilon(\bar{x})$ contains large amount of information about behavior of $f(x)$, its description is very complex in general case and determination of $d_\varepsilon(\bar{x})$ is a hard problem. Note that for any $\varepsilon > 0$ the nonimplemented method of ε-steepest descent: $x_{k+1}(h_k) = x_k + h_k d_\varepsilon(x_k), \bar{h}_k = \arg\min_{h_k} f(x_{k+1}(h_k))$, $x_{k+1} = x_{k+1}(\bar{h}_k)$ converges (see [Gol 75], [Gol 77]):

$$\lim_{k \to \infty} f(x_k) = f_\varepsilon^* \leq f^* + \varepsilon.$$

This is caused by the fact that for $\varepsilon > 0$ the vector-function $d_\varepsilon(x)$ is continious in x even for nonsmooth convex $f(x)$ ([Nu 77]).

We have the following relation for ε-directional derivative which generalizes the formula for the usual directional derivative:

$$f'_\varepsilon(\overline{x}, s) = \max_{g \in G^\varepsilon_f(\overline{x})} (s, g) = \inf_{\rho \geq 0} \left\{ \frac{f(\overline{x} + \rho s) - f(\overline{x}) + \varepsilon}{\rho} \right\} \tag{2.41}$$

Hence, if a direction $\overline{s}$ exists such that $f'_\varepsilon(\overline{x}, \overline{s}) < 0$, then one can find $\overline{\rho} > 0$ for which $f(\overline{x} + \overline{\rho s}) < f(\overline{x}) - \varepsilon$. If such a direction does not exist, then $0 \in G^\varepsilon_f(\overline{x})$ and $f(x) \leq f^* + \varepsilon$. Therefore, one may think of an algorithm for finding $\min f(x)$ with accuracy ε which would compute search directions s in current point $\overline{x}$ with negative $f'_\varepsilon(\overline{x}, s)$. The difficulty is that the set $G^\varepsilon_f(\overline{x})$ is not defined explicitly. This difficulty is overcome in Lemarechal's algorithm [Le 75] by approximating $G^\varepsilon_f(x)$ with a convex hull of subgradients collected at previously generated points belonging to a certain neighhorhood of x. This approximation is based on the following proposition:

Lemma 2 *Let $\overline{x}, y \in E^n$, $g_f(y)$ is the subgradient of f in the point y and*

$$(g_f(y), y - \overline{x}) \leq 0; \; f(\overline{x}) - f(y) \leq \varepsilon \tag{2.42}$$

Then $g_f(y) \in G^\varepsilon_f(\overline{x})$.

Proof. For any $x \in E^n$

$$f(x) - f(y) \geq (g_f(y), x - y) \tag{2.43}$$

Due to the conditions (2.42):

$$f(y) - f(\overline{x}) \geq -\varepsilon \tag{2.44}$$

$$0 \geq (g_f(y), y - \overline{x}) \tag{2.45}$$

By summing of right and left parts of inequalities (2.43)—(2.45) one can obtain:

$$f(x) - f(\overline{x}) \geq (g_f(y), x - \overline{x}) - \varepsilon,$$

so $g_f(y) \in G^\varepsilon_f(\overline{x})$.

Let us proceed to the description of the principal scheme of Lemareshal's algorithm. It is a two-level algorithm with an outer loop, generating a minimizing sequence (indexed by p) and an inner one determining search directions (indexed by k).

L-algorithm: Let $\varepsilon > 0$, $\eta \geq 0$, $x_0 \in E^n$, $g_0 \in G_f(x_0)$, $p = 0$.

1. Set $s_0 = -g_0$, $k = 0$.
2. Find $\rho_k = \arg\min_\rho f(x_p + \rho s_k)$ and $g_{k+1} \in G_f(x_p + \rho_k s_k)$ such that
$$(g_{k+1}, s_k) = 0 \tag{2.46}$$
3. If $f(x_p + \rho_k s_k) < f(x_p) - \varepsilon$ (descent step), set $x_{p+1} = x_p + \rho_k s_k$, $g_0 := g_{k+1}$, increase p by 1 and go to step 1; otherwise go to step 4 (null step).
4. Compute s_{k+1} by projecting the null vector onto the convex hull of $\{-g_0, \ldots, -g_{k+1}\}$, by solving the problem:
$$\min_\lambda \|\sum_{i=0}^{k+1} \lambda_i g_i\|^2, \ \lambda = (\lambda_1, \ldots, \lambda_{k+1}) \tag{2.47}$$
subject to constraints:
$$\lambda \geq 0; \ \sum_{i=1}^{k+1} \lambda_i = 1 \tag{2.48}$$
and setting $s_{k+1} = -\sum_{i=0}^{k+1} \lambda_i^* g_i$, where $\lambda^* = (\lambda_0^*, \ldots, \lambda_{k+1}^*)$ is the solution of the problem (2.47)–(2.48).
5. If $\|s_{k+1}\| \leq \eta$, then stop. Otherwise go to step 6.
6. Increase k by 1 and go to step 2. It is easy to observe that if $f(x)$ is bounded from below, then the number of descent steps is finite. Let $x_{\overline{p}}$ be the last minimizing point. Then $f(x_{\overline{p}}) - f^* \leq \varepsilon$. Note, that due to the property (2.46) in step 2 of L-algorithm for fixed p
$$\|s_{k+1}\| \leq \frac{\|s_k\| \|g_{k+1}\|}{\sqrt{\|s_k\|^2 + \|g_{k+1}\|^2}} \leq \|s_k\|, \tag{2.49}$$
so if $\|g_{k+1}\|$ does not tend to zero (i.e. $x(p;k) = x_p + \rho_k s_k$ does not tend to minimum point), then $\|s_{k+1}\|$ tends to zero, and for $\eta > 0$ after a finite number of inner steps the stop-criterion in step 5 will be fulfilled. Then we may continue the procedure with smaller ε and η.

One can see that property (2.49) and corresponding convergence results remain true if in step 4 for finding current search direction we will use the aggregate technique: replace the problem (2.47)–(2.48) by problem:
$$\min \|\sum_{i \in I_k} \lambda_i g_i + \overline{\lambda} s_k + \lambda_{k+1} g_{k+1}\|^2 \tag{2.50}$$

subject to constraints:

$$\sum_{i\in I_k} \lambda_k + \overline{\lambda} + \lambda_{k+1} = 1;\ \lambda_k, i \in I_k; \overline{\lambda}, \lambda_{k+1} \geq 0 \tag{2.51}$$

and set

$$s_{k+1} = -(\sum_{i\in I_k} \lambda_i^* g_i + \overline{\lambda}^* s_k + \lambda_{k+1}^* g_{k+1}) \tag{2.52}$$

Here: I_k is an arbitrary subset of indeces $i \in \{0, 1, \ldots, k\}$ (it may be empty), $\{\lambda_i^*, i \in I_k;\ \overline{\lambda}^*, \lambda_{k+1}^*\}$ is the solution of the problem (2.50)–(2.51).

The contracted subproblem (2.50)–(2.51) is used in order to prevent the unbounded growth of information that is needed to calculate the search directions. In addition, the size and the complexity of the subproblem is decreased, however the convergence $\{||s_k||\}$ to zero may slow down.

Consider the problem of unconstrained minimization of convex function $f(x)$ defined on E^n, $\lim_{||x||\to\infty} f(x) = +\infty$. After k steps of minimization process we obtain the point $x_k \in E^n$ and a "bundle" of the portions of information

$$\{y_i,\ f(y_i),\ g_f(y_i) = g_i\}_{i\in I_k} \tag{2.53}$$

where y_k is the subset of indeces of the points $y \in E^n$ in which we have calculated the values of the function f and corresponding subgradients. Using the bundle (2.53) one can obtain the piecewise affine lower approximation of $f(x)$:

$$\widehat{f}_k(x) = \max_{i\in I_k}[(g_i, x - y_i) + f(y_i)] \tag{2.54}$$

Let $d := x - x_k$. We obtain:

$$\widehat{f}_k(x) = \max_{i\in I_k}[(g_i, d) + f(y_i) + (g_i, x_k - y_i)] = \max_{i\in I_k}[(g_i, d) + \varphi_i(x_k)] = \overline{f}(x_k, d), \tag{2.55}$$

where

$$\varphi_i(x_k) = (g_i, x_k - y_i) + f(y_i),\ i \in I_k \tag{2.56}$$

The approximation $\widehat{f}_k(x)$ is not always bounded from below. Let us add to it a quadratic part $\frac{1}{2}t||d||^2$, $t > 0$, then we obtain the quadratic approximation:

$$\tilde{f}_k(d) = \max_{i\in I_k}[\frac{1}{2}t(d, d) + \overline{f}(x_k, d)] \tag{2.57}$$

Calculate the portion of information in the point x_k, and let $\overline{I}_k = I_k \cup \{\overline{k}\}$. Consider the problem of determination of descent direction d_k:

$$\min\{\frac{1}{2}t\|d\|^2 + u\} \tag{2.58}$$

subject to constraints:

$$(g_i, d) + \varphi_i(x_k) \le u, \quad i \in \overline{I}_k \tag{2.59}$$

Let d_k, u_k be an optimal solution of (2.58), (2.59). In [Ki 85] the following scheme of search in the direction is proposed. Let numbers $m \in (0,1]$; $\overline{s} \in (0,1]$ are given. From the set of values $\{1, \frac{1}{2}, \frac{1}{4}, \ldots\}$ try to find the largest one $s_L^k \ge \overline{s}$ such that

$$f(x_k + s_L^k d_k) \le f(x_k) + m s_L^k v_k,$$

where $v_k = u_k - f(x_k)$. If such s_L^k exists then set:

$$x_{k+1} := x_k + s_L^k d_k \text{ and } y_{k+1} := x_{k+1}$$

(a descent step). Otherwise set $x_{k+1} := x_k$ (the null step), and $y_{k+1} = x_k + s_R^k d_k$, where $s_R^k \in [\overline{t}, 1]$ and

$$f(x_k + s_R^k d_k) > f(x_k) + m s_R^k v_k.$$

In any case generate new point y_{k+1} and add the corresponding portion of information to the bundle (2.53).

Using the Kuhn-Takker theorem to the quadratic-type problem (2.58), (2.59) one can construct a dual problem in Lagrangean multiplies $\lambda = \{\lambda_i\}_{i \in \overline{I}_k}$:

$$\min(\|\sum_{i \in I_{\overline{k}}} \lambda_i g_i\|^2 - t \sum_{i \in \overline{I}_k} \lambda_i \varphi_i(x_k)) \tag{2.60}$$

subject to constraints:

$$\lambda \ge 0; \ \sum_{i \in \overline{I}_k} \lambda_i = 1 \tag{2.61}$$

One can see that when $t = 0$ the problem (2.60), (2.61) is converted into problem (2.47), (2.48). Note, that parameter t in (2.60) may be associated with "radius" parameter ρ in the so-called bundle trust region method in which the following model is used to choose the descent direction d:

$$minimize \overline{f}(x_k, d) \tag{2.62}$$

$$\|d\|^2 \le \rho_k^2 \tag{2.63}$$

Increasing of parameter t in (2.60) corresponds to decreasing of trust region radius ρ_k. For large t (small ρ) d is determined by those cutting hyperplanes which give "good" approximation of the graph of function f in the small neighborhood of point x_k. If $\lambda^* = \{\lambda_i^*\}_{i \in \overline{I}_k}$ is the optimal solution of dual problem (2.60), (2.61), then

$$d_k = -\frac{1}{t} \sum_{i \in I_k} \lambda_i^* g_i$$

is the optimal solution of the primal problem (2.57). In order to prevent the fast growth of the number of information portions that we use for calculating the descent directions d_k one can use the aggregation. Each i-th portion of information is associated with inequality:

$$(g_i, d) + \varphi_i(x_k) - f(x_k) \le 0, \ i \in \overline{I}_k.$$

We can choose the nonnegative weights d_i, $i \in \overline{I}_k$ and form the so-called surrogate constraint :

$$(\sum_{i \in I_k} \alpha_i g_i, d) + \sum_{i \in I_k} \alpha_i [\varphi_i(x_k) - f(x_k)] \le 0 \tag{2.64}$$

Let us write the surrogate constraint in the form:

$$z(x) = (\sum_{i \in I} \alpha_i g_i, x - \overline{x}) + b \le 0.$$

If $\alpha_i \ge 0$, $i \in I$ and $\sum_{i \in I} \alpha_i = 1$, then for any x

$$(\sum_{i=1}^{n} \alpha_i g_i, x - \overline{x}) + b \le f(x) \tag{2.65}$$

Due to expression (2.65) one may record the portion of information in form (2.53):

$$\{\overline{x}, \ b, \ \sum_{i \in I} \alpha_i g_i\}$$

and add it to the set of previous portions. If the Lagrangean multipliers α_i^* obtained by solving dual problem (2.60), (2.61) play the role of α_i the aggregate portion of information equals

$$\{\overline{x}_k,\ b = \sum_{i\in\overline{I}_k} \alpha_i^* \varphi_i(x_k) - s_k,\ \sum_{i\in\overline{I}_k} \alpha_i^* g_i\}.$$

The bundle and aggregate procedures similar to ε-subgradient processes of the type (2.50),(2.51), (2.52) were investigated, in which dual problems (2.60),(2.61) are written in contracted form: (which doesn't essentially affect the convergence):

$$\min(\|\sum_{i\in I_k} \lambda_i \overline{g}_i t\overline{\lambda} d_k + \lambda_{k+1} g_{k+1}\|^2 + t[\sum_{i\in I} \overline{\lambda}_i \overline{b}_i^{(k)} + \overline{\lambda b} + \lambda_{k+1}\varphi_{k+1}(x_k)),$$

subject to constraints:

$$\lambda_i, i \in I_k,\ \overline{\lambda}, \lambda_{k+1} \geq 0,\ \sum_{i\in I_k} \lambda_i + \overline{\lambda} + \lambda_{k+1} = 1$$

where I is a subset of indeces of information portions (it may be empty), $\overline{g}_i, \overline{b}_i^{(k)}$ are parameters, standing for "gradient" and constant parts of information portions attached to the current point x_k.

In the last twenty years dozens of different modifications of bundle and aggregate methods were proposed (including local algorithms for nonconvex case). Systematic description of theoretical aspects of such methods one can find in Kiwiel's monography [Ki 85] and in the works [Le 78a], [Le 78b], [LNN 95], [Aus 77], [Aus 87], [LSB 81], [Ki 83a], [Ki 83b], [Ki 90], [Ki 95] and so on. As a rule, the proposed methods are two level. The "inner" level corresponds to null-steps and relates to the choice of a "good" search direction for moving from current point. At this phase the subroutines of convex quadratic programming (or LP) are used. When appropriate direction is chosen, we switch to the subroutine of approximate minimum search along this direction and accumulate new portions of information. Then we make a "null-step" or pass to a new current point and construct a new approximating model for minimized function, attached to this point.

The speed of convergence of bundle processes is not sufficiently investigated in general. But there are many publications where results of numerical experiments (with test and real life problems) of function minimization by bundle methods are described ([Le 81], [Le 82], [LNN 95], [Zo 85]).

But we may use the inner convex approximation of convex set $G_f^\varepsilon(\overline{x})$ and find the **descent** directions $\overline{d}_\varepsilon(\overline{x})$ as approximation of $d_\varepsilon(\overline{x})$. The appropriate search direction $\overline{d}_\varepsilon(\overline{x})$ is found based on information provided by the subgradients calculated in a certain neighborhood of the current point.

Let

$$f'_\varepsilon(x,d) := \inf_{t>0} \frac{1}{t}[f(x+td) - f(x) + \varepsilon] \tag{2.66}$$

be an ε-directional derivative in direction d that is a support function of ε-subgradient set $G_f^\varepsilon(x)$:

$$f'_\varepsilon(x,d) = \max_{g \in G_f^\varepsilon(x)} (g,d) \tag{2.67}$$

The ε-steepest direction $d_\varepsilon(\overline{x})$ corresponds to the solution of the problem:

$$\min_{\|d\| \leq 1} f'_\varepsilon(\overline{x},d) = d_\varepsilon(\overline{x}). \tag{2.68}$$

Formula (2.66) implies that a move from $\overline{x}$ along $d_\varepsilon(\overline{x})$ guarantees a decrease of at least ε if $0 \notin G_f^\varepsilon(\overline{x})$.

One can propose ε-subgradient method analogous to the steepest descent method:

$$\begin{aligned} f(x_k + t_k d_k) &= \min_{t \geq 0} f(x_k + t d_k), \text{ where } d_k = d_\varepsilon(x_k); \\ x_{k+1} &= x_k + t_k d_k. \end{aligned} \tag{2.69}$$

It is easy to prove the following.

Theorem 45 . *Let $f^* := \inf f(x)$ and let$\{x_k\}$ be the sequence generated by (2.69) for an arbitrary starting point $x_0, \varepsilon > 0$.*

(a) *If $f^* = -\infty$ then $\lim_{k\to\infty} f(x_k) = -\infty$;*

(b) *If $f^* > -\infty$ then there exists k such that $f(x_k) \leq f^* + \overline{\varepsilon}$ for arbitrary $\overline{\varepsilon} > \varepsilon$.*

The reason of "ε-converging" of the method (2.69) is that the mapping $x \to d_\varepsilon(x)$ used in (2.69) is continious for $\varepsilon > 0$ in contrast to the case of steepest descent method ($\varepsilon = 0$).

The direct implementation of (2.69) is impossible. That is why we have to use the "bundle" idea for approximation $d_\varepsilon(x)$.

Let J_k be a set of points $x^{(j)}$ and corresponding subgradients g_j already computed and stored in memory after k steps of algorithm

$$J_k = \{x^{(j)};\ g_j = g_f(x^{(j)}) \in G_f(x^{(j)})\}_{j \in I_k}, \tag{2.70}$$

where I_k is the corresponding index subset.

In the point x_k, obtained on the k-th step of algorithm, one may calculate

$$\varepsilon_j^{(k)} := f(x_k) - f(x^{(j)}) - (g_j, x_k - x^{(j)}),\ j \in I_k \tag{2.71}$$

We may construct a polytope P_k for any given $\varepsilon > 0$ as a convex hull of $g_f(x_k)$ and $\{g_j\}_{j \in I_k}$:

$$P_k = \{\lambda_0 g_f(x_k) + \sum_{j \in I_k} \lambda_j g_j / \lambda_0, \lambda_j \geq 0; \lambda_0 + \sum_{j \in I_k} \lambda_j = 1; \sum_{j \in I_k} \lambda_j \varepsilon_j^{(k)} \leq \varepsilon\} \tag{2.72}$$

It is easy to show that $G_f^{\varepsilon}(x_k) \supseteq P_k$.

If we replace $G_f^{\varepsilon}(x_k)$ by P_k in (2.69), the determination of the search direction reduces to a quadratic programming problem:

$$d_k = -\arg\min\{||g||^2 | g \in P_k\} \tag{2.73}$$

If the search direction is obtained this way one can build the schematic bundle algorithm in the following form [Le 78a]:

0-step. Start from $x_1 \in E^n, g^{(1)} \in G_f(x_1),\ I_1 = \emptyset$.

After k steps we have sets J_k, I_k and the point $x \in E^n$.

At the ***k*-th step**:

(a) compute the search direction d_k by using (2.73);

(b) check for stopping: if $||d_k||$ is small enough then quit or reduce ε, and go to **(a)**; otherwise

(c) do the line search: find $\bar{t} \geq 0$ such, that $(g_{k+1}, d_k) = 0$, where $g_{k+1} \in G_f(x_k + \bar{t} d_k)$, yielding either a **descent step**:

$$x_{k+1} = x_k + \bar{t} d_k,$$

if $\Delta_k = f(x_k) - f(x_k - \bar{t} d_k)$ is large ehough; or a **null-step** if Δ_k is small or null. Then $x_{k+1} = x_k$.

(d) In the case of descent step update the $\varepsilon_j^{(k)}$:

$$\varepsilon_j^{(k+1)} = \varepsilon_j^{(k)} + [f(x_{k+1}) - f(x_k) + t(g_j, d_k)].$$

In both cases update the list J_k by adding a new portion of information in the point x_{k+1}:

$$J_{k+1} = \{J_k \cup \{x_{k+1}, g_{k+1}\}\}$$

(e) loop to **(a)**, replacing k by $(k+1)$.

The more general concept of bundle algorithms is based on piecewise linear lower approximations of $f(x)$ in the neighborhood of point x_k:

$$\widehat{f}_k(x_k + d) = f(x_k) + \max_{\{j:j \le k\}} \{-p_j^{(k)} + (g_j, d)\}, \tag{2.74}$$

where

$$g_j = g_f(x_j), \quad p_j^{(k)} = f(x_k) - f(x_j) + (g_j, x_j - x_k).$$

Such approximations are built by supporting hyperplanes to the graph of function f, received at the points $\{x_j, f(x_j)\}_{j \le k}$.

The minimization of $\widehat{f}_k(x_k + d)$ corresponds to one step of cutting plane method [Ke 60] which as we know may give us slow convergence. In the "bundle" algorithms the quadratic regularization is used.

Consider the problem:

$$\min_{\{v,d\}} [v + \frac{1}{2} s \|d\|^2] \tag{2.75}$$

$$v \ge -p_j^{(k)} + (g_j, d), \ \forall j, j \le k. \tag{2.76}$$

where $s > 0$ is given parameter.

The dual problem has the following form:

$$\min \frac{1}{2} \sum_{i=1}^{k} \|\lambda_i g_i\|^2 + s \sum_{i=1}^{k} \lambda_i p_j^{(k)} \tag{2.77}$$

subject to constraints:

$$\lambda \ge 0; \ \sum_{i=1}^{k} \lambda_i = 1; \ \lambda = \{\lambda_1, \dots, \lambda_k\}. \tag{2.78}$$

Let $\lambda^*(k)$ be the solution of the problem (2.77),(2.78). Then the optimal solution $d^*(k)$ of a primal problem is given by the formula:

$$\begin{aligned} d^*(k) &= -\frac{1}{s}\sum_{i=1}^{k} \lambda_i^*(k) g_i, \\ \text{and} & \\ x_{k+1} &= x_k + d^*(k). \end{aligned}$$

If $f^* = \inf f(x) > -\infty$, then $f(x_k)$ tends to f^*.

Some versions of bundle methods, known as conjugate subgradient methods, were based on the idea: when we compute the direction of descent, take in the "bundle" only those portions of information $\{f(x_j), g_j\}$ which give a good approximation for $f(x_k)$. For a given $\varepsilon > 0$ let

$$I_k = \{i : p_i^{(k)} \leq \varepsilon, \ i \leq k\} \tag{2.79}$$

Then the corresponding dual problem is:

$$\text{minimize } \frac{1}{2}\sum_{i \in I_k} \|\lambda_i g_i\|^2, \lambda \geq 0; \sum_{i \in I_k} \lambda_i = 1,$$

and

$$x_{k+1} = x_k - \frac{1}{s}\sum_{i \in I_k} \lambda_i g_i \tag{2.80}$$

For this simplified version of bundle method the convergence results have the following form (see [Le 78b]):

Let $f(x)$ be a convex function defined on E^n, $\{x_k\}_{k=1}^{\infty}$ is generated by algorithm (2.79), (2.80). Then:

— either $f(x_k) \to -\infty$ for $k \to +\infty$;

— or the line-search detects at some iteration k that f is unbounded from below;

— or the stopping criterion is satisfied for some finite k, at which there holds

$$f(x_{\overline{k}}) \leq f(y) + \overline{\varepsilon} + \delta\|y - x_k\| \text{ for all } y \in E^n, \tag{2.81}$$

where $\overline{\varepsilon} > 0, \delta > 0$ are parametres included in stopping conditions: the value $\overline{\varepsilon}$ is the final tolerance, or the lowest useful value for $\varepsilon = \varepsilon_k$, that is used in (2.79), (2.80) when we find direction of descent d_k; δ determines the condition for decreasing ε: ε changes to $q\varepsilon$, $0 < q < 1$, when $\|d_k\| \leq \delta$.

The detail description and justification of bundle methods and their aggregate generalizations (some portions of information in the form of cutting plane inequalities are replacedby an aggregate inequality that is a linear combination of given inequalities with positive coefficients) are given in the book, published by Kiwiel [Ki 85]. In general case the speed of convergence of bundle methods was not studied theoretically in detail. Now we have hundreds publications devoted to the theory of bundle methods and numerical experiments with "bundle" algorithms (see [Ki 85], [Le 78a], [Le 78b], [LNN 95], [LSB 81], [Ki 83a], [Ki 83b], [Ki 90], [Le 81], [Le 82], [LNN 95], [Zo 85]).

Remark. Some versions of bundle method are similar to the methods of linearization used for the solution of minimax problems and general smooth problems of nonlinear programming (see [Psh 93]).

2.4 THE STOCHASTIC SUBGRADIENT METHOD

In [ES 68] the notion of **stochastic subgradient** was introduced and used for building a random search method which is a stochastic analogue of the subgradient method. First used for a solution of two-stage convex problems of stochastic programming [ES 68] and further developed in the works of Yu.M.Ermoliev and his research group, R.Wets, and others (see [Erm 76], [Gup 79], [Nu 79], [Nu 82], [EW 87]), the stochastic subgradient method became an efficient tool for solving manifold stochastic programming problems. The method belongs to the broad class of random search methods,i.e. iterative algorithms that proceed along directions resulting from random (pseudorandom) events, contrary to "deterministic" procedures when various numerical errors are neglected.

The **stochastic subgradient method** for minimizing a convex function $f(x), x \in E^n$, is defined by the formula:

$$x_{k+1} = x_k - h_k(x_k) g_w(x_k), \; k = 0, 1, 2, \ldots, \tag{2.82}$$

where $h_k(x_k)$ is a stepsize coefficient at the k-th iteration, and $g_w(x_k)$ is a random vector with mathematical expectation equal to subgradient $g_f(x_k)$ of f at x_k. For simplicity we assume that probabilistic characteristics of the vector $g_w(x_k)$ are fully determined by the point x_k and do not depend on the previous points. In reality this requirement is not so essential, and convergence theorems were proven in more general cases.

Let x^* be a unique minimum point for convex function f defined on E^n. Then

Theorem 46 *(see [Erm 76]). Let the following conditions be satisfied:*

$$\text{(i)} \quad \sum_{k=0}^{\infty} h_k(x_k) = +\infty; \; h_k(x_k) > 0 \textit{ for all } k; \tag{2.83}$$

$$\text{(ii)} \quad \sum_{k=0}^{\infty} h_k^2(x_k) < \infty; \tag{2.84}$$

$$\text{(iii)} \quad E\{\|g_w(x_k)\|^2\} \leq c, \textit{ for all } k; \; c > 0, \tag{2.85}$$

where E denotes the mathematical expectation.

Then with probability one the sequence $\{x_k\}$, generated by stochastic subgradient method (2.82) converges to the minimal point, i.e. for arbitrary $\delta > 0$

$$\lim_{k\to\infty} P\{\|x_k - x^*\| > \delta\} = 0.$$

Proof. The proof of convergence is based on properties of random sequences called supermartingales. A sequence of random variables $\{y_k\}_{k=1}^{\infty}$ is a **supermartingale**, if

$$E\{y_n/y_{n-1},\dots,y_0\} \leq y_{n-1},$$

where $E\{y_n/y_{n-1},\dots,y_1\}$ denotes the conditional expectation of y_n with respect to $y_{n-1},\dots,y_0$. This concept generalizes the notion of monotonic sequences.

The following convergence theorem holds for supermartingales [Do 83]: if $E\{|y_n|\} \leq c < +\infty$ then with probability one the limit

$$\lim_{n\to\infty} y_n = y^*,$$

exists, and $E\{|y^*|\} < +\infty$ (see [Do 83]).

Let us study the sequence $\{E\{\|x_k - x\|^2\}\}_{k=0}^{\infty}$. Note that

$$\|x_{k+1} - x^*\|^2 = \|x_k - h_k(x_k)g_w(x_k) - x^*\|^2 =$$
$$= \|x_k - x^*\|^2 - 2h_k(x_k)(g_w(x_k), x_k - x^*) + h_k^2(x_k)\|g_w(x_k)\|^2,$$
$$E\{\|x_{k+1} - x^*\|^2/x_k\} =$$
$$= \|x_k - x^*\|^2 - 2h_k(x_k)(E\{g_w(x_k)/x_k\}, x_k - x^*) +$$
$$+h_k^2 E\{\|g_w(x_k)\|^2/x_k\}.$$

Since $E\{g_w(x_k)/x_k\} = g_f(x_k)$ and $(g_f(x_k), x_k - x^*) \geq 0$, we have

$$E\{\|x_k - x^*\|^2/x_k\} \leq \|x_k - x^*\|^2 + ch_k^2(x_k). \tag{2.86}$$

Consider the random variable $z_k = \|x_k - x^*\|^2 + c\sum_{s=k}^{\infty} h_s^2(x_s)$. Inequality (2.86) is equivalent to the inequality $E\{z_{k+1}/z_k, \ldots, z_1\} \leq z_k$. Consequently, $\{z_k\}_{k=1}^{\infty}$ is a supermartingale and converges almost surely to a certain limit z^*. Since $\lim_{k\to\infty} \sum_{s=k}^{\infty} h_s^2(x_s) = 0$, the sequence $\{\|x_k - x^*\|^2\}$ also converges to z^* wth probability one. We shall prove from the opposite that the limit z^* is equal to zero. If this were not true, then one could find $\varepsilon > 0$ and $\delta > 0$ such that with probability $\delta > 0$ $\|x_k - x^*\| \geq \varepsilon$ for all sufficiently large k, whence

$$\sum_{k=0}^{\infty} h_k(x_k)(E\{g_w(x_k)/x_k\}, x_k - x^*) = +\infty$$

with probability δ, which in turn would yield

$$E\left\{\sum_{k=0}^{\infty} h_k(x_k)(g_w(x_k), x_k - x^*)\right\} = +\infty.$$

However, this would contradict the relation

$$E\{\|x_{k+1} - x^*\|^2\} =$$
$$= \|x_0 - x^*\|^2 - 2E\left\{\sum_{s=0}^{k} h_s(x_s)(g_w(x_s), x_s - x^*)\right\} +$$
$$+E\left\{\sum_{s=0}^{k} h_s^2(x_s)\|g_w(x_s)\|^2\right\}.$$

The proof is complete.

3

SUBGRADIENT-TYPE METHODS WITH SPACE DILATION

3.1 HEURISTICS OF METHODS WITH SPACE DILATION

The analysis of the subgradient methods has shown that improvements only in the stepsize rules cannot, in general, significantly accelerate convergence if at each iteration the algorithm proceeds in the direction opposite to that of the subgradient. Indeed, slow convergence is due to the fact that the subgradient is often almost perpendicular to the direction towards the minimum. In such circumstances the reduction of the distance to the minimum is much smaller than the stepsize, and therefore the stepsizes cannot diminish too rapidly, if we want to guarantee the convergence to a minimum.

On the other hand, there is a simple way of changing the angles between the subgradient and the direction towards the minimum , namely to use linear nonorthogonal space transformation. An idea is to construct at each iteration a certain linear transformation changing the metric of the space, and to use the direction opposite to that of the subgradient in the space with transformed metric. Such a direction may significantly differ from the direction opposite to subgradient in the original space, but it forms an acute angle with the direction from the given point to the point of minimum for convex functions.

How should the operators of space transformation be constructed? Recalling well-known quasi-Newton algorithms for minimizing twice continuously differentiable functions, we see that the basic idea of the variable metric methods consist in obtaining in one way or another, a matrix close to the inverse of the Hessian (or its positive multiple) at the minimum point, i.e. one uses a quadratic approximation of the objective function and imitates the Newton-Raphson method without calculating the second derivatives (see [Min 86], [Fle 87]). This approach does not apply to nonsmooth functions for fundamental reasons. For instance, the Hessian of a piecewise linear function vanishes almost everywhere, and its inverse simply does not exist. Therefore, other ideas should be used for constructing gradient-type methods with space transformation for a sufficiently broad class of nonsmooth functions.

As indicated above, the basic factor that slows down the subgradient method with stepsizes determined off line, is that the cosine of the angle between the subgradient and the direction towards the minimum is close to zero. If it was possible to substantially reduce the component of the subgradient orthogonal to the direction towards the minimum while leaving the component parallel to that direction almost unchanged, then the resulting vector would define a better search direction than the subgradient. In this case one could expect to see the acceleration of convergence, provided an appropriate stepsize rule is used. But how can we find a direction providing a significant decrease of the function, or directions orthogonal to it? Let us note that in most difficult cases the direction of the subgradient is often almost orthogonal to the direction towards the minimum. Assuming such an orthogonality, we should try to reduce at subsequent iterations the components of the subgradient that are parallel to the last subgradient. This can be done by performing space dilation in the direction of the subgradient. So, from rather nonformal heuristic considerations we have derived the idea of the subgradient method with space dilation along the subgradient. A detailed description and justification of the method will be given below.

Let a vector $\xi \in E^n$,$||\xi|| = 1$, and a number $\alpha \geq 0$ be fixed. Every vector $x \in E^n$ may be represented as follows:

$$x = \gamma_\xi(x)\xi + d_\xi(x), \tag{3.1}$$

where

$$(x, d_\xi(x)) = 0 \tag{3.2}$$

From (3.1) and (3.2) we obtain $\gamma_\xi(x) = (x, \xi)$ and $d_\xi(x) = x - (x, \xi)\xi$.

Definition. *An operator $R_\alpha(\xi)$ which transforms a vector x of the form (3.1) into*

$$R_\alpha(\xi)x = \alpha\gamma_\xi(x)\xi + d_\xi(x)$$

is called an operator of space dilation along direction ξ with coefficient α

It follows from the above definition that the following statements are true:

1)
$$R_\alpha(\xi)x = \alpha\gamma_\xi(x)\xi + [x - (x,\xi)\xi] = (\alpha-1)(x,\xi)\xi + x. \qquad (3.3)$$

2) The operator $R_\alpha(\xi)$ is linear and symmetric,

$$(R_\alpha(\xi)x, y) = (\alpha-1)(x,\xi)(y,\xi) + (x,y) = (x, R_\alpha(\xi)y).$$

3) $R_{\alpha\beta}(\xi) = R_\alpha(\xi)R_\beta(\xi)$,

4) For $\alpha > 0$ one has $R_\alpha(\xi)R_{\frac{1}{\alpha}}(\xi) = R_1(\xi) = I$, I is the identity matrix.

5) The operator $R_0(\xi)$ is operator of projection on the subspace orthogonal to ξ,
$$R_0(\xi)x = d_\xi(x).$$

6) The operator $R_\alpha(\xi)$ has for $n \geq 2$ two eigenvalues $\lambda_1 = \alpha$, and $\lambda_1 = 1$, the first one corresponding to the subspace of eigenvectors generated by ξ, and the second one to the subspace of eigenvectors orthogonal to ξ .

7) If the coordinates of the vector ξ in a certain orthonormal basis $s = \{e_1, e_2, \ldots, e_n\}$ are equal to $\{\xi_1, \xi_2, \ldots, \xi_n\}$, then it follows from (3.3) that the operator $R_\alpha(\xi)$ is represented in this coordinate system by the matrix $R_\alpha(\xi)$ with entries $\{r_{ij}\}$ defined as follows

$$\{r_{ij}\} = (R_\alpha(\xi)e_i, e_j) = \begin{cases} (\alpha-1)\xi_i\xi_j & \text{for } i \neq j \\ (\alpha-1)\xi_i^2 & \text{for } i = j \end{cases}$$

8) By (3.3), the computation of a vector $R_\alpha(\xi)x$ requires $(2n+1)$ multiplications, and the computation of matrices of the form $R_\alpha(\xi)A$ or $AR_\alpha(\xi)$ for a given matrix A requires $n(2n+1)$ multiplications.

9) For any vector $x \in E_n$ one has

$$\|R_\alpha(\xi)x\| = \sqrt{\|x\|^2 + (\alpha^2-1)(x,\xi)^2}, \qquad (3.4)$$

since from (3.3) we obtain

$$\begin{aligned} \|R_\alpha(\xi)x)\|^2 &= (R_\alpha(\xi)x, R_\alpha(\xi)x) \\ &= (x + (\alpha-1)(x,\xi)\xi, x + (\alpha-1)(x,\xi)\xi) \\ &= \|x\|^2 + 2(\alpha-1)(x,\xi)^2 + (\alpha-1)^2(x,\xi)^2 \\ &= \|x\|^2 + (\alpha^2-1)(x,\xi)^2. \end{aligned}$$

10) $R_\alpha(\xi)$ has the matrix representation

$$R_\alpha(\xi) = I + (\alpha - 1)\xi\xi^T.$$

Indeed,

$$(I + (\alpha - 1)\xi\xi^T)x = x + (\alpha - 1)\xi(\xi, x), \tag{3.5}$$

which is exactly the defining relation (3.3).

3.2 THE SUBGRADIENT METHOD WITH SPACE DILATION IN THE DIRECTION OF SUBGRADIENT

We shall consider a class of algorithms for minimizing convex functions, in which movement in the direction of the subgradient is combined with space dilation along this direction. Algorithms from this class will be called subgradient methods with space dilation along the subgradient (SDG algorithms, SDG for "Space Dilations along the Gradient"). Assume that we have a finite process for computing a subgradient $g_f(x)$ of a convex objective function f at any point $x \in E_n$, rules for determining sequences of positive reals $\{h_k\}$ and $\{\alpha_k\}$, $k = 1, 2, \ldots$, (stepsizes and space dilation coefficients), a starting point x_0 , and an initial nonsingular matrix $B_0 = A_0^{-1}$ (e.g. $B_0 = I$). Under these assumptions we shall describe an infinite iterative algorithm, which has the $(k+1)$ -st iteration, $k = 1, 2, \ldots$, defined as follows: let after k steps we obtained the point x_k and the matrix B_k^*

1) Evaluate $g_f(x_k)$ (if $g_f(x_k) = 0$ the computations stop, since then x_k is the optimal point).

2) Set

$$g_{\varphi_k}(y_k) = B_k^* g_f(x_k) = \tilde{g}_k, \tag{3.6}$$

where $\varphi_k(y_k) = f(B_k y)$, $y_k = A_k x_k$, $A_k = B_k^{-1}$ and B_k^* is the conjugate operator to B_k . By formula (3.6) we calculate a subgradient of the function $\varphi_k(y_k) = f(A_k^{-1} y)$, which is obtained from f by the linear transformation of variables $y = A_k x$. This formula is a special case of a general formula for the supporting functional of a convex functional defined on a Banach space which is a linear transformation of another Banach space [Psh 69]. The formula may be also easily derived from the definition of the subgradient. At those points at which $g_f(x)$ is not unique, formula (3.6) should be read as a single-valued mapping from the set $G_f(x)$ into the set $G_{\varphi_k}(y)$ (if the operator B_k is nonsingular then this mapping is subjective and invertible).

Next compute

3)

$$\xi_{k+1} = \frac{g_{\varphi_k}(y_k)}{\|g_{\varphi_k}(y_k)\|} = \frac{\tilde{g_k}}{\|\tilde{g_k}\|} \tag{3.7}$$

4) h_{k+1}

5) α_{k+1}

6)

$$x_{k+1} = x_k - h_{k+1} B_k \xi_{k+1} \tag{3.8}$$

Relation (3.8) is obtained from the formula $A_k x_{k+1} = y_k - h_{k+1}\xi_{k+1}$, which describes one iteration of the subgradient method for minimization φ_k in the transformed space, by premultiplying it with the operator B_k to map both sides into the original space.

7) Calculate

$$B_{k+1} = A_{k+1}^{-1} = B_k R_{1/\alpha_{k+1}}(\xi_{k+1}) \tag{3.9}$$

Formula (3.9) defines the operator B_{k+1}, which is the inverse of the space transformation operator

$$A_{k+1} = R_{\alpha_{k+1}}(\xi_{k+1}) \ldots R_{\alpha_1}(\xi_1) A_0$$

resulting from succesive space dilations in the directions of the normed subgradients $\xi_1, \xi_2, \ldots, \xi_{k+1}$ with coefficients $\alpha_1, \alpha_2, \ldots, \alpha_{k+1}$. Thus

$$B_{k+1} = A_{k+1}^{-1} = A_k^{-1} R_{\alpha_{k+1}}(\xi_{k+1})^{-1} = B_k R_{1/\alpha_{k+1}}(\xi_{k+1})$$

8) Proceed to the $(k+2)$ -nd iteration.

At each iteration the SDG method involves more operations than the subgradient method. The most laborious ones are the operations 2),6) and 7), each of them requiring a number of arithmetic operations of order n^2, including about $4n^2$ multiplications. Moreover, additional memory is required for storing matrices B_k.

Let us observe that if $B_0 = I$ and $\alpha_{k+1} = 1$ for $k = 1, 2, \ldots$, then the SDG algorithm becomes equivalent to the subgradient method. But the SDG method is more flexible since its operation depends on two sequences $\{\alpha_k\}$ and $\{h_k\}$. Using various rules for determining those sequences, we can modify algorithm in many ways and choose the most efficient version for a given class of problems.

We shall show that certain SDG algorithm improves the function values with a speed of geometrical progression with a ratio depending on properties of the objective function that are invariant to nonsingular linear transformations. Moreover, for a special SDG algorithm for convex minimization this ratio depends only on the dimension of the domain of the function. This algorithm is described further and named "ellipsoid method".

The proofs of the results to follow are based on the analysis of the behavior of the norm of the gradient of the objective function in the transformed space.

Theorem 47 *Let the SDG method (3.6)–(3.9) be applied to the minimization of an almost differentiable function f and let it generate a sequence $\{x_k\}_{k=o}^{\infty}$. Suppose that there exist positive numbers d, α^* and δ such that for $k = 0, 1, \ldots$ the following condidions are satisfied:*

1) $\|g_f(x_k)\| \leq d$,

2) $1 + \delta \leq \alpha_k \leq \alpha^*$.

Then one can find a subsequence $\{x_{k_p}\}_{p=o}^{\infty}$, $k_p < k_{p+1}$, and a constant $c > 0$ such that

$$\|\tilde{g}_{k_p}\| < c \left(\prod_{j=1}^{k_p} \alpha_j \right)^{-1/n} \quad \textit{for } p = 0, 1, \ldots.$$

(see [Shor 70a], [Shor 85]).

Proof. Let us express the matrix A_k as a product of an ortogonal matrix O_k and a symmetric positive definite matrix S_k (the polar factorization):

$$A_k = O_k S_k; \; A_0 = I.$$

Since $A_k = R_{\alpha_k}(\xi_k) \ldots R_{\alpha_1}(\xi_1)$ and the product of the eigenvalues of matrix is equal to its determinant, the product of the eigenvalues of the matrix S_k is equal to

$$\det S_k = \det A_k = \prod_{j=1}^{k} \alpha_j$$

Let $\left\{\lambda_i^{(k)}\right\}_{i=1}^n$ be the eigenvalues of S_k, $\left\{\lambda_1^{(k)}\right\}_{i=1}^n \leq \left\{\lambda_2^{(k)}\right\}_{i=1}^n \leq \ldots \leq \left\{\lambda_n^{(k)}\right\}_{i=1}^n$, and let $\left\{e_i^{(k)}\right\}_{i=1}^n$ be the corresponding orthonormal system of eigenvectors. We have

$$S_k e_i^{(k)} = \lambda_i^{(k)} e_i^{(k)}; \ \prod_{j=1}^n \lambda_i^{(k)} = \prod_{j=1}^k \alpha_j; \ \lambda_n^{(k)} \geq \left(\prod_{j=1}^k \alpha_j\right)^{1/n}. \tag{3.10}$$

Moreover, $\lambda_i^{(k)} \geq 1$, since $\alpha_j > 1$, $j = 1, \ldots, k$, and thus

$$\|S_k e_i^{(k)}\| = \|A_k e_i^{(k)}\| \geq 1.$$

Let $O_k e_i^{(k)} = \tilde{e}_i^{(k)}$. The system of vectors $\left\{\tilde{e}_i^{(k)}\right\}_{i=1}^n$ is also orthonormal, because O_k is orthogonal operator.

Let us write $g_f(x_k)$ as follows:

$$g_f(x_k) = \sum_{i=1}^n g_i^{(k)} e_i^{(k)}.$$

Since $\|g_f(x_k)\| \leq d$,

$$\|g_i^{(k)}\| \leq d; \ i = 1, 2, \ldots, n; \ k = 1, 2 \ldots. \tag{3.11}$$

We shall prove the theorem by contradiction.

Suppose that the assertion is false. Then for arbitrarily large c one can find $k(c)$ such that for all $k > k(c)$

$$\| \tilde{g}_k\| = \|B_k^* g_f(x_k)\| = \| \sum_{i=1}^n \frac{g_i^{(k)}}{\lambda_i^{(k)}} \tilde{e}_i^{(k)}\| \geq c \left(\prod_{j=1}^k \alpha_j\right)^{-1/n}. \tag{3.12}$$

Thus for any $k > k(c)$ there exists i_k^* , $1 \leq i_k^* \leq n$, such that

$$\frac{\|g_{i_k^*}^{(k)}\|}{\lambda_{i_k^*}^{(k)}} \geq \frac{c\left(\prod_{j=1}^k \alpha_j\right)^{-1/n}}{\sqrt{n}}$$

whence

$$\lambda_{i_k^*}^{(k)} \leq \frac{\sqrt{n}d}{c} \left(\prod_{j=1}^k \alpha_j\right)^{1/n}.$$

Next

$$\lambda_n^{(k)} \max_{1\le i\le n} \lambda_i^{(k)} \ge \left(\frac{\prod_{j=1}^k \alpha_j}{\lambda_{i_k^*}^{(k)}}\right)^{1/(n-1)} \ge \frac{c}{\sqrt{n}d}^{1/(n-1)} \left(\prod_{j=1}^k \alpha_j\right)^{1/n}$$
$$= \tilde{c}\left(\prod_{j=1}^k \alpha_j\right)^{1/n},$$

where

$$\tilde{c} = \left(\frac{c}{\sqrt{n}d}\right)^{1/(n-1)}.$$

Let

$$e_n^{(k+1)} = \sum_{i=1}^n d_i^{(k)} e_i^{(k)}; \quad \sum_{i=1}^n (d_i^{(k)})^2 = 1.$$

Then

$$A_{k+1} e_n^{(k+1)} = \lambda_n^{(k+1)}\, \overline{e}^{(k+1)} = R_{\alpha_{k+1}}(\xi_{k+1}) \left(\sum_{i=1}^n \lambda_i^{(k)} d_i^{(k)}\, \overline{e}_i^{(k)}\right).$$

Set $r_k = \min \left\{\lambda_n^{(k)}/\lambda_r^{(k)} \le \alpha_{k+1} c,\ 1 \le r \le n\right\}$ and express $y = \sum_{i=1}^n \lambda_i^{(k)} d_i^{(k)}\, \overline{e}_i^{(k)}$ as $y = y_1 + y_2$, where

$$y_1 = \sum_{i=1}^{r_k - 1} \lambda_i^{(k)} d_i^{(k)}\, \overline{e}_i^{(k)},$$

$$y_2 = \sum_{r_k}^{n} \lambda_i^{(k)} d_i^{(k)}\, \overline{e}_i^{(k)}.$$

Let us estimate $\|R_{\alpha_{k+1}}(\xi_{k+1})y\|$. Observe that

$$\xi_{k+1} = \frac{\tilde{g}_k}{\|\tilde{g}_k\|} = \sum_{i=1}^n \mu_i^{(k)}\, \overline{e}_i^{(k)},$$

where $\mu_i^{(k)} = g_i^{(k)}/\lambda_i^{(k)}\|\tilde{g}_k\|$. From (3.11) and (3.12) we obtain

$$\|\mu_i^{(k)}\| = \frac{d\left(\prod_{j=1}^k \alpha_j\right)^{1/n}}{\lambda_i^{(k)} c}.$$

For $i \geq r_k$ we have

$$\|\mu_i^{(k)}\| = \frac{d\left(\prod_{j=1}^k \alpha_j\right)^{1/n} \alpha_{k+1}}{\lambda_n^{(k)} c} \leq \frac{d\alpha^*}{\tilde{c}}.$$

Thus

$$\|R_{\alpha_{k+1}}(\xi_{k+1})y_1\| \leq \alpha_{k+1}\|y_1\| \leq \lambda_n^{(k)}\frac{1}{c}\|\sum_{i=1}^{r_k-1} d_i^{(k)}\, \bar{e}_i^{(k)}\| \leq \frac{\lambda_n^{(k)}}{c}$$

$$\leq \lambda_n^{(k)}\left[1 + (\alpha^* - 1)\frac{d\alpha^*}{\tilde{c}}\right],$$

$$\|R_{\alpha_{k+1}}(\xi_{k+1})y_2\| \leq \|y_2\| + (\alpha_{k+1} - 1)|(y_2, \xi_{k+1})|$$

$$\lambda_n^{(k+1)} = \|R_{\alpha_{k+1}}(\xi_{k+1})y\| \leq \lambda_n^{(k)}\left[1 + \frac{1}{c} + (\alpha^* - 1)\frac{d\alpha^*}{\tilde{c}}\right]. \tag{3.13}$$

By (3.13), for any ε_0 one can find c so large that $\lambda_n^{(k+1)} \leq (1+\varepsilon_0)\lambda_n^{(k)}$ for all $k > k(c)$. But if we choose $\varepsilon_0 < (1+\delta)^{1/n} - 1$ then for sufficiently large k we obtain $\lambda_n^{(k)} < (1+\varepsilon_0)^k < (1+\delta)^{k/n} \leq \left(\prod_{j=1}^k \alpha_j\right)^{1/n}$, which contradicts (3.10). This proves the theorem. □

A similar result may be obtained for the rate of convergence of the sequence $v_k = \min_{1 \leq r \leq k} \|\tilde{g}_r\|$. This result will be used later to estimate the rate of convergence of the function value "records " in SDG algorithm (the "record " of the k-th iteration equals $\min_{1 \leq r \leq k} f(x_r)$).

Theorem 48 *Under the conditions of the Theorem 47 assume that $\alpha_k = \alpha > 1$ for all k. Then*

$$v_k \leq \frac{d\sqrt{k(\alpha^2 - 1)}}{\sqrt{\alpha^{2k/n} - 1}} \quad \textit{for } k = 1, 2, \ldots$$

Proof. We shall use the notation introduced in the proof of the Theorem 47.

We have

$$\xi_{k+1} = \frac{\tilde{g}_k}{\|\tilde{g}_k\|} = \sum_{i=1}^n \mu_i^{(k)}\, \bar{e}_i^{(k)},$$

$$\mu_i^{(k)} = \frac{g_i^{(k)}}{\lambda_i^{(k)}\|\tilde{g}_k\|}, \quad i = 1, 2, \ldots, n, \quad \sum_{i=1}^n \left(g_i^{(k)}\right)^2 \leq d^2. \tag{3.14}$$

Suppose that

$$v_k = \min_{1\le r\le k} \|\tilde{g}_k\| > \frac{d\sqrt{k(\alpha^2-1)}}{\sqrt{\alpha^{2k/n}-1}}$$

for some k . Then for $1 \le r \le k$ and $1 \le i \le n$ (3.14) yields

$$|\mu_i^{(r)}| \le \frac{|g_i^{(r)}|\sqrt{\alpha^{2k/n}-1}}{\lambda_i^{(r)} d\sqrt{k(\alpha^2-1)}} \tag{3.15}$$

Using this relation we shall estimate $(\lambda_n^{(r+1)})^2 - (\lambda_n^{(r)})^2$ for $1 \le r \le k-1$ ($\lambda_n^{(r)}$ is the largest eigenvalue of the matrix S_r). For any vector a such that $\|a\| = 1$ we have

$$\begin{aligned}
a &= \sum_{i=1}^{n} a_i e_i^r, \sum_{i=1}^{n} a_i^2 = 1, \\
A_r a &= \sum_{i=1}^{n} a_i \lambda_i^{(r)} \, \bar{e}_i^{(r)}, & (3.16)\\
\lambda_i^{(r+1)} &= \max_{\|a\|=1} \|A_{r+1} a\| = \max_{\|a\|=1} \|R_\alpha(\xi_{r+1}) A_r a\|. & (3.17)
\end{aligned}$$

Recall formula (3.4):

$$\|R_\alpha(\xi)x\| = \sqrt{\|x\|_2 + (\alpha^2-1)(x,\xi)^2}.$$

Setting $x = A_r a$, we obtain from (3.15)–(3.17) the inequality

$$\begin{aligned}
\left[\lambda_n^{(r+1)}\right]^2 &= \max_{\|a\|=1} \left[\sum_{i=1}^{n} a_i^2 (\lambda_i^{(r)})^2 + (\alpha^2-1)\left(\sum_{i=1}^{n} a_i \lambda_i^{(r)} \mu_i^{(r)}\right)^2\right] \\
&\le \max_{\|a\|=1} \left[(\lambda_n^{(r)})^2 + (\alpha^2-1)\left(\sum_{i=1}^{n} |a_i g_i^{(r)}|\right)^2 \frac{\alpha^{2k/n}-1}{kd^2(\alpha^2-1)}\right] \\
&< [\lambda_n^{(r)}]^2 + \frac{\alpha^{2k/n}-1}{k}, \qquad r = 0, 1, \ldots, k.
\end{aligned}$$

Since $\lambda_n^{(0)} = 1$, one has $[\lambda_n^{(k)}]^2 < \alpha^{2k/n}$, which contradicts the inequalitiy $[\lambda_n^{(k)}] \ge \alpha^{k/n}$. The proof is complete. □

To estimate the rate of convergence of function values in the SDG method we shall need estimates of the distance in the transformed space from the current point to the optimal set. Such estimates can be easily derived for some version of the SDG method.

Theorem 49 *Let f be an almost differentiable function defined on a certain sphere $S_d = \{x : \|x - x^*\| \le d\}$ around the local minimum point x^* of f, and let for all $x \in S_d$ the almost-gradient $g(x)$ satisfies the inequalities*

$$N[f(x) - f(x^*)] \le (g(x), x - x^*) \le M[f(x) - f(x^*)], \tag{3.18}$$

where $M > N > 0$. If in the SDG method we set

1) $x_0 \in S_d$;

2)

$$h_{k+1} = \frac{2MN}{M+N} \frac{[f(x_k) - f(x^*)]}{\|\tilde{g}_k\|} \tag{3.19}$$

3)

$$1 < \alpha_{k+1} = \alpha \le \frac{M+N}{M-N}, k = 0, 1, \ldots, \tag{3.20}$$

then

$$\|A_k(x - x^*)\| \le d \qquad for \quad k = 0, 1, \ldots.$$

(see [Shor 70a]).

Proof. We shall prove the theorem by induction. Let

$$\|A_p(x_p - x^*)\| \le d.$$

for some p. We shall estimate

$$\|A_{p+1}(x_{p+1} - x^*)\| \le d.$$

Denote $\|A_k(x_k - x^*)\| = z_k$. We have

$$\begin{aligned}
\|z_{p+1}\|^2 &= \|R_{\alpha_{p+1}}(\xi_{p+1})(z_p - h_{p+1}\xi_{p+1})\|^2 \\
&= \|R_{\alpha_{p+1}}(\xi_p)z_p\|^2 - \\
&- 2h_{p+1}(R_{\alpha_{p+1}}(\xi_{p+1})z_p, R_{\alpha_{p+1}}(\xi_{p+1}))\xi_{p+1}) + \alpha_{p+1}^2 h_{p+1}^2 \\
&= \left[1 + ({\alpha_{p+1}}^2 - 1)\frac{(z_p, \xi_{p+1})^2}{\|z_p\|^2}\right] \|z_p\|^2 - \\
&- 2h_{p+1}{\alpha_{p+1}}^2(z_p, \xi_{p+1}) + {\alpha_{p+1}}^2{h_{p+1}}^2 \\
&= \|z_p\|^2 + {\alpha_{p+1}}^2[(z_p, \xi_{p+1}) - h_{p+1}]^2 - (z_p, \xi_p)^2.
\end{aligned} \tag{3.21}$$

Observe that

$$(z_p, \xi_{p+1}) = (A_p(x_p - x^*), \xi_{p+1}) = (x_p - x^*, A_p^* \xi_{p+1})$$
$$= \frac{1}{\tilde{g_p}}(g_f(x_p), x_p - x^*).$$

Let us consider two cases.

Case (a): $(z_p, \xi_{p+1}) \le h_{p+1}$.

Using the inequality

$$h_{p+1} = \frac{[f(x_p) - f(x^*)]2MN}{\|\tilde{g_p}\|(M+N)} \le \frac{(g_f(x_p), x_p) - x^*)2M}{\|\tilde{g_p}\|(M+N)}$$
$$= (z_p, \xi_{p+1})\frac{2M}{(M+N)},$$

which results from (3.18) and (3.19) , we see that

$$\alpha_{p+1}{}^2[(z_p, \xi_{p+1}) - h_{p+1}]^2 - (z_p, \xi_{p+1})^2$$
$$\le \alpha_{p+1}{}^2 \left[(z_p, \xi_{p+1}) - (z_p, \xi_{p+1})\frac{2M}{(M+N)}\right]^2 - (z_p, \xi_{p+1})^2$$
$$\le \left[\left(\frac{M+N}{(M-N)}\right)^2 \left(1 - \frac{2M}{(M+N)}\right)^2 - 1\right] (z_p, \xi_{p+1})^2 = 0.$$

Case (b): $(z_p, \xi_{p+1}) \ge h_{p+1}$.

In a similar way we obtain

$$h_{p+1} = \frac{[f(x_p) - f(x^*)]2MN}{\|\tilde{g_p}\|(M+N)} \ge \frac{(g_f(x_p), x_p) - x^*)2N}{\|\tilde{g_p}\|(M+N)}$$
$$= (z_p, \xi_{p+1})\frac{2N}{M+N},$$

and

$$\alpha_{p+1}{}^2[(z_p, \xi_{p+1}) - h_{p+1}]^2 - (z_p, \xi_{p+1})^2$$
$$\le \alpha_{p+1}{}^2 \left[(z_p, \xi_{p+1}) - (z_p, \xi_{p+1})\frac{2N}{(M+N)}\right]^2 - (z_p, \xi_{p+1})^2$$
$$\le \left[\left(\frac{M+N}{(M-N)}\right)^2 \left(1 - \frac{2N}{(M+N)}\right)^2 - 1\right] (z_p, \xi_{p+1})^2 = 0.$$

Hence in view of inequality 3.2 $||z_{p+1}|| \leq ||z_p||$. Consequently, $||A_k(x_k - x^*) \leq d||$ for all $k = 0, 1, \ldots,$ as required. □

From Theorems 47 – 49 we can deduce the following result.

Theorem 50 *If the assumptions of Theorem 49 are satisfied then there exists a subsequence of indeces $k_1, k_2, \ldots$ and a positive constant c such that*

$$f(x_{k_p}) - f(x^*) \leq c\alpha^{-k/n}, p = 1, 2, \ldots.$$

Moreover,

$$\min_{1 \leq i \leq k} [f(x_i) - f(x^*)] \leq \frac{G\sqrt{k(\alpha^2 - 1)}d}{\sqrt{\alpha^{2k/n} - 1}}$$

where $G = \max_{x \in S_d} ||g_f(x)||$ (see [Shor 70b]).

Proof. Since the minimizing sequence $\{x_k\}_{k=0}^{\infty}$ remains in the bounded set S_d , the almost gradients are uniformly bounded and thus the conditions of Theorem 47 and 48 are satisfied .

Let $\{k_p\}_{p=1}^{\infty}$ be a sequence for which, according to the Theorem 47 , one has $|| \tilde{g}_{k_p}|| \leq \tilde{c}\alpha^{-k_p/n}$. From (3.8) we obtain

$$\begin{aligned} N[f(x_{k_p}) - f(x^*)] &\leq (g_f(x_{k_p}), x_{k_p} - x^*) \\ &= (A^*_{k_p}\tilde{g}_{k_p}, x_{k_p} - x^*) \\ &= (\tilde{g}_{k_p}, A_{k_p}(x_{k_p} - x^*)) \\ &\leq ||(\tilde{g}_{k_p}||d \leq \tilde{c}d\alpha^{-k/n}. \end{aligned}$$

Setting $\frac{\tilde{c}d}{N} = c$ we get

$$f(x_{k_p}) - f(x^*)] \leq \frac{\tilde{c}d}{N}\alpha^{-k/n} = c\alpha^{-k/n}.$$

In similar way, using Theorem 48

$$\min_{1 \leq r \leq k} [f(x_r) - f(x^*)] \leq \min_{1 \leq r \leq k} ||\tilde{g}_r|| \frac{d}{N},$$

whence

$$\min_{1 \leq r \leq k} [f(x_r) - f(x^*)] \leq \frac{G\sqrt{k(\alpha^2 - 1)}d}{N\sqrt{\alpha^{2k/n} - 1}}$$

This proves the result. □

One can extend Theorem 49 to the case of nonunique minimum points of the function f. Suppose that the set M^* of all minimum points of a differentiable function f is bounded and the value of f on M^* is equal to f^*. Let us denote

$$\varrho_k(x_k) = \min_{x^* \in M^*} \|A_k(x_k - x^*)\| = \|A_k(x_k - x_k^*(x_k))\|; x_k^*(x_k) \in M^*.$$

Theorem 51 *Assume that for all x and x^* satisfying the conditions $\varrho_0(x) \leq d, x \notin M^*$ the following inequalities hold:*

$$N[f(x) - f^*] \leq (g_f(x), x - x^{**}) \leq M[f(x) - f^*],$$

where x^{**} is the point nearest to x in $M^* \cap \{ y : y = x + t(x^* - x), t \leq 0\}$.

If in the SDG algorithm

1) $\varrho_0(x_0) \leq d$,

2) $h_{k+1} = \frac{2MN}{M+N} \frac{f(x_k) - f(x^*)}{\|g_{\varphi_k}(y_k)\|}$,

3) $1 < \alpha_{k+1} = \alpha \leq \frac{M+N}{M-N}$,

then $\varrho_{k+1}(x_{k+1}) \leq d$ for $k = 1, 2, \ldots$ (see [Shor 70a]).

Proof. We shall show that the sequence $\{\varrho_k(x_k)\}$ is nonincreasing:

$$\begin{aligned} \varrho_{k+1}^2(x_{k+1}) &= \min_{y \in M^*} \|R_{\alpha_{k+1}}(\xi_{k+1}) A_k(x_{k+1} - y)\|^2 \\ &= \min_{y \in M^*} \|R_{\alpha_{k+1}}(\xi_{k+1}) A_k(x_k - y) - R_{\alpha_{k+1}}(\xi_{k+1}) h_{k+1} \xi_{k+1}\|^2 \\ &\leq \|R_{\alpha_{k+1}}(\xi_{k+1}) A_k(x_k - x_k^*(x_k)) - \alpha_{k+1} h_{k+1} \xi_{k+1}\|^2. \end{aligned}$$

Let $z_k = A_k(x_k - x_k^*(x_k))$. The remaining part of the proof is similar to that of the Theorem 49. We obtain $\|z_{k+1}\| \leq \|z_k\| \leq \|z_0\| \leq d,\ k = 0, 1, 2, \ldots$, as required. □

Theorem 52 *Let f be a convex function posessing the following property: there exists a constant $M > 1$ such that if the function*

$$\varphi(\alpha) = f((1 - \alpha)x_1 + \alpha x_2)$$

is strictly decreasing on $[0,1]$ *then*

$$(g_f(x_1), x_1 - x_2) \leq M[f(x_1) - f(x_2)]. \tag{3.22}$$

Furthermore, suppose that $\lim_{\|x\|\to\infty} f(x) = +\infty$. *Then the SDG method with*

$$\alpha_{k+1} = \frac{M+N}{M-N},$$

$$h_{k+1} = \frac{2M}{M+N} \frac{f(x_k) - \overline{f}}{\|\tilde{g}_k\|}$$

has the following properties: if $\overline{f} \geq f^*$ *then the sequence* $\{h_k\}$ *is bounded and for any* $\varepsilon > 0$ *one can find* $\overline{k}$ *such that* $f(x_{\overline{k}}) < \overline{f} + \varepsilon$ *(we assume that if* $f(x_k) < \overline{f}$ *at a certain step then the algorithm terminates); if* $\overline{f} < f^*$ *then the sequence* $\{h_k\}$ *is unbounded (see [Shor 70b]).*

Proof. Suppose that $\overline{f} \geq f^*$ and consider the function

$$\overline{f}(x) = \begin{cases} f(x) & \text{if } f(x) \geq \overline{f}, \\ \overline{f} & \text{if } f(x) < \overline{f}. \end{cases}$$

The assumption of Theorem 51 is satisfied for this function, and so the sequence $\{\varrho_k(x_k)\}$ is bounded, which implies the convergence of function values by virtue of Theorem 29. The convexity of $\overline{f}(x)$ yields

$$\frac{f(x_k) - \overline{f}}{\|\tilde{g}_k\|} \leq \varrho(x_k).$$

Hence $h_k \leq \frac{2M}{M+1}\varrho_k(x_k)$, i.e. $\{h_k\}$ is bounded.

If $\overline{f} < f^*$ then two cases are possible:

(a) $\sup_k f(x_k) < \infty$;

(b) $\sup_k f(x_k) = +\infty$.

In case (a) the conditions of Theorem 49 are satisfied, which implies then existence of a sequence $\{\|\tilde{g}_{k_p}\|\}_{p=0}^{\infty}$ converging to zero, and so

$$\sup_k h_k = \sup_k \frac{2M}{M+1} \frac{f(x_k) - \tilde{f}}{\|\tilde{g}_k\|} \geq \sup_k \frac{f^* - \tilde{f}}{\|\tilde{g}_k\|} = +\infty.$$

In case (b) we shall consider a ray originating at x_k and passing through the set of minima M^*. Let x_k^* be the first minimum point on this ray. By (3.22)

$$\frac{f(x_k) - f^*}{\|g_f(x_k)\|} \geq \frac{1}{M} \frac{(g_f(x_k), x_k - x_k^*)}{\|g_f(x_k)\|} = \frac{R_k}{M},$$

where R_k is the distance from x_k^* to the hyperplane supporting the level set $\{x : f(x) \leq f(x_k)\}$ at x_k. Since the set M^* is bounded and $\{f(x_k)\}$ is unbounded, $\sup_k R_k = +\infty$. Next,

$$\sup_k h_k = \sup_k \frac{2M}{M+1} \frac{f(x_k) - f}{\|\tilde{g}_k\|} \geq \sup_k \frac{2M}{M+1} \frac{f(x_k) - \tilde{f}}{\|\tilde{g}_f(x_k)\|} \geq$$
$$\geq \frac{2}{M+1} \sup_k R_k = +\infty.$$

The proof is complete. □

We shall use the above result to construct an algorithm for minimizing a function f with unknown f^*, assuming f satisfies the conditions of Theorem 52. The algorithm consists of a sequence of stages at which the SDG method of Theorem 52 is used.

At the beginning we choose $x_0^{(1)} \in E_n, h > 0, \Delta_0 > 0, \tilde{f}_0 = f(x_0^{(1)})$. Assume that r stages of the algorithm have been performed. At the beginning of the $(r+1)$-st stage we have

$$x_0^{(r+1)} \in E_n, h > 0, \Delta_r; \tilde{f}_r = f(x_0^{(r+1)}),$$

where $x_0^{(r+1)}$ is the point with the smallest objective value $\tilde{f}_r$ obtained so far. We set

$$\Delta_{r+1} = \begin{cases} \Delta_r & \text{if } f(x_0^{(r+1)}) \leq f(x_0^{(r)}) - \frac{\Delta_r}{2}, \\ \frac{\Delta_r}{2} & \text{if } f(x_0^{(r+1)}) > f(x_0^{(r)}) - \frac{\Delta_r}{2}, \end{cases}$$

and $\tilde{f}_{r+1} = f(x_0^{(r+1)}) - \Delta_{r+1}$. Setting $x_0 = x_0^{(r+1)}$ and $\tilde{f} = \tilde{f}_{r+1}$ we use the SDG method of the form described in Theorem 52. After a finite number of steps, for some $k = \tilde{k}_{r+1}$ one of the following cases must occur:
(a) $f(x_{\tilde{k}_{r+1}}) \leq f(x_0^{(r+1)}) - \frac{\Delta_{r+1}}{2}$;
(b) $h_{\tilde{k}_{r+1}+1} > h$.
This completes the $(r+1)$-st stage. The "record" value $\tilde{f}_{r+1} = \min_{0 \leq k \leq \tilde{k}_{r+1}} f(x_k^{(r+1)})$ and the corresponding point $x_{k^*}^{(r+1)} = x_0^{(r+2)}$ are stored and the $(k+2)$-nd stage is initialized.

We shall prove that $\lim_{r\to\infty} f(x_0^{(r)}) = f^*$. Since $\{f(x_0)^{(r)}\}$ is nonincreasing and bounded from below, $\lim_{r\to\infty} f(x_0^{(r)})$ exists. It is clear from the description of the algorithm and the proof of Theorem 52 that for a sufficiently small Δ_r case (b) may occur only if $\tilde{f}_{r+1} < f^*$. On the other hand once case (b) happens, case (a) may occur for only a finite number of subsequent stages. Therefore $\lim_{r\to\infty} \Delta_r = 0$. Since for sufficiently small Δ_r (with h fixed) case (b) implies that $f(x_0^{(r)}) - \Delta_r$ is a lower bound for f^*, we have

$$\lim_{r\to\infty} f(x_0^{(r)}) = f^*,$$

as required.

Under certain circumstances, e.g. when solving a system of nonlinear equations, we know f^*, but difficulties may arise in the estimation of M and N. If the constants M and N are chosen improperly, the (M, N)–**algorithm with space dilation** described in the Theorem 49 may converge too slowly or even diverge. Therefore it is important to have tests for detecting such "patalogical" situations. The simplest test is based on the analysis of the behavior of the sequence $\{h_k\}$.

Theorem 53 *If the parameters M and N in the (M, N)- algorithm are chosen correctly then the sequence $\{h_k\}$ is bounded. If the values of these parameters are wrong and the algorithm does not converge, then either $\{h_k\}$ or $\{f(x_k)\}$ is unbounded (see [Shor 70a], [Shor 85]).*

The proof is similar to Theorem 52.

To sum up, if h_k or $f(x_k)$ exceed sufficiently large bounds $h^{\max}$ or $f^{\max}$, one should increase constant M, decrease N and restart the SDG algorithm of Theorem 49 either from the initial point or from the best point obtained so far.

The operators of space dilation can be used for converting the ellipsoidal approximations of a region of minimum localization in initial space to the "ball" approximations in transformed space. This idea is used in so-called ellipsoid method that may by interpreted as a partial case of SDG methods.

3.3 THE ELLIPSOID METHOD AND ITS GENERALIZATIONS

There are some methods of centered sections for solution of n -dimensional convex programming problems based on the common ideas:

a) approximation of the set of localization of minimum points by convex bodies taken from the given class $\mathcal{K}$ of "simple' bodies.

b) finding of the "center" (in a certain sense) $\overline{c}(K)$ of approximating body $K \in \mathcal{K}$ and constructing cutting hyperplane $P(\overline{c}, g)$, crossing the central point orthogonally to the subgradient direction $g(\overline{c})$.

c) by hyperplane $P(\overline{c}, g)$ the space E^n is divided into two halfspaces: $H_1(\overline{c}, g) = \{x : (g(\overline{c}), x - \overline{c}) \leq 0\}$ and $H_2(\overline{c}, g) = \{x : (g(\overline{c}), x - \overline{c}) \geq 0\}$ and it is guaranted that minimum points are in the $H_1(\overline{c}, g)$.

d) the intersection $K \bigcap H_1(\overline{c}, g) = \overline{K_1}$ gives us a new region of minimum points localization. Then we approximate $\overline{K}_1$ by $K_1 \in \mathcal{K}$, go iteratively to (b), and the procedure is repeated.

One of the first methods of such type was proposed by A.Levin in 1965 [Lev 65] (see also [New 65]) for minimizing convex programming problems. This method was called **the method of centered sections**. At each iteration of the method, knowing a compact convex polyhedron locating an optimal point, one finds the center of gravity of this polyhedron. By calculating a subgradient at this center one can locate the optimal point in the halfspace defined by the corresponding supporting hyperplane to the level set, and thus one can cut off the part of the polyhedron that lies on the other halfspace. It can be shown (see [Mit 68]) that at each iteration at least an $(n/(n+1))^n$ - th part of the volume polyhedron is cut off. This leads to a decrease in the location volumes with a speed not less than the speed of decreasing geometrical progression with a ratio

$$q_n = \left(1 - \frac{n}{n+1}\right)^n < 1 - \frac{1}{e} \quad (q_n \to 1 - \frac{1}{e} \text{ when } n \text{ tends to infinity }) \quad (3.23)$$

However for $n > 3$ this algorithm is impractical, since calculating centers of gravity of multidimensional polyhedra requires much effort. D.B.Yudin and

A.S.Nemirovskii [YN 76] proposed a modified method of centered sections for solving convex programming problems:

$$f(x) \to \min, x \in E^n, \tag{3.24}$$

$$\varphi(x) \leq 0, \tag{3.25}$$

where f and φ are convex functions. (If the number of constraint inequalities is more than one, for example, $f_i(x) \leq 0, i = 1, \ldots, m$, we can set: $\varphi(x) = \max_{\{i: i \in \{1,\ldots,m\}\}} f_i(x)$ and obtain one equivalent inequality). Let the optimum point x^* be located in a ball $S(x_0, R) = \{x : \|x - x_0\| \leq R\}$.

Define the hyperplane $P(x_0) = \{x : (g(x_0), x - x_0) = 0\}$, and halfspace $H(x_0) = \{x : (g(x_0), x - x_0) \leq 0\}$, where

$$g(x_0) = \begin{cases} g_f(x_0) & \text{if } \varphi(x_0) \leq 0; \\ g_\varphi(x_0) & \text{if } \varphi(x_0) > 0. \end{cases} \tag{3.26}$$

The location area of x^* reduces to a halfball $S_H(x_0, R) = S(x_0, R) \cap H(x_0)$. Let us construct an ellipsoid of minimum volume containing this halfball. The center of the ellipsoid lies on the ray $x_0 - hg(x_0), h > 0$, and the ellipsoid is flattered in the direction $g(x_0)$.

To complete the recursion, one has to transform the ellipsiod into a sphere, which can be done by space dilation in the direction $\xi = \frac{g(x_0)}{\|g(x_0)\|}$ with a certain coefficient. Thus modified centered sections lead in a natural way to a subgradient type algorithm with space dilation in the direction of the gradient (see [Shor 77]).

Consider the procedure of finding parameters of n-dimensional minimum volume ellipsoid circumscribed around the half-sphere. For simplicity let $x_0 = 0, R = 1$ and $P_H(0) = \{x : x_n \leq 0\}$, half-sphere

$$S_H(0,1) = S(0,1) \cap P_H(0)$$

The desired optimal ellipsoid has the form:

$$E^* = \{x : \frac{\|\overline{x}\|^2}{a^2} + \frac{(x_n - h)^2}{b^2} \leq 1\}$$

where $\overline{x} = \{x_1 \ldots, x_{n-1}, 0\}$, a is the length of maximal half-axis, b is the length of minimal half-axis, h is the n-th coordinate of the center of E^*. Set $y^2 = \|\overline{x}\|^2$.

The points that satisfy the equation $\frac{1}{y^2} + \frac{h^2}{b^2} = 1$ and the point $(\overbrace{0\ldots,0}^{(n-1)\text{ time}}, 1)$

lies on the surface of E^*. We obtain: $(1+h)^2 = b^2$, hence

$$b = 1 + h$$

$$\frac{1}{a^2} = 1 - \frac{h^2}{b^2} = 1 - \frac{h^2}{1+h^2};\, a^2 = \frac{(1+h)^2}{1+2h}$$

The square of volume E^*

$$V^2(E^*) = \omega_n^2 (a^2)^{n-1} b^2 = \omega_n^2 \cdot \frac{(1+h)^{2n}}{(1+2h)^{n-1}} \tag{3.27}$$

(ω_n is the volume of a ball $S(0,1)$).

To find the optimal value $\bar{h}$ of h we have to minimize $\varphi(h) = \ln[\frac{1}{\omega_n^2} V^2(E^*)]$.

$$\begin{aligned} \varphi(h) &= (n-1)\ln\frac{(1+h)^2}{1+2h} + 2\ln(1+h) \\ &= 2n\ln(1+h) - (n-1)\ln(1+2h). \end{aligned}$$

$$\frac{d\varphi}{dh}(\bar{h}) = \frac{2n}{1+\bar{h}} - \frac{2(n-1)}{1+2\bar{h}} = 0 \tag{3.28}$$

$$\bar{h} = -\frac{1}{n+1}. \tag{3.29}$$

From (3.29) one can deduce:

$$a = \frac{n}{\sqrt{n^2-1}}; \quad b = \frac{n}{n+1} \tag{3.30}$$

$$\alpha = \frac{a}{b} = \sqrt{\frac{n+1}{n-1}}; \quad \frac{V^2(E^*)}{V^2(S(0,1))} = \left(\frac{n^2}{n^2-1}\right)^n \frac{n-1}{n+1} \tag{3.31}$$

For the arbitrary initial sphere $S(x_0, R)$ and vector $g(x_0) \neq 0$, let $\xi = \frac{g(x_0)}{\|g(x_0)\|}$.

Consider the minimum volume ellipsoid $E^*(x_0, R, \xi)$ containing

$$S_H(x_0, R) = S(x_0, R) \cap H(x_0).$$

Due to the formulas (3.29)–(3.31) the center of $E^*(x_0, R, \xi)$ lies in the point $x_0 - \frac{R}{n+1}\xi$, the length of a small half-axis is $\frac{Rn}{n+1}$, and of large ones is $\frac{Rn}{\sqrt{n^2-1}}$. The direction of the small axis is collinear with ξ.

Due to (3.31) one can obtain the ratio ϱ of volume of ellipsoid $E^*(x_0, R, \xi)$ to the volume of sphere $S(x_0, R)$: $\varrho^2 = \frac{V^2(E^*_{x_0,R,\xi})}{V^2(S(x_0,R))}$ equal to the right part of the formula (3.31). For large n one may use an approximate formula:

$$\varrho^2 \approx 1 - \frac{1}{n}.$$

Thus, we have

$$\varrho = \frac{V(E^*_{a_0,R,\xi})}{V(S(a_0,r))} = \frac{n^n}{\sqrt{n^2-1}^n}\sqrt{\frac{n-1}{n+1}} = 1 - \frac{1}{2n} + O\left(\frac{1}{n^2}\right). \qquad (3.32)$$

Use the dilation of the space in the direction ξ with coefficient $\alpha = \sqrt{\frac{n+1}{n-1}}$; i.e. linear transformation $E \to E^n : y = R_\alpha(\xi)x$.

After this transformation the ellipsoid $E^*_{x_0,R,\xi}$ converts into a sphere $S(y, R_1)$ in transformed space with center in the point $y_1 = R_\alpha(\xi)(x_0 - \bar{h}\xi)$ and $R_1 = \frac{n}{\sqrt{n^2-1}}R$. So we can repeat the described above procedure in the transformed space and so on. Corresponding algorithm was independently developed in the work [Shor 77] as a partial case of SDG-method, and in work [YN 76] as a modified method of centered sections.

Note that we use only the following property of $g(x)$ and x^*, namely, $(g(x), x - x^*) \leq 0$ for all $x \in E^n$. So we can formulate the problem in a more general form.

Assume we have a vector field g on E^n, which is not necessarily continuous; $g \in E^n$ for all $x \in E_n$. We shall consider the problem of finding a point x^* such that $(g(x), x - x^*) \geq 0$ for all $x \in E^n$. We assume that this problem has a solution and that it is known a priori that $x^* \in S(x_0, R)$, where $S(x_0, R)$ denotes the closed ball with center x_0 and radius R.

Consider the following iterative algorithm for solving this problem for $n > 1$. In the description of the algorithm we assume that $g \neq 0$ if $x \neq x^*$.

Before starting the computation we have $x_0 \in E_n$, $B_0 = I$ (an identity matrix) and $h_0 = R/(n+1)$. Suppose that at the k-th iteration we obtained: $x_k \in E^n$ and $n \times n$ matrix B_k. At $(k+1)$-st iteration calculate:

(1)

$$g(x_k) \text{ (if } g(x_k) = 0\text{, then } x_k \text{ is a solution)}, \qquad (3.33)$$

(2)

$$\xi_k = \frac{B_k^* g(x_k)}{||B_k^* g(x_k)||}, \tag{3.34}$$

(3)

$$x_{k+1} = x_k - h_k B_k \xi_k \tag{3.35}$$

(4)

$$B_{k+1} = B_k R_\beta(\xi_k), \quad \beta = \sqrt{\frac{n-1}{n+1}}, \tag{3.36}$$

where $R_\beta(\xi_k)$ is the operator of space dilation in the direction ξ_k with the coefficient β,

(5)

$$h_{k+1} = rh_k, \quad \text{with} \quad r = \frac{n}{\sqrt{n^2-1}}. \tag{3.37}$$

The values h_0, β and r are choosen in accordance with formulas (3.29), (3.30).

Theorem 54 *A sequence $\{x_k\}_{k=0}^{\infty}$ generated by the algorithm satisfies the inequalities*

$$||A_k(x_k - x^*)|| \leq h_k(n+1) \quad \textit{for all } k = 0, 1, \ldots \tag{3.38}$$

where $A_k = B_k^{-1}$.

Proof. We shall prove the theorem by induction. For $k = 0$ inequality reads $||x_0 - x^*|| \leq R$ and holds by assumption. Suppose that it satisfies for $k = \bar{k}$. We shall show that it also holds for $k = \bar{k} + 1$.

For brevity let $z_k = A_k(x_k - x^*)$ and $\alpha = 1/\beta$. Then

$$\begin{aligned}
||z_{k+1}||^2 &= ||R_\alpha(\xi_{\bar{k}})(z_{\bar{k}} - h_{\bar{k}}\xi_{\bar{k}})||^2 \\
&= ||z_{\bar{k}} - h_{\bar{k}}\xi_{\bar{k}} + (\alpha - 1)[(z_{\bar{k}} - h_{\bar{k}}\xi_{\bar{k}})]||^2 \\
&= ||z_{\bar{k}} + [-\alpha h_{\bar{k}} + (\alpha - 1)(z_{\bar{k}}, \xi_{\bar{k}})]\xi_{\bar{k}}||^2 \\
&= ||z_{\bar{k}}||^2 - 2h_{\bar{k}}(z_{\bar{k}}, \xi_{\bar{k}})\alpha^2 + (\alpha^2 - 1)(z_{\bar{k}} - \xi_{\bar{k}})^2 + \alpha^2 h_{\bar{k}}^2 \\
&= ||z_{\bar{k}}||^2 - 2h_{\bar{k}}(z_{\bar{k}}, \xi_{\bar{k}})\tfrac{n+1}{n-1} + \tfrac{2}{n-1}(z_{\bar{k}}, \xi_{\bar{k}})^2 + \tfrac{n+1}{n-1} h_{\bar{k}}^2.
\end{aligned}$$

Observe that

$$\begin{aligned}(z_{\bar{k}}, \xi_{\bar{k}}) &= \frac{1}{\|\tilde{g}(x_{\bar{k}})\|}(A_{\bar{k}}(x_{\bar{k}} - x^*)), B^*_{\bar{k}} g(x_{\bar{k}})) \\ &= \frac{1}{\|\tilde{g}(x_{\bar{k}})\|}(x_{\bar{k}} - x^*, g(x_{\bar{k}})) \geq 0,\end{aligned}$$

where $\tilde{g}(x_{\bar{k}}) = B^*_{\bar{k}} g(x_{\bar{k}})$, so $(z_{\bar{k}}, \xi_{\bar{k}}) \leq \|z_{\bar{k}}\| \leq h_{\bar{k}}(n+1)$. Therefore

$$\frac{2}{n-1}(z_{\bar{k}}, \xi_{\bar{k}})^2 - 2h_{\bar{k}}\frac{n+1}{n-1}(z_{\bar{k}}, \xi_{\bar{k}}) \leq \frac{2}{n-1}h^2_{\bar{k}} \leq \frac{2}{n-1}[(z_{\bar{k}}, \xi_{\bar{k}})^2 - (z_{\bar{k}}, \xi_{\bar{k}})^2] = 0.$$

Hence

$$\begin{aligned}\|z_{k+1}\|^2 \leq \|z_{\bar{k}}\|^2 + \frac{n+1}{n-1}h^2_{\bar{k}} &\leq (n+1)^2 h^2_{\bar{k}} + \frac{n+1}{n-1}h^2_{\bar{k}} \\ &= (n+1)^2 \frac{n^2}{n^2-1} h^2_{\bar{k}} = (n+1)^2 h^2_{\bar{k}+1}.\end{aligned}$$

and the desired conclusion follows. □

The set of all points x satisfying the inequality

$$\|A_k(x - x^*)\| \leq (n+1)h_k = R\left(\frac{n}{\sqrt{n^2-1}}\right)^k$$

is an ellipsoid Φ_k with volume $v(\Phi_k)$ equal to $v_0 R^n \left(\frac{n}{\sqrt{n^2-1}}\right)^{nk} \Big/ \det A_k$ where v_0 is the volume of the unit n-dimensional ball. We obtain

$$\begin{aligned}\frac{v(\Phi_{k+1})}{v(\Phi_k)} &= \frac{\left(\frac{n}{\sqrt{n^2-1}}\right)^2 \det A_k}{\det A_{k+1}} = \frac{\left(\frac{n}{\sqrt{n^2-1}}\right) \det A_k}{\det R_\alpha(\xi_k) \det A_{k+1}} \\ &= \frac{1}{\alpha}\left(\frac{n}{\sqrt{n^2-1}}\right)^n = \frac{n-1}{n+1}\left(\frac{n}{\sqrt{n^2-1}}\right)^n = q_n < 1.\end{aligned}$$

The volume of the ellipsoid locating the solution x^* according to inequality (3.38) decreases with the speed of a geometrical progression with a ratio q_n.

Let us consider possible applications of the above procedure to the solution of various mathematical programming problems.

1. The Problem of Minimizing a Convex Function on a Ball

Let f be a convex function defined on E^n. We seek its minimum on the ball $S(x_0 R) = \{x : \|x - x_0\| \leq R\}$.

Let x^* be a minimum point of f on $S(x_0, R.)$ Then for all $x \in S(x_0 R)$ we have

$$(g_f(x), x - x^*) \geq f(x) - f(x^*) \geq 0.$$

For any $x \notin S(x_0, R)$

$$\begin{aligned}(x - x_0, x - x^*) &= \|x - x_0\|^2 - (x - x_0, x^* - x_0) \\ &\geq \|x - x_0\|^2(\|x - x_0\| + \|x^* - x_0\|) \geq 0.\end{aligned}$$

Thus we may construct a vector field g, satisfying $(g(x), x - x^*) \geq 0$ for all $x \in E^n$, as folows

$$g(x) = \begin{cases} g_f(x) & \text{if } x \in S(x_0 R), \\ \frac{x - x_0}{\|x - x_0\|} & \text{if } x \notin S(x_0 R) \end{cases} \tag{3.39}$$

Let us apply the algorithm (3.33)-(3.37) with g defined by (3.39) to minimize a convex function on a ball. From Theorem 54 we obtain that the volume of the area locating the solution decreases with speed of geometrical progression with a ratio

$$q_n = \sqrt{\frac{n-1}{n+1}\left(\frac{n}{\sqrt{n^2 - 1}}\right)^n}.$$

Let us consider the case when the unconstrained minimum of f is attained at interior point x^* of the ball $S(x_0, R)$. Since the algorithm (3.33)–(3.37), when applied to the problem of minimizing f, in fact realizes the SDG algorithm with minor modifications, one may use Theorem 50 for estimating the speed of convergence of the function values. We obtain the following estimate

$$\begin{aligned} \varrho_k &= \min_{i \in \overline{1k}} \|B_i^* g(x_i)\| \leq \frac{d_k \sqrt{k(\alpha^2 - 1)}}{\sqrt{\alpha^{2k/n} - 1}}, \\ d_k &= \max_{i \in \overline{1k}} \|g(x_i)\|. \end{aligned} \tag{3.40}$$

Let the minimum in (3.40) be attained at $i = i(k)$. Then

$$\begin{aligned} \mu_k &= (x_{i(k)} - x^*, g(x_{i(k)})) = (A_{i(k)}(x_{i(k)} - x^*) B_{i(k)}^* g(x_{i(k)})) \\ &\leq \|A_{i(k)}(x_{i(k)} - x^*)\| \frac{d_k \sqrt{k(\alpha^2 - 1)}}{\sqrt{\alpha^{2k/n} - 1}} \\ &\leq h_{i(k)}(n+1) \frac{d_k \sqrt{k(\alpha^2 - 1)}}{\sqrt{\alpha^{2k/n} - 1}} \\ &\leq h_{i(k)}(n+1) d_k \frac{\sqrt{k(\alpha^2 - 1)}}{\sqrt{\alpha^{2k/n} - 1)}} = \frac{R d_k \sqrt{k(\alpha^2 - 1)}}{\sqrt{1 - \beta^{2k/n}}} q_n^{k/n}. \end{aligned}$$

We observe that if $x_{i(k)} \notin S(x_0, R)$ then $g(x_{i(k)}) = (x - x_0)/\|x - x_0\|$ and since x^* lies in the interior of $S(x_0, R)$, $(x_{i(k)} - x^*, (x_{i(k)} - x_0)/\|x_{i(k)} - x_0\|) \geq \delta > 0$.

Therefore $\mu_k \to 0$ implies that $x_{i(k)} \in S(x_0, R)$ for all sufficiently large k. But if $x_{i(k)} \in S(x_0, R)$ then $g_{i(k)} = g_f(x_{i(k)})$ and

$$\begin{aligned} f(x_{i(k)}) - f(x^*) &\leq (g_f(x_{i(k)}), x_{i(k)} - x^*) \\ &\leq \frac{Rd_k\sqrt{k(\alpha^2-1)}}{\sqrt{1-\beta^{2k/n}}} q_n^{k/n}. \end{aligned}$$

Consequently if f attains its minimum at the interior point x^* of the ball $S(x_0, R)$, then in the algorithm (3.33)–(3.37) the deviation of the "record" function values from $f(x^*)$ satisfies

$$r_k = \min_{i \in \overline{1,k}} [f(x_i) - f(x^*)] \leq \frac{Rd_k\sqrt{k(\alpha^2-1)}}{\sqrt{1-\beta^{2k/n}}} q_n^{k/n}.$$

If $x_k \in S(x_0, R)$ for $k = 0, 1, \ldots$, then one can obtain from Theorem 50 the following more accurate estimate

$$f(x_k) - f(x^*) \leq c\, q_n^{k/n}$$

for some subsequence $\{k_i\}_{i=1}^{\infty}$. For large n, q_n is well approximated by the following asymtotic formula

$$q_n \approx 1 - \frac{1}{2n}.$$

It follows that although r_k converges essentially at the speed of a geometrical progression (if one neglects the slow inner multiplier $\sqrt{k}$), the coefficient of this progression is close to one for large n i.e. the convergence may be slow in practice.

An advantage of the algorithm (3.33)-(3.37) is that the guaranteed speed of convergence depends only on the dimension of the space and does not require the knowledge of specific properties of f.

2. The General Convex Programming Problem

The problem is to

$$\begin{array}{l} \text{minimize } f_0(x), \\ \text{subject to } f_i(x) \leq 0, \quad i = 1 \ldots, m \quad x \in E_n \quad n > 1, \end{array} \tag{3.41}$$

where f_v are convex functions defined on E_n with subgradients $g_v(x) v = 0, 1, \ldots, m$. We assume that it is known a priori that an optimal point x^* exists in a ball $S(x_0, R)$ (formally, one may append the constraint $\|x - x_0\| \leq R$ to (3.41)).

Consider the vector field g

$$g(x) = \begin{cases} g_0(x) & \text{if } \max_{i\in[1,m]} f_i(x) \le 0, \\ g_{i^*}(x) & \text{if } \max_{i\in[1,m]} f_i(x) = f_{i^*}(x) > 0. \end{cases} \tag{3.42}$$

We shall show that $(g(x), x - x^*) \ge 0$ for all $x \in E^n$. If $\max_{i\in\overline{1,m}} f_i(x) \le 0$, then $g(x) = g_0(x)$ and

$$(g(x), x - x^*) = (g_0(x), x - x^*) \ge f_0(x) - f_0(x^*) \ge 0$$

If $\max_{i\in\overline{1,m}} f_i(x) > 0$, then $g(x) = g_{i^*}(x)$ with $f_{i^*}(x^*) > 0$, and $f_{i^*}(x^*) \le 0$, so $(g(x), x - x^*) = (g_{i^*}(x), x - x^*) \ge f_{i_*}(x) - f_{i^*}(x^*) \ge 0$. Therefore, $(g(x), x - x^*) \ge 0$ for all $x \in E^n$. Due to this inequality, we can use the algorithm (3.33)-(3.37) for locating x^*, with $g(x)$ calculated by (3.42).

By Theorem 54, after k iterations of the algorithm (3.33)-(3.37) x^* will be located in an ellipsoid Φ_k centered at x_k with volume $v(\Phi_k) = v_0 q_n^k$ (v_0 denotes the volume of the ball $S(x_0, R)$).

Let us observe that the result remains valid if we use $g_{\bar{i}}(x)$ in (3.42) instead of $g_{i^*}(x)$, where $\bar{i}$ is an arbitrary index satisfying $f_{\bar{i}}(x) > 0$.

3. The Saddle Point Problem

Let f be a convex-concave function of two vector variables $x \in E^n$ and $y \in E^m$, $\{x, y\} = z \in E^n \times E^m = E^{n+m}$, z^* be a saddle point of f, z_0 be an initial approximation to a solution, and suppose that one knows *a priori* that $\|z_0 - z^*\| \le R$.

Consider the pseudo-gradient set

$$G(z) = G_f^x(x, y) \times (-G_f^y(x, y),$$

where $G_f^x(x, y)$ is the set of partial subgradients of the function $f(x, y)$, considered as a function of x for fixed y, and $-G_f^y(x, y)$ is the set of subgradients of the function $-F(x, y)$ of y for fixed x.

Define the vector field g as follows

$$g(z) = \{g_f^x(z), -g_f^y(z)\} \quad g_f^x(z) \in G_f^x(z), \quad g_f^y(z) \in G_f^y(z).$$

We shall show that $(g(z), z - z^*) \ge 0$.

By the definition of a saddle point, we have $f(x, y^*) \geq f(x^*, y^*) \geq f(x^*, y)$, so

$$\begin{aligned} 0 &\leq f(x,y^*) - f(x^*,y) = f(x,y^*) - f(x,y) + f(x,y) - f(x^*,y) \\ &\leq (g_f^x(z), x - x^*) - (g_f^y(z), y - y^*) \\ &= (g(z), z - z^*). \end{aligned}$$

Consequently, for locating x^* one can apply the algorithm (3.33)-(3.37), using the pseudo-gradients $\{g_f^x(z); -g_f^y(z)\}$ instead of $g(z)$. Other subgragient-type methods for saddle point seeking may be found in [Gols 72].

Ellipsoid method for minimization of a convex function converges (in a certain sense) to the minimum value of the objective function with the speed of geometrical progression with ratio $q_n^{\frac{1}{n}}$ but for large n $q_n^{\frac{1}{n}} \approx 1 - \frac{1}{2n^2}$ and very close to 1, i.e. the convergence may be slow in practice. The number of iterations necessary to guarantee the 10 times decrease of the divergence of the record obtained at previous iteration from the optimal value, equals approximately $1.6n^2$. Therefore the efforts were made to find accelerated algorithms similar to ellipsoid method. We used two main strategies for this purpose.

a) Constructing "deeper" sections.

Consider the convex programming problem.

$$f(x) \to \min, x \in E^n, \tag{3.43}$$

$$\varphi(x) \leq 0, \tag{3.44}$$

and at the point $\bar{x} \in E^n$ $\varphi(\bar{x}) > 0$.

The hyperplane

$$P(\bar{x}) = \{x : (g_\varphi(\bar{x}), x - \bar{x}) + \varphi(\bar{x}) = 0\}$$

can be considered as a true cutting hyperplane in the sense that the half-space

$$H(\bar{x}) = \{x : (g_\varphi(\bar{x}), x - \bar{x}) + \varphi(\bar{x}) \leq 0\}$$

contains the optimal point set.

Similarly, if $\varphi(\bar{x}) \leq 0$ and we know some value $\bar{f} \geq f^*$, $\bar{f} < f(\bar{x})$, then one may construct a trully "deep" section by the cutting hyperplane

$$P_f(\bar{x}) = \left\{x : (g_f(\bar{x}), x - \bar{x}) + [f - \bar{f}] = 0\right\} \tag{3.45}$$

In this case we use the half-space

$$H_f(\overline{x}) = \{x : (g_f(\overline{x}), x - \overline{x}) + [f - \overline{f}] \leq 0\} \tag{3.46}$$

as the region of localization of optimal set for problem (3.43), (3.44).

The cutting hyperplanes mentioned in (3.45) are different from the usual central cutting hyperplanes $P(c, g)$ by parallel shift towards the more essential section of region containing the optimal points.

b) Constructing of a new ellipsoid of minimum volume circumscribed around the region of localization of optimal points obtained after one, two or more true sections of a given ellipsoid. In this case the volume of the ellipsoid containing the optimum may decrease faster per each section.

Based on the mentioned above techniques new versions of the algorithm were developed (see [Ge 77], [Ge 80], [GTo 82], [Go 83], [SGe 79]). The most complex stage of iteration in generalized ellipsoid methods is the calculation of parameters of minimum volume ellipsoids circumscribed around the bodies which arise from a ball after two or more deep sections. Note, that after r - multiple sections of a ball ($r < n$) the optimal circumscribed ellipsoid saves due to the symmetry $(n - r)$ equal principal axes. Therefore to convert it in a ball no more than r transformation of space dilation type are needed.

V.J.Gershovich [Ge 81b] proved the theorem about the speed of convergence for different modifications of ellipsoid method which were used to minimize convex function $f(x)$ defined on E^n. Let X^* be the minimum set of $f(x)$. $S_0 = \{x : \|x - x_0\| \leq r_0\}$ is a ball, containning points from X^*.

Omitting some nonessential details the generalized scheme of ellipsoid method can be described as following.

Before the $(k+1)$-step, $(k = 1, 2, \ldots)$ we have a point $x_k \in E^n$, nondegenerate $n \times n$ matrix B_k and number $r_0 > 0$ which determine an ellipsoid $E_k = \{x : \|B_k^{-1}(x - x_k)\| \leq r_k\}$, such that $\overline{X} = X^* \cap E_k$ is nonempty ($B_0 = I_n; E_0 = S(x_0, R)$). $(k+1)$-st step consists of two stages:

a) *Refinement of minimum point location.*

According to some procedure choose the pairs $\{x_k^l, f_{kl}\}$ ($\{x_k^l \in E^n, f_{kl}$ is the value for "shifting") $l = 0, 1, \ldots, l_k$.

Each pair $\{x_k^l, f_{kl}\}$ defines the cutting hyperplane

$$P(x_k^l, f_{kl}) = \{x : (g(x_k^l), x - x_k^l) + (f(x_k^l) - f_k^l) = 0\},$$

and the correspoding half-space

$$H(x_k^l, f_{kl}) = \{x : (g(x_k^l), x - x_k^l) + (f(x_k^l) - f_k^l) \leq 0\},$$

such that $\overline{X} \subseteq H(x_k^l, f_{kl})$. The body

$$\mathcal{Z}_k = E_k \cap \left(\cap_{l=o}^{l_k} H(x_k^l, f_{kl})\right)$$

contains $\overline{X} \subseteq X^*$. The main property that the procedure of constructing $\mathcal{Z}_k$ has to possess is the following: the ellipsoid $E(\mathcal{Z}_k) \supset \mathcal{Z}_k$ exists, such, that

$$q(\mathcal{Z}) = \frac{V(E(\mathcal{Z}))}{V(E_k)} \leq q < 1,$$

where q is some number, independent on k.

b) *Second stage: recalculation of parameters.*

Find $E_{k+1} = E(\mathcal{Z}_k)$, i.e. calculate the vector $x_{k+1} \in E^n$, nondegenerate $n \times n$ matrix B_{k+1} and number $r_{k+1} > 0$, such that

$$E_{k+1} = \{x \| \|B_{k+1}^{-1}(x - x_{k+1})\| \leq r_{k+1}\} = E(\mathcal{Z}_k).$$

Consider the sequence of records $\{f_k^r\}_{k=0}^{\infty}$ generated by generalized ellipsoid algorithm.

$$f_k^r = \left\{f_{k-1}^r, \min_l f(x_k^l)\right\}, \quad l = 0, 1, \ldots, l_k; \quad k = 1, 2, \ldots \tag{3.47}$$

$$f_0^r = f(x_0).$$

Theorem 55 *Let $x^* \in X^* \cap \mathrm{int}\ S_0$, $\|g(x)\| \leq L$ for all $x \in S_0$ and "shift" values for f_{kl} satisfy the condition:*

$$f_{kl} = \min\left\{f_{k-1}^r, \min_p \{f(x_k^p)\}, \quad 0 \leq p \leq l\right\}, \ l = 0, 1, \ldots, l_k; \ k = 0, 1, \ldots$$

Then the sequences of matrices $\{B_k\}_{k=0}^{\infty}$, *numbers* $\{r_k\}_{k=0}^{\infty}$ *and* $\{f_k^r\}_{k=0}^{\infty}$ *generated by a generalized ellipsoid algorithm satisfy the inequalities :*

$$f_k^r - f(x^*) \le 2Lr_{k+1}\sqrt{\lambda_n(B_{k+1}B_{k+1}^{(t)})}, \quad k = 0, 1, \ldots \tag{3.48}$$

and the following inequalities are true:

$$f_k^r - f(x^*) \le 2Lr_0 \left(\prod_{i=0}^{k} q(Z_i)\right)^{\frac{1}{n}} \le 2Lr_0 q^{\frac{k}{n}}, \quad k = 0, 1, \ldots \tag{3.49}$$

Proof. Note that all previous cuttings do not cut the Lebesque set $\{x \in E^n : f(x) \le f_k^r\}$, and the constant L can be used as Lipschitz's constant of function f in a ball $\{x : ||x|| \le r_0\}$.

Consider the optimal point $x^* \in E_{k+1}$ and a straight line $S = \{x \in R^n : x = x^* + vt\}$, where $t \in R, v \in R^n$ is the vector which is collinear to the eigenvector associated with minimal eigenvalue of matrix $B_{k+1}B_{k+1}^{(t)} = H_k$. This line intersects with the level surface $\{x : f(x) = f_k^r\}$ in some point $y \in E_{k+1}$. The length of the smallest principal axis of ellipsoid E_{k+1} equals $2r\sqrt{\lambda_n(H_k)}$. So, $||y - x^*|| \le 2r\lambda_n(H_k)$. Further, $f_k^r - f(x^*) \le f(y) - f(x^*) \le 2Lr_{k+1}\sqrt{\lambda_n(B_{k+1} \cdot B_{k+1}^{(t)})}$. Since $\sqrt{\lambda_n(B_{k+1}B_{k+1}^{(t)})} \le \frac{r_0}{r_{k+1}} \left(\prod_{i=0}^{k} q(Z_i)\right)^{\frac{1}{n}}$ we have:

$$f_k^r - f(x^*) \le 2Lr_0 \left(\prod_{i=0}^{k} q(Z_i)\right)^{\frac{1}{n}} \le 2Lr_0 q^{\frac{k}{n}}, \; k = 0, 1, \ldots$$

This completes the proof.□

3.4 METHODS WITH SPACE DILATION IN DIRECTION OF THE DIFFERENCE OF TWO SUCCESSIVE SUBGRADIENTS IN TRANSFORMED SPACE

Experience of application of the algorithms with space dilation in the direction of the gradient revealed the possibility of significant acceleration of the subgradient process with employment of linear operators that change the space metric. Difficulties inherent in the selection of step multipliers when f^* is unknown gave an impetus to the search for new methods of nonsmooth optimization with variable metric where the selection of a step multiplier is associated witn an approximate search for the minimum along the chosen direction. In 1971 the author suggested in his doctoral thesis and in cooperation with N.G. Zhurbenko, studied experimentally a family of gradient-type algorithms with space dilatation in the direction of the difference of two successive gradients ([Shor 70b], [SZ 71]).

In a certain sense this family of algorithms can be considered as generalization of the conjugate gradient algorithms for minimization of smooth functions.

Let $f(x), x \in E^n$, be a convex quadratic function: $f(x) = \frac{1}{2}(Ax, x)$, $x^* = 0$ is a minimum point of this function. For any given point $x_0 \neq 0$, $g_f(x_0) = Ax_0$, and one can made iteration of the steepest descent method:

$$x_1 = x_0 - h_0 g_f(x_0),$$

where

$$h_0 = \arg\min_t f(x_0 - t g_f(x_0)),$$

i.e.

$$h_0 = \frac{(Ax_0, x_0)}{(A^2 x_0, x_0)}.$$

Consider the difference of gradients of $f(x)$ taken in the points x_1 and x_0: $r = g_f(x_1) - g_f(x_0)$ and one can see that

$$(x_1, g_f(x_1) - g_f(x_0)) = (Ax_1, x_1 - x_0) = (g_f(x_1), x_1 - x_0) = 0,$$

so r is orthogonal to the direction $-x_1$ from x_1 to the point of minimum $x^* = 0$. Therefore, if one uses the transformation of argument space: $y = R_\alpha(\xi)x$ with any $\alpha \geq 0$ and $\xi = \frac{r}{\|r\|}$, the $R_\alpha(\xi)x_1 = x_1$.

It is a matter of experience that on minimization of positive definite quadratic functions by the method of conjugate gradients a sequence of gradients $\{g(x_k)\}_{k=0}^{n-1}$ taken at the point of the minimizing sequence has the following property: $g(x_{k+1}) - g(x_k), k = 0, \ldots, n-2$, is orthogonal to the direction from the point x_{k+1} to the minimum point.

Let $f(x)$ be the convex piecewise linear function, x the point lying in the first domain near the boundary of two domains of linearity $(a) f_1(x) = (g_1, x) + b_1$;

$(b) f_2(x) = (g_2, x) + b_2$ and $(g_1, g_2) < 0$. After one step of the subgradient descent $x_1 = x_0 - hg_1$ we obtain for a specific h the point x_1 lying on the boundary of two domains. Putting $g(x_1) = g_2$ we derive that $g_2 - g_1$ is orthogonal to the affine manifold $\{x : f_1(x) = f_2(x)\}$ in which the direction of the steepest descent from the point x_1 lies.

Thus, both in the first and second examples, the space dilation in the direction of the difference of two successive gradients (subgradients) decreases the components of the gradients orthogonal to the direction towards the minimum point (direction of steepest descent), that is, in a sense, it is "useful". These heuristic considerations are at the essence of the family of algorithms with space dilation in the direction of the difference of two successive subgradients, or for short, the "r-algorithms".

Now we shall give a short description of r-algorithm.

Let $f(x)$ be a convex function defined on n-dimensional Euclidean space E^n, and $f(x) \to +\infty$, when $\|x\| \to \infty$. By $g_f(x)$ we denote a subgradient of function f in the point x.

Consider a general scheme of r-algorithm.

0-step. Initial point $x_0 \in E^n$ and nonsingular matrix $n \times n$ B_0 are given (for example, $B_0 = I_n$ (identity matrix)).

1-st step. Calculate:

$$g_0 = g_f(x_0);\ \tilde{g}_0 = B_0^T g_0;\ \xi_0 = \frac{\tilde{g}_0}{\|\tilde{g}_0\|};$$

$$x_1 = x_0 - h_0 B_0 \xi;\ (h_0 > 0 \text{ is the step multiplier }). \tag{3.50}$$

Remember x_1, $g_f(x_0)$, B_0. After k-steps we have:

$$x_k \in E^n,\ g_f(x_{k-1}) \in E^n,\ n \times n \text{ matrix } B_{k-1}.$$

(k+1)-st step. Calculate:

$$(a)\quad g_f(x_k);\ d_k = g_f(x_k) - g_f(x_{k-1}); \tag{3.51}$$

$$(b)\quad r_k = B_k^T d_k;\ \xi_k = \frac{r_k}{\|r_k\|}; \tag{3.52}$$

$$(c)\quad B_k = B_{k-1} \cdot R_{\beta_k}(\xi_k),\ 0 < \beta_k < 1; \tag{3.53}$$

$$(d)\quad x_{k+1} = x_k - h_k B_k p_k, \tag{3.54}$$

where

$$p_k = \frac{\tilde{g}_k}{\|\tilde{g}_k\|};\ \tilde{g}_k = B_k^T g_f(x_k),$$

$h_k > 0$ is k-th step multiplier.

Remember x_{k+1}, $g_f(x_k)$, B_k. Stop if stopping criteria are satisfied; otherwise go to the next step.

Comments. After k steps of r-algorithm let A_{k-1} be the resulting matrix of space transformation: $y = A_{k-1}x$, or $x = B_{k-1}y$, where $B_{k-1} = A_{k-1}^{-1}$, and $\varphi_k(y) = f(B_{k-1}y)$. Since $g_{\varphi_k}(y) = B_{k-1}^T g_f(x)$, r_k is the difference of two subgradients of function φ_k taken in the points $y_k = A_{k-1}x_k$ and $\tilde{y}_{k-1} = A_{k-1}x_{k-1}$. So ξ_k is the normed direction of the difference of two successive subgradients of transformed function $\varphi_k(y)$. In this direction we must make current dilation of transformed space and obtain resulting matrix $A_k = A_{k-1}R_{\alpha_k}(\xi_k)$. The inverse matrix $B_k = R_{\beta_k}(\xi_k)B_{k-1}$, $\beta_k = 1/\alpha_k$. Consider $\varphi_{k+1}(y) = f(B_k y)$, and use subgradient step for function $\varphi_{k+1}(y)$ from $\tilde{y}_k = B_k x_k$:

$$y_{k+1} = \tilde{y}_k - h_k \frac{B_k^T g_f(x_k)}{\|B_k^T g_f(x_k)\|}.$$

In original space the point $x_{k+1} = B_k y_{k+1}$ corresponds to the point x_{k+1}, and we obtain

$$x_{k+1} = x_k - h_k B_k \frac{B_k^T g_f(x_k)}{\|B_k^T g_f(x_k)\|}.$$

Thus we calculate x_{k+1}, $g_f(x_k)$, B_k, and are ready to make next iteration.

In the [Sk 74] the so-called H-form of r-algorithms was considered. One can transform the main formula of r-algorithm (3.50)–(3.54) by the following way.

Let $H_{k-1} = B_{k-1}B_{k-1}^T$.

Then

$$\begin{aligned} H_k &= B_{k-1}R_{\beta_k}(\xi_k)R_{\beta_k}(\xi_k)B_{k-1}^T \\ &= B_{k-1}(I - (1-\beta_k^2)\xi_k\xi_k^T)B_{k-1}^T \\ &= H_{k-1} - (1-\beta_k^2)B_{k-1}\frac{B_{k-1}^T d_k}{\|B_{k-1}^T d_k\|}\cdot\frac{d_k^T B_{k-1}}{\|B_{k-1}^T d_k\|}B_{k-1}^T \\ &= H_{k-1} - (1-\beta_k^2)H_{k-1}\frac{d_k d_k^T}{(H_k d_k, d_k)}H_{k-1}^T, \qquad (3.55) \end{aligned}$$

where

$$d_k = g_f(x_k) - g_f(x_{k-1}),$$

$$x_{k+1} = x_k - h_k \frac{H_k g_f(x_k)}{\sqrt{H_k g_f(x_k), g_f(x_k)}}.$$

These formulas are similar with formulas of iterations in quasi-Newton methods with variable metric (see [PD 75], [Fle 87]). The implementation of r-algorithm in H-form demandes less arithmetical operations for vector and matrix transformations in comparison with B-form, but H-form is not so stable relatively the rounding errors. Theoretically matrices B_k $(k = 1, 2, \ldots)$ must be positive definite, but accumulation of rounding errors for large k may distort the entries of H_k so that distorted image $\overline{H}_k$ of H_k will have negative eigenvalues and the normal convergence of r-algorithm may be break for large k.

The family of r-algorithms (3.50) – (3.54) has two sequences of parameters: $\{\beta_k\}_{k=1}^{\infty}$ and $\{h_k\}_{k=1}^{\infty}$. Naturally, we must think about rational choice of these sequences to obtain "good" convergence to the optimal point of $f(x)$.

For minimization of nonsmooth convex functions defined on E^n we recomend to use the following specifications of r-algorithm:

The space dilation coefficients α_k equal to α, where $\alpha \in [2; 4]$. For determination of a step multiplier h_k we use adaptive technique of regulating step length (see [SSt 89]), which determined by parameters: $h_0^{(0)}$ (initial step-length), natural number $\overline{m} > 1$, and coefficients $q_1 < 1$ and $q_2 > 1$ for decreasing (increasing) of step-multiplier. After k iterations of r-algorithm we obtain step constant h_k^0. On $(k+1)$-st iteration we choose the direction of descent due to r-algorithm and move in this direction with step multiplier h_k^0 until the condition of stopping the search along the direction is fulfiled or the number of steps would be equal $\overline{m}$. In the last case we continue descent along the same direction with new step constant $h_k^1 = q_2 h_k^0$. If after current $\overline{m}$ steps the condition of interrupting of search direction is not fulfiled we set $h_k^2 = q_2 h_k^1$ and so on.

We suppose that $\lim_{\|x\| \to +\infty} f(x) = +\infty$, so after finite number of steps the stopping condition for directional search will be fulfiled.

The step constant $h_k^{p_k} = q_2^{p_k} h_k^0$ $(p \in \{0, 1, 2, \ldots\})$, obtained at last stage of k-th iteration, is used as h_{k+1}^0.

The parameter $q_1 < 1$ may be used in the case when after the first move along the direction the stopping criteria are fulfiled. In this case $h_{k+1}^0 = q_1 h_k^0$.

When we use r-algorithm for solving smooth minimization problems we may hope for attaining more fast speed of convergence. In such cases we may permit to use more large values of α_k in combination with more exact technique of obtaining minimum along direction of descent.

Let one use r-algorithm for minimization of a convex quadratic function $f(x) = \frac{1}{2}(Ax, x) + (c, x)$, $x \in E^n$ with constant coefficient $\alpha > 1$ and exact minimization along the direction. As we have seen, at each iteration the direction of space dilation is orthogonal to the direction from current point to minimum, so the sequence $||A_{k-1}(x_k - x^*)||_{k=1}^{\infty}$ monotonically tends to zero (or attains 0 after finite number of steps). This fact ensures the convergence of $f(x_k)$ to f^* with the speed of geometrical progression:

$$f(x_k) - f^* \leq c\alpha^{\frac{k}{n}}$$

for any $\alpha > 1$. Let us tend α to $+\infty$. The limit version of mentioned algorithm ($\beta_k = \beta = 0$) gives optimal point x^* for the number of iteration not exceedong n, when we use it for minimization of convex quadratic function.

Let us consider in a more detailed way an extreme variant of the r-algorithm with resetting, in which β_k, $k = 1, 2, \ldots$, are chosen equal to zero and h_{k+1} is determined by minimizing $f(x_k - hB_{k+1}\tilde{g}_{k+1})$. As one may easily observe, one has then either $\tilde{g}_{k^*+1} = 0$ for some $k^* < n$, or $B_{n+1} = 0$.

Indeed, $R_0(\eta)$ is the operator of projection on the subspace orthogonal to the vector η. The product of such operators, $\prod_{i=1}^{k} R_0(\eta_i)$, does not depend on the order of factors, is a selfadjoint operator, and realizes the orthogonal projection on the supspace being the orthogonal complement of the linear span of the vectors η_i, $i = 1, \ldots, k$.

In the variant of the r-algorithm under consideration, one has to apply the resetting when $B_k^* g_f(x_{k-1}) = 0$. This variant may be regarded as a modification of the conjugate gradient method. It is easy to prove that for a nonnegative definite quadratic form the solution is found in no more than n iterations. For any sufficiently smooth function the n-step quadratic rateof convergence may be proved.

Theorem 56 *Let f be a function defined on E^n and twice continuously differentiable in a certain neighborhood S of the optimal point x^*, and let the matrix of the second derivatives (the Hessian) $H(x)$ be Lipschitz in that neighborhood:*

$$||H(x) - H(x')|| \leq L||x - x'|| \text{ for all } x,\ x' \in S. \tag{3.56}$$

Furthermore, suppose that $H(x^*)$ is a positive definite matrix. Then there exists a neighborhood S' of x^*, $S' \subset S$, such that if $x_0 \in S'$ then one can find a constant c such that

$$\|x_n - x^*\| \leq c\|x_0 - x^*\|^2,$$

where x_n is the point obtained after n iterations of the algorithm described above (if $g_{\overline{k}} = 0$ for some $\overline{k} < n$, then we set $x_n = x_{\overline{k}}$).

Some years ago the competition was carried out for determination of the best codes for unconstrained minimization of convex smooth, but badly conditional functions. Our program based on certain version of r-algorithm solved all proposed instances. It was one the best among the some of proposed codes on execution time, number of iterations and attained accuracy of solution. So, the r-algorithm may be successively used for minimization of bad conditional convex smooth function not yilding in practical effectivness to such methods as quasi-Newton's with variable metric, conjugate direction algorithms and so on.

We proved in [Shor 75] the theorem of convergence of so-called $r_\mu(\alpha)$-algorithm, used for continuous piecewise smooth functions (nonconvex, in general case).

In formulas (3.50) – (3.54) we use almost gradients instead of subgradients. $r_\mu(\alpha)$-algorithm is the idealized version of r-algorithm, in which $\alpha_k = \alpha > 1$ and h_k is chosen from the condition of obtaining the local minimum nearest to x_k along the chosen direction of descent and if the set of almost-gradients $\{g_f(x_k)\}$ is nonunique we must choose $g_f(x_k)$ so, that $(p_{k-1}, g_f(x_k)) \geq 0$. ($p_{k-1}$ is the direction of descent).

Theorem 57 *[Shor 75], [Shor 85] If $f(x)$ is a continuous piecewise smooth function, $\lim_{\|x\|\to\infty} f(x) = +\infty$ and the sequence $\{x_k\}$ generated by the $r(\alpha)$-algorithm possesses the property that $\lim_{k\to\infty} \|x_{k+1} - x_k\| = 0$, then for any $\alpha > 1$ there exists a limit point $\overline{x}$ of the sequence $\{x_k\}$ whose set of almost-gradients forms a linearly dependent family of vectors. By virtue of the monotonicity of the descent process here $\lim_{k\to\infty} f(x_k) = f(\overline{x})$.*

Corollary 3 *By the above Theorem, if x^* is an isolated local minimum, if x_0 is a point such that the connected component of the set $\{x : f(x^*) \leq f(x) \leq f(x_0)\}$ containing x^* and x_0 does not contain other points z aside from x^* where the family of almost-gradients is linearly dependent, then the sequence $\{x_k\}_{k=0}^{\infty}$ generated by the $r(\alpha)$-algorithm converges to x^*.*

As far as the convergence rate of the r-algorithms is concerned, this problem is studied insufficiently from theoretical point of view. At the same time wide experience has been gained in solving various sophisticated problems, which were reduced to minimization of nonsmooth functions by mentioned above implementable versions of r-algorithm. This experience made it possible to find rational parameters of r-algorithms ensuring a fairly high efficiency of this family.

In this book many examples of using implemented versions of r-algorithms are given. In Chapter 5 the application of r-algorithms for realization of decomposition schemes are given. The solution of minimax problems by using r-algorithm is described in Chapter 6. In Chapters 5–9 we describe the use of r-algorithms for obtaining Lagrange dual estimates in quadratic-type problems. In this applications the additional difficulties arise: the *dom f* is the part of E^n, and optimal x^* can be at the boundary of *dom f*. For these problems we worked out special versions of r-algorithm (see [Shor 95], [SSt 97]).

For example, if one intend to solve a matrix optimization problem of the type

$$find \quad \inf_{x \in E^m} f(A(x)) \mid A(x) \succeq 0,$$

where $A(x)$ are $n \times n$ symmetric matrices which enties affinely dependentofparametricvector x, often the domain of $\varphi(x) = f(A(x))$ has no intersection with the set $\{x : A(x)$ is not positive semifedinite$\}$. In this case we construct a function:

$$\psi(x) = \begin{cases} f(A(x)) & , \text{ if } A(x) \succ 0 \\ -\lambda_n(A(x)), & \text{if } A(x) \text{ is not positive semidefinite} \end{cases}$$

(where λ_n denotes the minimal eigenvector of A), and use described above version of r-algorithm to function $\psi(x)$.

Let $\{x_k\}_{k=0}^{\infty}$ be the sequence of points generated by r-algorithms. If for $k = \overline{k}$ the point $x_{\overline{k}}$ is not feasible, i.e. $\lambda_n(A(x)) < 0$, then we pass to the process of minimization of convex function $(-\lambda_n(A(x))$ for obtaining a feasible point (with $\lambda_n(A(x))0)$. The details of such approach and corresponding numerical results are described in Chapters 8–9.

A "careful" user of a nonsmooth minimization program based on described above r-algorithm would like to have in the process of calculation the estimates for $\Delta_k = f_k^r - f^*$, where $f_k^r = \min_{0 \leq j \leq k} f(x_j)$, and $F^* = \inf f(x)$. One may use a heuristic supposition that for large enough $\overline{k}$ $\Delta^r_{\overline{k+n}} \leq \frac{1}{2}\Delta^r_{\overline{k+n}}$. If this supposition is true then $\Delta^r_{\overline{k+n}} - f^* \leq \Delta^r_{\overline{k}} - \Delta^r_{\overline{k+n}}$. One can use the stoppng criteria of the same form: "to stop" the r-algorithm if $\bar{k} \geq N$ and $f^r_{\bar{k}+n} - f^r_{\bar{k}} \leq \delta$, where $\delta > 0$ is given parameter.

To be absolutely sure in attained accuracy one may combine the r-algorithm with some cutting plane method giving a lower bound for f^*. For instance, one may use at the last stage of performing r-algorithm the information about subgradients $g_f(x_{\bar{k}}), g_f(x_{\bar{k}+1}), \ldots, g_f(x_{\bar{k}+s})$ can be used for generating cutting plane linear inequalities:

$$(g_f(x_k), x_k - x) \geq f(x_k) - f, \;\; \bar{k} \leq k \leq \bar{k} + s \tag{3.57}$$

Let $\bar{f}$ be an optimal value of the problem $\max f$ under the constraints 3.57. $\bar{f}$ may be used as lower estimate for f^*.

The convergence of r-algorithms with fixed or even adaptively increased step-size in transformed space can be explained by behavior of sequence $\{B_k\}_{k=1}^{\infty}$. As a rule B_k tends to zero matrix when k tends to infinity. Only in very rare cases the trajectory of descent lies (or approximately lies) in some linear manifold of dimension $m < n$. In this case the sequence B_k may be not convergent to zero, but $B_k \tilde{g}_k / \|\tilde{g}_k\|$ tends to zero.

Mentioned above way of stepsize regulating is based on the following heuristic idea. Let one uses r-algorithms with constant coefficient of step dilation α. Then after n iterations randomly chosen direction in initial space will be contracted in α times in average . It should be expected that the step-size in initial space would be decreased in α times. "in average" for each n iterations.

If the distance to the point of minimum would be decreased with the same average speed, then the number of steps along current direction of descent will be limited. But if the temp of convergence would be not so fast then the number of steps along current direction of descent may increase unlimitedly. The adaptive procedure of increasing h just is introduced for stabilization of the number of steps along the directions by one or several steps.

Numerous test and real life applications of r-algorithms showed that if we set $\alpha \in [2, 4]$ and use described above technique of step-size regulation , the average number of steps along current direction is not exceeded 2 as a rule, besides the deviation $f(x_k) - f^*$ decreases in 2–3 times for n iterations. It should be noted that in typical problems of nonsmooth minimization the complexity of calculations the value of minimized function and its subgradient in given point are differ small, therefore it should be taken into special consideration the decreasing of the number of search steps along direction.

When we use r-algorithm for solving of smooth minimization problems we may hope for attaining more fast speed of convergence. In this case we can permit themself to use more large values of α_k in combination with more exact ways of determination of minimum along direction.

Some mathematicians used other implementations of r-algorithm (see, for example, [Le 82], [Mai 91], [DMi 95]), in which they applied distinctive from our methods of step-size regulating. It may affect essentially on convergence of the algorithm, and as a rule, nonpositively.

My student P.I. Stetsjuk worked out a standard implementation of r-algorithm with adaptive rule of step-length regulation for PC-computers (see [SSt 97]), and performed numerous test experiments with his program. Some of the results of these tests are represented in the tables 3.1, 3.2, 3.3, 3.4.

We use the following test functions:

(1) bad conditional convex quadratic functions (*Quad*):

$$f(x) = \sum_{i=1}^{n} 10^{i-1} x_i^2, \quad x \in E^5, \quad x \in E^{10}, \quad x_0 = (1, \ldots, 1),$$

(2) convex piece-wise linear functions (*Sabs*):

$$f(x) = \sum_{i=1}^{n} 10^{i-1} |x_i|, \quad x \in E^5, \quad x \in E^{10}, \quad x_0 = (1, \ldots, 1),$$

(3) convex piece-wise quadratic functions (*Shor*) (see [Shor 85], p.137):

$$f(x) = \max_{i=1}^{10} \varphi_i(x), \quad x \in E^5, \quad x_0 = (0, 0, 0, 0, 1),$$

(4) convex piece-wise quadratic functions (*Maxquad*) (see [Le 78a], p.151):

$$f(x) = \max_{i=1}^{5} \varphi_i(x), \quad x \in E^{10}, \quad x_0 = (1, \ldots, 1),$$

(4) nonconvex "generalized Rozenbrock functions":

$$f(x) = (x_1 - 1)^2 + 100 \sum_{i=2}^{n} (x_i - x_{i-1}^2)^2, \qquad x \in E^2, \quad x \in E^3, \quad \ldots, \quad x \in E^6,$$

$$x_0 = (x_1^0, x_2^0, \ldots, x_n^0), \quad x_1^0 = -1.0, x_i^0 = (x_{i-1}^0)^2 - 0.2, i = 2, \ldots, n.$$

The simplified condition of interrupting the iterations of $r(\alpha)$-algorrithm (stopping criterion) is the following: $||x_{k+1} - x_k|| \leq \varepsilon_x$, for given $\varepsilon_x > 0$. In our test examples $\varepsilon_x \sim 10^{-5}$–$10^{-6}$.

We use the following notations in the tables:

$iter$ – the number of r-algorithm iterations;

n_f – the number of calculations of $f(x)$ and $g_f(x)$;

x_{iter} – the approximate minimum point, obtained by r-algorithm on the iteration where stopping criterion is fulfiled.

In table 3.1 the results of numerical testing obtained by $r(\alpha)$-algorithm with parameters $\alpha = 2, h_0 = 1.0, q_1 = 1.0, \overline{m} = 3, q_2 = 1.1$ are described.

In table 3.2 we described the results of application of r-algorithm with exchanged value $q_1 (q_1 = 0.9)$ for the minimization of the same family of functions.

For the functions MAXQUAD we used $r(\alpha)$-algorithm with different $\alpha, \alpha = \{2, 3, 4\}$ and $q_1 = \{0.9; 1.0\}$. The values of other parameters are: $q_2 = 1.1; \overline{m} = 3; h_0 = 1.0$

In the Table 3.3 the following results are described: the number of iterations of r-algorithms (the number of calculations of $f(x)$ and $g_f(x)$ in brackets) necessary for obtaining the point x_k in which $f(x_k) - f^* \leq \varepsilon$ (the optimal value $f^* = -0.841408334596$).

In the table 3.4 the results of applications of $r(\alpha)$-algorithm with parameters $\alpha = 2, q_2 = 1.1, q_1 = \{0.9; 1\}; \overline{m} = 3; h_0 = 1$ for minimization of nonconvex smooth "generalized Rosenbrock" functions are described.

All obtained by $r(\alpha)$-algorithm results demonstrated in tables 3.1–3.4 show relatively fast convergence to minimum value of minimized functions and high stability of calculations even for very bad-behaved functions. In subsequent Chapters 5–9 we represent many other real-life and test instances of solving various optimization problems by r-algorithm.

Table 3.1

Test	n	$\varepsilon_x = 10^{-6}$		$\varepsilon_x = 10^{-5}$	
		$f(x_{iter})$	$iter(n_f)$	$f(x_{iter})$	$iter(n_f)$
Shor	5	22.6001620960	145(156)	22.6001621426	109(119)
Maxquad	10	-.841408284867	175(195)	-.841407416044	148(164)
Quad	5	.114774125059E-10	106(144)	.207932545789E-09	93(128)
Sabs	5	.313658856047E-04	115(143)	.149241240487E-03	103(130)
Quad	10	.310359728383E-10	239(307)	.273742459477E-09	207(265)
Sabs	10	.868752157795E-04	294(337)	.185796463607E-03	278(317)

Table 3.2

Test	n	$\varepsilon_x = 10^{-6}$		$\varepsilon_x = 10^{-5}$	
		$f(x_{iter})$	$iter(n_f)$	$f(x_{iter})$	$iter(n_f)$
Shor	5	22.6001620975	91(125)	22.6001620996	85(116)
Maxquad	10	-.841408290372	86(109)	-.841408254265	80(101)
Quad	5	.137271876048E-11	76(109)	.321988639576E-09	64(92)
Sabs	5	.841899676020E-05	87(124)	.143213610435E-03	72(103)
Quad	10	.376028664155E-10	133(191)	.205212211807E-08	115(165)
Sabs	10	.101302789721E-04	228(398)	.366580679542E-03	207(376)

Table 3.3

ε	$q_1 = 1.0$			$q_1 = 0.9$		
	$\alpha = 2$	$\alpha = 3$	$\alpha = 4$	$\alpha = 2$	$\alpha = 3$	$\alpha = 4$
1.	19(23)	17(23)	24(43)	14(21)	13(20)	17(28)
0.1	37(42)	29(37)	32(55)	24(32)	21(33)	22(36)
10^{-2}	59(65)	45(62)	39(63)	32(44)	33(55)	31(52)
10^{-3}	76(82)	59(82)	51(80)	37(49)	39(63)	36(61)
10^{-4}	102(112)	70(97)	58(91)	49(63)	49(77)	44(73)
10^{-5}	117(128)	82(113)	73(110)	60(77)	59(91)	53(88)
10^{-6}	142(157)	97(135)	86(125)	68(86)	67(104)	61(103)
10^{-7}	164(182)	112(158)	97(141)	79(99)	78(123)	70(116)
10^{-8}	183(205)	123(174)	107(157)	88(113)	87(135)	79(131)

Table 3.4

n	$f(x_0)$	q_1	$f(x_{iter})$	$\|x_{iter} - x^*\|$	$iter$	n_f
2	8.0	1.0	.13188E-11	.11040E-05	51	101
		0.9	.13070E-10	.21658E-06	39	82
3	12.0	1.0	.33276E-13	.64373E-06	83	141
		0.9	.23136E-12	.14724E-06	70	149
4	16.0	1.0	.25197E-14	.10690E-06	137	281
		0.9	.53802E-13	.13781E-06	117	277
5	20.0	1.0	.81394E-13	.17662E-05	212	486
		0.9	.54224E-13	.12596E-06	184	455
6	24.0	1.0	.12920E-11	.11666E-04	303	749
		0.9	.10480E-13	.15137E-05	284	734

4

ELEMENTS OF INFORMATION AND NUMERICAL COMPLEXITY OF POLYNOMIAL EXTREMAL PROBLEMS

4.1 INTRODUCTION TO THE COMPLEXITY THEORY OF OPTIMIZATION PROBLEMS

After the famous works of Church, Godel and Turing (see [Ta 51], [AHU 76]) in 1930-th one can clearly define what the notion of potential solvability of some class of problems means. They established the existense of non-solvable (non-resolvable) problems that arise in logic and in the theory of computation in a natural way. Problem is considered algorithmicaly solvable if it can be solved by appropriate Turing machine. But the partition of all problems in two classes (solvable or nonsolvable) is a very rough classification. Further more detailed separation of solvable problems into subclasses can be done by establishing of upper bounds on computational resources necessary for obtaining their solution. Two main resources used by an algorithm is time (the number of elementary steps of the algorithm) and the memory (working memory area). We can obtain the upper bounds for the necessary resources studying an arbitrary algorithm that solves the given problem. But to get an answer for the question of how good a particular algorithm is we must find the lower bounds for computational resources, the limits that cannot be improved by the "best" algorithm among the potentially possible ones. Obtaining the nontrivial lower bounds is a very hard job in many important cases. But even in the case with no chance for lower and upper bounds to be close enough, one has a possibility to make a comparative classification of the problem by associating it with other problems from a wide enough class. So we have two types of estimates of computational complexity: absolute and relative. As we deal with polynomial functions it is natural for us to consider examples linked with the first order theory of solvability for real numbers.

Let us consider formulae of predicate calculus of the first order with equality, containing binary operations $+, \times$ and "standard" relations $>, \geq, =, \neq, \leq, <$. We can interpret these symbols in a usual sense for real variables. The sentences, which are bounded by quantifiers may be true or false. For such calculus $Th(R,t,x)$ Tarsky [Ta 51] proved solvability. If we exclude the operation $\times$, we obtain more simple calculus $Th(R,t)$. Suppose that length $|s|$ of the sentence s is determined by the number of all symbols, contained in the sentence. Fisher and Rabin proved the following

Theorem 58 *The constant $c > 1$ exists such that each Turing machine that can resolve infinite number of sentences from $Th(R,t)$ calculus must do no less than $c^{|s|}$ steps to work with entry s.*

This is an example of absolute lower exponential estimates for the complexity of solving simple arithmetical sentences. Exponential lower bound shows that even for relatively short sentences from $Th(R,t)$ arbitrary solving procedure may demand tremendous number of steps, so in a certain sense $Th(R,t)$ is "practically nonsolvable".

Renegar found more explicit estimates for the complexity of certain problems of the first-order theory of the reals [Re 92a], [Re 92b]. Note that the most general results of the complexity of polynomial programming are described in the mentioned above articles.

A formula (in the first-order theory of the reals) is an expression of the form

$$(SF)\ \ (Q_1 x^{[1]} \in R^{n_1}) \dots (Q_w x^{[w]} \in R^{n_w}) P(y, x^{[1]} \dots x^{[w]}),$$

where:

- $y = (y_1, \dots, y_k) \in R^k$ is the vector of free variables;

- each Q_i, $i = 1, \dots, w$ is one of the quantifiers $\exists$ or $\forall$;

- $P(y, x^{[1]} \dots x^{[w]})$ is a quantifier free Boolean formula with r "atomic predicates" of the form

$$g_i(y, x^{[1]} \dots x^{[w]}) \Delta_i 0, \ \ i = 1, \dots, r,$$

where $\Delta_i$is one of the "standard relations" $>,<,\geq,\leq,=,\neq$ and the g_i's are real polynomials of degree at most $d \geq 2$.

Note that the above formula is in prenex form: all quantifiers in (SF) occur in front. Formulas without free variables are called sentences. We say that $y \in R^k$ is a solution of (SF) if the sentence obtained by substituting y into (SF) is true.

Proposition 11 *[Re 92a] If formula (SF) has only integer coefficients, each of bit length at most B, then every connected component of the set of its solutions intersects the ball $\{y \in R^k | \ \|y\| \leq R\}$, where R satisfies*

$$\log R \leq B(rd)^{2^{O(w)} k \prod_i n_i}.$$

A quantifier elimination method for the first-order theory of the reals constructs for any input formula (SF) an equivalent quantifier free formula.

Proposition 12 *[Re 92b] There is an algorithm which, given a formula (SF), finds an equivalent quantifier free formula of the form*

$$\bigvee_{i=1}^{I} \bigwedge_{j=1}^{J_i} (h_{ij}(y) \Delta_{ij} 0),$$

where: - $I \leq (rd)^{2^{O(w)} k \prod_i n_i}$,

- $J_i \leq (rd)^{2^{O(w)} k \prod_i n_i}$,

- *the degree of* $h_{ij}(y) \leq (rd)^{2^{O(w)} k \prod_i n_i}$,

- Δ_{ij} *is one of the standard relations* $>,<,\geq,\leq,=,\neq$.

The algorithm requires $(rd)^{2^{O(w)} k \prod_i n_i}$ *operations and* $(rd)^{O(k+\sum_i n_i)}$ *evaluations of the input formula. If the coefficients of the atomic polynomials* g_i, $i = 1,\ldots,r$ *are integers of bit length at most* B, *then the algorithm works with numbers of binary length*

$$(B+k)(rd)^{2^{O(w)} k \prod_i n_i}.$$

This bound also holds for the binary length of the coefficients of the polynomials h_{ij}.

The following special case of the above results deals with the decision problem for the first-order theory of the reals: determine whether a sentence (SF) is true or false.

Proposition 13 *[Re 92b] There is an algorithm for the decision problem of the first-order theory of the reals that requires*

$$(rd)^{2^{O(w)}k\prod_i n_i}$$

operations and $(rd)^{O(\sum_i n_i)}$ *evaluations of the input formula. When restricted to sentences involving only polynomials with integer coefficients of bit length at most* B*, the procedure works with numbers of binary length* $B(rd)^{2^{O(w)}k\prod_i n_i}$.

The following inequality is a well-known bound on nonzero roots of univariate polynomials (see, [Mig 82]).

Proposition 14 *Let* $p(x) = a_0x^d + a_1x^{d-1} + \ldots + a_{d-1}x + a_d$ *be a univariate polynomial with integer coefficients, and let* α *be a nonzero root of* $p(x)$*. Then* $|\alpha| \geq 1/(1+h)$*, where* $h = \max\{|a_0|, |a_1|, \ldots, |a_d|\}$ *is the height of* $p(x)$.

In the theory of algoritmic complexity the so-called "decision problems" with binary output "yes" or "no" are represented separately. Other problems which suppose more complex output are reduced to the series of decision problems.

An **instance** of a problem is a single occurence of such a problem, which is determined by certain values of parameters entered in logic relations describing given problem. The **size** of an instance is the number of binary bits required to represent the instance, i.e. to code the list of parameters with their values. The problems equivalent in mathematical sense may have different systems of parameters and the corresponding relations for their description. So in the theory of algorithmic complexity the notion "problem" is characterized not only by its mathematical model, but also by a certain way of its description (coding), which determines the sizes of instances or lengths of input for given algorithm.

We may suppose that execution of an algorithm on an instance of a problem consists of performing a sequence of elementary steps, with execution time independent of the size of the particular instance. In order to investigate the comparative efficiency of different algorithms solving a given decision problem one needs to estimate the number of elementary steps of each algorithms necessary to set a proper output. The class P is the set of all decision problems which can be solved "polynomially", i.e. for each problem $\Pi \in P$ there must exist an algorithm and a polynomial $p(l)$ such that an instance of Π whose input is of length l can be solved by the algorithm in at most $p(l)$ elementary steps.

After the series of the works ([Ed 65a],[Ed 65b],[Coo 71]) class of the "practically solvable" problems is identified with class P. Though apriori we do not know what algorithm is "practically" worse: one with execution time estimate l^{100} or $O(2^l)$, class P proved its usefullness in mathematical sense because it is invariant to large variety of calculation models if "elementary step"is understood as subroutine with polynomial in l upper bound for required number of bit operations.

We say that problem Π_1 is polynomially reduceable to problem Π_2 if there exist a natural number s and $c > 0$ exist such that for each instance I_1 of Π_1 with input length l_1 one can construct an instance I_2 of Π_2 with the same output by algorithm of polynomial in l_1 complexity with input length $l_2 < cl_1^s$. Using the technique of polynomial reducing Cook [Coo 71] showed that problems can be concentrated into polynomially equivalent subclasses so that either all representatives are in P or no one is.

We say that a decision problem Π is in class NP if a "yes" answer for it can be verified in polynomial time, provided that some extra information is given. This extra information is called a certificate and for each instance, its length must be uniformly polynomially bounded by the length of the corresponding input l, besides, the time , required to obtain the certificate is not counted.

Trivially $P \subseteq NP$. Unfortunately, there are many very important combinatorial and optimization NP-problems for which the belonging to class P was not proved. A NP-problem Π is called **NP-complete** if each NP-problem is polynomially reduceable to the problem Π. Cook [Coo 71] showed that NP-complete problem existed. Karp [Karp 72], [Karp 75] showed that many known hard combinatorial and optimization problems are NP-complete. Since then thousands of problems have be shown to been NP-complete.

A problem Π is **NP-hard** if an arbitrary NP-complete problem is polynomially reduceable to this problem. Besides the problem Π itself doesn't need to lie in NP and we permit more than one call to the subroutine.

Suppose that for some problem Π, I^* is a set of instances which are valid, and for which the answer is "yes". The complement of I^* is the set of all other possible inputs. We say that problem Π is in *co-NP* if an invalid input or an input for which the answer is "no" can be verified in time polynomial in the length of the input, given appropriate certificate. Naturally, $P \subseteq co\text{-}NP$, and although it is not known whether $P = NP \cap co\text{-}NP$, for the most of known relatively hard problems for which the fact that they lie in $NP \cap co\text{-}NP$ was proven, the polinomial-time algorithms were found.

4.2 POLYNOMIAL-TIME ALGORITHMS FOR LINEAR PROGRAMMING

For a long time the problem of finding a polynomial-time algorithm for linear programming(LP) problems was very actual. It was observed that simplex procedures are effective in most cases of solving the real-world LP problems. But Klee, Minty [KM 72] and others showed by construction of special simple examples that certain variants of the simplex method demand exponentially growing number of iterations for obtaining exact solutions when length of input grows linearly. So there was a large gap between the worst case bounds for known simplex procedures and hopes that LP problem has polynomial time complexity. These hopes were stimulated by the conjecture that if a problem belongs to the class NP and co-NP it can be solved by polynomial-time algorithm. It is very easy to prove that LP problem belongs to both classes: NP and co-NP.

Most investigators tried to find a polynomial time algorithm for LP as a finite number steps procedure. Unexpectedly the first polynomial-time algorithm for LP proposed by L.G. Khachian in 1979 [Kh 79] proved to be slight modification of the mentioned above ellipsoid method demanding infinite number of steps in general for convergence to the minimum value of convex function.

L.G. Khachian considered a LP problem of finding an exact solution of a system of linear inequalities with integer coefficients given in the following form:

$$l_i(x) = \sum_{j=1}^{n} a_{ij}x_j - b_i \leq 0, \ i = 1, \ldots, m, \tag{4.1}$$

where $x = (x_1, \ldots, x_n)$.

The length of input may be estimated by

$$L = \sum_{i=1}^{m}\sum_{j=1}^{n} [\log(|a_{ij}|+1)+1] + \sum_{i=1}^{m} [\log(|b_i|+1)+1] \tag{4.2}$$

Let

$$\begin{aligned} \varphi(x) &= \max_{i\in\{1,\ldots,m\}} l_i(x), \\ \varphi^* &= \inf_{x\in E^n} \varphi(x) \end{aligned}$$

If $\varphi^* > 0$, then the system (4.1) is infeasible. Otherwise it is feasible.

Using elementary facts of linear algebra Khachian proved [Kh 79]:

(i) if the system (4.1) is feasible, then there exists a solution in the ball $S(0; 2^L)$.

(ii) if the system (4.1) is infeasible, then $\varphi^* > 2^{-L}$.

Now we are ready to use ellipsoid algorithm to determine the feasibility of (4.1). For this aim it is enough to find φ^* with precision 2^{-L}.

One can use the ball $S(0; 2^L)$ as an initial set of localization for solutions of (4.1). It is easy to deduce that

$$\max_{x\in S(0;2^L)} \varphi(x) \leq 2^L \cdot \max_{i\in\{1,\ldots,m\}} \sqrt{\sum_{j=1}^{n} a_{ij}^2} < \sqrt{n}\cdot L\cdot 2^L$$

From the theory of the ellipsoid method one can deduce that it is sufficient to make $O\left(\left[n^2 \log \frac{\sqrt{n}L\cdot 2^L}{2^{-L}}\right]\right)$ steps of the ellipsoid method for attaining required precision of determining φ^*. We must remember that in strong theory of computational complexity all arithmetical operations are made with fixed number of bits. So one must account for the possible rounding errors in realization of the ellipsoid method. The rounding errors can be taken in account by small increasing of ellipsoid, containing system solutions, at each step of the algorithm. It can be made by using "blow-up" coefficient $\xi > 1$ (for example, $\xi = 1 + \frac{1}{4(n+1)^2}$) for increasing the linear sizes of localizing ellipsoids (see [GLS 88]).

Let

$$E_k = \{x : (A_k(x-a_k), x-a_k) \leq 1\}$$

be a n-dimensional ellipsoid, containing solutions of the system (4.1) and generated after k steps of the usual ellipsoid method applied to the function $\varphi(x)$.

Let subgradient $g_\varphi(a_k) = c_k$. Then one iteration of the central ellipsoid method reduces to the following formulas:

$$\begin{aligned} a_{k+1} &= a_k - \rho\frac{1}{\sqrt{(A_k c_k, c_k)}}A_k c_k, \\ A_{k+1} &= \xi\cdot\sigma\left(A_k - \tau\frac{1}{(A_k c_k, c_k)}\right)A_k c_k c_k^T A_k, \end{aligned}$$

where $\rho = \frac{1}{n+1}$; $\sigma = \frac{n^2}{n^2-1}$; $\tau = \frac{2}{n+1}$; $\xi = 1$ (ρ is a step parameter; σ is a dilatation parameter; τ is an expansion parameter; ξ is a "blow-up" parameter).

The basic iteration of modificated ellipsoid method in which the rounding errors are taken into consideration has the following form:

$$\begin{aligned} a_{k+1} &:\approx a_k - \rho\frac{1}{\sqrt{(A_k c_k, c_k)}}A_k c_k \\ A_{k+1} &:\approx \bar{\xi}\cdot\sigma\left(A_k - \tau\frac{1}{(A_k c_k, c_k)}\right)A_k c_k c_k^T A_k, \end{aligned}$$

where "$\approx$" means cutting off after the p-th digit behind the decimal point in the binary representation of the numbers at the right hand side; $\bar{\xi} = 1 + \frac{1}{4(n+1)^2}$. The blow-up factor $\bar{\xi}$ is chosen for compensating of rounding errors, if $p = 8N$; $N = 50(n+1)^2 < A, b >$, where $< A, b >$ denotes the encoding length of A and b.

One can deduce (see [GLS 88]) that if initial ellipsoid is $E_0 = S\left(0; \left[\sqrt{n}2^{<A,b>-n^2}\right]\right)$, then after N steps of using the modified central ellipsoid method we obtain $\varphi(x_N)$, approximating φ^* with required precision.

As a result we can obtain the following theorem:

Theorem 59 *The ellipsoid method can be adopted to find a feasible solution to (4.1), if one exists, in time bounded by a polynomial of the length of its input, given by (4.2).*

The detailed analysis shows (see [Kh 79]) that the number of bit operations to determine the feasibility is bounded by $O(n^4L^2)$.

One can show that if the system (4.1) is feasible, then there exist rational solutions $\overline{x}$ of (4.1) such that $\overline{x}_i = \frac{a_i}{\Delta}(a_i$ and Δ are integers) with $|a_i| \leq 2^L$ and $\Delta \geq 1 (i = 1, \ldots, n)$. Thus, $-2^L \leq \overline{x}_i \leq 2^L$.

Let one knows that $b_1 \leq \overline{x}_i \leq b_2$. By using the additional inequality $x_i \leq \frac{b_1+b_2}{2}$ for the system (4.1) and performing the mentioned above ellipsoid procedure for the new system one can recognize whether there exists a feasible point $\overline{x}$ with $\overline{x}_i \in [b_1, \frac{b_1+b_2}{2}]$ in the case of positive answer ($\overline{x}_i \in (\frac{b_1+b_2}{2}, b_2]$ – in the case of negative answer). So we can use the binary search for localization $\overline{x}_1$ with necessary precision, then for $\overline{x}_2$ and so on. It is clear that after $nO(L)$ applications of the ellipsoid procedure one can localize the coordinates of a feasible solution in rational intervals of the length $\frac{1}{2^{2L}}$. Further we can determine the exact values of solution components using the technique of continued fraction approximation (see [Kh 80]).

It is known that solving LP problem can be reduced to the solving of the system of linear inequalities by simultaniously consideration of primal and dual problems. Thus one can obtain the following result [Kh 80]:

Theorem 60 *There exists a polynomial-time algorithm for solving LP problems with integer coefficients (based on the ellipsoid methods). The number of bit arithmetical operations to determine the feasibility of LP problem is bounded by $O(n^4L^2)$. A rational optimal solution for LP problem (if it exists) can be obtained by $O(p^5L^3)$ bit operations, where $p = m + n$.*

4.3 INTERIOR POINT ALGORITHMS FOR LINEAR PROGRAMMING AND SPECIAL CONVEX PROGRAMMING PROBLEMS

The title "interior point methods" initially was linked with using **barrier penalty functions** for solving constrained nonlinear programming problems. The corresponding **interior penalty methods** were studied in the classical monography of Fiacco and McCormick [FM 68]. But after the famous paper of Karmarkar [Kar 84], in which the polynomial-time projective interior point method was developed for LP problems, hundreds of publications appeared devoted to similar methods for solving LP and special convex programming problems. One can find the concluding (in certain sense) results in this area of research activity in the bright book of Nesterov and Nemirovskii [NN 94].

Dikin [Di 67], [Di 74], [DZ 80], [DPop 95] used the title "interior point method" in another sense. Let A be a matrix m by n $(m \leq n)$ with entries a_{ij}, $i = 1, \ldots, m$; $j = 1, \ldots, n$; $x = (x_1, \ldots, x_n)$ be a n-dimensional vector of variables, $b = (b_1, \ldots, b_m)$ be a m-dimensional vector, $c = (c_1, \ldots, c_n)$ be a vector of cost coefficients. Consider the LP problem in canonical form:

$$\text{minimize}\,(c, x) \tag{4.3}$$

subject to the constraints

$$Ax = b, \tag{4.4}$$

$$x \geq 0. \tag{4.5}$$

Dual problem has the form:

$$\text{minimize}\,(b, \lambda) \tag{4.6}$$

subject to

$$A^T \lambda \leq c. \tag{4.7}$$

Assume for simplicity, that the prime and the dual LP problems have unique optimal solutions x^* and λ^*. The feasible region G of the prime problem (4.3)-(4.5) can be represented as intersection of affine manifold P given by (4.4), and nonnegative orthant R^+, given by (4.5). Let $\overline{x} = (\overline{x}_1, \ldots, \overline{x}_n) \in P$ and all $\overline{x}_i > 0$.

Consider an ellipsoid

$$E(\overline{x}) = \{x \in R^n : \sum_{i=1}^{n} \frac{(x_i - \overline{x}_i)^2}{\overline{x}_i^2} \leq 1\} \tag{4.8}$$

Since $E(\overline{x}) \subset R^+$, the solutions $x^*(\overline{x})$ of anxiliary problem:

$$\text{minimize}\,(c, x) \tag{4.9}$$

subject to

$$Ax - b = 0 \tag{4.10}$$

$$\sum_{i=1}^{n} \frac{(x_i - \overline{x}_i)^2}{\overline{x}_i^2} \leq 1 \tag{4.11}$$

will be feasible for the LP problem (4.3)–(4.5). Set $y_i = x_i - \overline{x}_i$, $y = (y_1, \ldots, y_n)$. Then the auxiliary problem (4.9)–(4.11) is equivalent to the problem:

$$\text{minimize } [(c, y) + (c, \overline{x})] \tag{4.12}$$

subject to constraints

$$Ay = 0, \tag{4.13}$$

$$\sum_{i=1}^{n} \frac{y_i^2}{\overline{x}_i^2} \leq 1 \tag{4.14}$$

Let $\lambda(\overline{x})$ be the vector of Lagrange multipliers corresponding to the constraints (4.13), $\lambda(\overline{x}) = (\lambda_1(\overline{x}), \ldots, \lambda_m(\overline{x}))$, and $\frac{v}{2}$ be the Lagrange multiplier for the constraint (4.14). One can form the Lagrange function $L_{\overline{x}}(y, \lambda)$ for the problem (4.12)–(4.14):

$$L_{\overline{x}}(y, \lambda) = (c, y + \overline{x}) + (\lambda(\overline{x}), Ay) + \frac{v}{2}\left(\sum_{j=1}^{n} \frac{y_j^2}{\overline{x}_j^2} - 1\right) =$$

$$= (c, y + \overline{x}) - (A^T \lambda(x), y) + \frac{v}{2}\left(\sum_{j=1}^{n} \frac{y_j^2}{\overline{x}_j^2} - 1\right).$$

By using the Lagrange function method we obtain the following necessary and sufficient conditions of optimum for the problem (4.12)–(4.14):

$$c_j - \sum_{i=1}^{m} a_{ij} \lambda_i(\overline{x}) + v \frac{y_j}{\overline{x}_j^2} = 0, \ j = \overline{1, n} \tag{4.15}$$

$$\sum_{j=1}^{n} a_{ij} y_j = 0, i = \overline{1, m} \tag{4.16}$$

$$\sum_{j=1}^{n} \frac{y_j^2}{\overline{x}_j^2} = 1 \tag{4.17}$$

From (4.15)–(4.17) one can obtain:

$$y_j(\overline{x}) = \overline{x}_j^2 \left(\sum_{i=1}^{m} a_{ij} \lambda_i(\overline{x}) - c_j\right) / v. \tag{4.18}$$

By substituting $y_j(\overline{x})$ (see (4.18)) in $L_{\overline{x}}(y, \lambda)$, we obtain:

$$\varphi(\lambda) = -\frac{1}{v} \Phi_{\overline{x}}(\lambda),$$

where

$$\Phi_{\overline{x}}(\lambda) = \sum_{j=1}^{m} \overline{x}_j^2 \delta_j^2,$$

$$\delta_j = c_j - \sum_{i=1}^{m} a_i \lambda_i(\overline{x}).$$

Let y^* be optimal solution of the problem (4.12)–(4.14), $\lambda^*(\overline{x}), v^*(\overline{x})$ be optimal vector of Lagrange multipliers. Due to the Kuhn-Tucker theorem, $(c, y^* + \overline{x}) = \max_{\lambda^*(\overline{x}), v^*(\overline{x})} \varphi(\lambda^*)$, so $\lambda^*(\overline{x})$ can be received as a result of the minimization of function

$$\Phi_{\overline{x}}(\lambda) = \sum_{j=1}^{n} (\overline{x}_j)^2 \left(\sum_{i=1}^{m} a_{ij} \lambda_i(\overline{x}) - c_j \right)^2, \tag{4.19}$$

besides $v^* = \sqrt{\Phi_{\overline{x}}(\lambda^*)}$.

The minimization of $\Phi(\lambda)$ is equivalent to solving the system of linear equations:

$$\frac{\partial \Phi_{\overline{x}}}{\partial \lambda_i}(\lambda) = 0, \; i = 1, \ldots, m. \tag{4.20}$$

When we calculate the solution $\lambda^*(\overline{x})$ of (4.20) we may use it for obtaining direction of descent $y(\overline{x})$ by formula (4.18).

Now we can formulate Dikin's interior point algorithm. We have an initial feasible point $x(0)$, $x_j(0) > 0$, $j = 1, \ldots, n$. After k steps we obtain a feasible interior point $x(k)$, $x_j(k) > 0$, $j = 1, \ldots, n$.

$(k+1)$-st step. Set $\overline{x} = x(k)$ and solve the problem of minimizing $\Phi_{\overline{x}}(\lambda)$ (see (4.19)). Let $\lambda_k = \lambda(\overline{x})$ be the solution of this problem. Use (4.18) for calculating descent direction $y(k) = \{y_j(\overline{x})\}_{j=1}^n$ and set

$$x(k+1) = x(k) + h_k y(k), \; 0 < h_k < 1. \tag{4.21}$$

Note, that if $h_k = \frac{1}{\sqrt{\Phi(\overline{x})}}$, (see (4.19)) $x(k+1)$ is the solution of the auxiliary problem (4.9)–(4.11). In this case we move in direction $y(k)$ to the boundary of ellipsoid $E(\overline{x})$. Denote the value $\Phi(x(k))$ by Φ_k.

It is easy to calculate the decreasing of objective function at $(k+1)$-st step:

$$(c, x(k) - x(k+1)) = h_k \Phi_k \tag{4.22}$$

If h_k is a "complete" step, i.e. $h_k = \frac{1}{\sqrt{\Phi_k}}$,

$$(c, x(k) - x(k+1)) = \sqrt{\Phi_k}.$$

But if we use $h_k = \frac{\alpha_k}{\sqrt{\Phi_k}}$, where $0 < \overline{\alpha} \leq \alpha_k \leq 1$, one can guarantee, that sequence $\{x(k)\}_{k=0}^{\infty}$ is the sequence of interior feasible points, and the system of linear equations (4.20) is uniquelly solvable at each iteration. Assume that prime problem (4.3)–(4.5) has unique solution x^*, $(c, x^*) = f^*$. Set, for example, $\alpha_k = \overline{\alpha} > 0$. Then, due to (4.22), (4.19)

$$\lim_{k\to\infty} \Phi_k = 0 \tag{4.23}$$

$$\lim_{k\to\infty} x_j(k) \left(\sum_{i=1}^{m} a_{ij}\lambda_i(k) - c_j \right) = 0, \; j = 1, \ldots, n \tag{4.24}$$

so $\lim_{k\to\infty} x(k) = x^*$. Let $f_k = (c, x(k)) - f^*$. Dikin [Di 67] showed that f_k assymptotically tends to zero with the speed of geometrical progression. Let $p_k = \{p_k^{(j)}\}_{j=1}^{n}$ denote the vector with coordinates

$$p_k^{(j)} = (x_j(k) - x_j^*)^2 / x_j(k).$$

J be the set if indeces, for which $x_j^* > 0$. Then $\lim_{k\to\infty} p_k^{(j)} = 0$ for $j \in J$; $\lim_{k\to\infty} p_k^{(j)} = 1$ for $j \notin J$. Thus $\lim_{k\to\infty} ||p_k|| = \sqrt{|J|}$. Further, using the inequality $|(a,b)| \leq ||a|| \cdot ||b||$, we obtain:

$$\begin{aligned} f_k &= \sum_{j=1}^{n} x_j(k) \left(\sum_{i=1}^{m} a_{ij}\lambda_i(k) - c_j \right) \frac{x_j(k) - x_j^*}{x_j(k)} \leq \\ &\leq \sqrt{\sum_{j=1}^{n} (x_j(k))^2 \cdot \left(\sum_{i=1}^{m} a_{ij}\lambda_i(k) - c_j \right)^2} \, ||p_k|| = \\ &= \sqrt{\Phi_k} ||p_k|| \end{aligned}$$

$$\frac{f_k - f_{k+1}}{f_k} \leq]1 - \frac{h_k \Phi_k}{\sqrt{\Phi_k} ||p_k||} = 1 - \frac{\overline{\alpha}}{||p_k||}$$

Thus

$$\lim_{k\to\infty} \sup_{x(0)} \frac{f_k - f_{k+1}}{f_k} \leq 1 - \frac{\overline{\alpha}}{\sqrt{|J|}} \leq 1 - \frac{\overline{\alpha}}{\sqrt{n-m}}.$$

Let us recall that this result was obtained in 1967 [Di 67]. After 17 years in 1984 N. Karmarkar proposed his famous projective polynomial time algorithm for LP which may be considered as interior point algorithm also. The first publications on this algorithm provoked a vast number of research papers on interior point methods in convex programming. We recommend a very good book of Nemirovskii and Nesterov [NN 94] for detailed study of this area of optimization. Below we give only a brief review on that subject.

In [Kar 84] Karmarkar consider the specific model of LP:

$$\text{minimize}\,(c, x) \tag{4.25}$$

$$Ax = 0, \tag{4.26}$$

$$\sum_{j=1}^{n} x_j = 1, \tag{4.27}$$

$$x_j \geq 0,\ \forall j,\ j = 1, \ldots, n \tag{4.28}$$

where c, x are n-dimensional vectors, $\frac{e}{n} = \left\{\frac{1}{n}, \frac{1}{n}, \cdots, \frac{1}{n}\right\}$ is the feasible vector, A is the $m \times n$ matrix of rank m, for optimal solution x^*

$$(c, x^*) = 0 \tag{4.29}$$

The constraints (4.27), (4.28) define a simplex Δ in R^n. In the algorithm the following projective transformation $\Delta \to \Delta$ is used:

$$x \to \widehat{x} = \frac{\overline{X^{-1}}x}{(e, \overline{X^{-1}}x} \tag{4.30}$$

with inverse

$$\widehat{x} \to x = \frac{\overline{X}\widehat{x}}{(e, \overline{X}\widehat{x}}, \tag{4.31}$$

where $\overline{X}$ is diagonal matrix with entries of vector $\overline{x}$. Let B be the matrix $\begin{bmatrix} A \\ e \end{bmatrix}$. One may begin with the initial feasible solution $x_0 = \frac{e}{n}$, and find the direction of descent by projection of vector c on affine manifold P defined by $Ax = 0$.

$$d = -(I - B^t(BB^t)^{-1}B)c,\ B = \begin{bmatrix} A \\ e \end{bmatrix} \tag{4.32}$$

Set $r = (n(n-1))^{-\frac{1}{2}}$ (the radius of ball inscribed in the simplex Δ) and choose $\alpha,\ 0 < \alpha < 1,\ \rho = \frac{\alpha}{n} < \alpha r$. Set

$$x_1 = \frac{e}{n} + \rho\frac{d}{\|d\|} \tag{4.33}$$

It is easy to verify that x_1 is a feasible point. This new feasible interior point is projected to $\frac{e}{n}$ by projective transformation (4.30), where $\bar{x} = \bar{x}_1$, and we obtain the problem in new variables equivalent to the original problem (4.25)–(4.28):

$$\begin{array}{ll} \text{minimize} & \dfrac{(c, \bar{X}_1\widehat{x})}{(e, \bar{X}_1\widehat{x})} \\ \text{subject to} & (A\bar{X}_1, \widehat{x}) = 0 \\ & \widehat{x} \in \Delta, \\ \text{besides,} & \bar{x}_1 \rightarrow \dfrac{e}{n}. \end{array}$$

Defining

$$\widehat{A} = A\bar{X}_1; \quad \widehat{c} = \bar{X}_1 c,$$

we get LP problem

$$\text{minimize} \quad (\widehat{c}, \widehat{x}) \tag{4.34}$$

$$\text{subject to} \quad \widehat{A}\widehat{x} = 0 \tag{4.35}$$

$$\widehat{x} \in \Delta \tag{4.36}$$

One can observe that the problem (4.34)–(4.36) is of the same type as the problem (4.25)–(4.28), so we can make a new step, substituting $\widehat{A}$ and $\widehat{c}$ for A and c in formulas (4.32) , (4.33) . Hence, we receive a new point $\widehat{x}_2$. In similar way we may produce unlimited number of iterations and get the sequence of points $x_0, x_1, \ldots, x_k, \ldots$.

Let us define the potential function:

$$f(x) = \sum_{j=1}^{n} \ln(c, x)/x_j$$

and set $\alpha = \frac{1}{3}(\rho = \frac{1}{3n})$ in formula (4.33). Karmarkar proved the following:

Theorem 61 *If the sequence $\{x_k\}_{k=0}^{\infty}$ is defined by the described above algorithm, then*

(i) $f(x_k) \leq f(x_0) - \frac{1}{5}k$

(ii) $(c, x_k) \leq (c, x_0)\exp(-k/5n)$

So, Karmarkar's algorithm guarantees the convergence of the sequence (c, x_k) to the optimal value 0 with the speed of geometrical progression with ratio $q = e^{-\frac{1}{5}}$ that is independent of input date. This algorithm was generalized for the general class of LP problems in canonical form (4.3)–(4.5).

Let L be the length of input of given LP problem. Since at most $O(L \max\{m, n\})$ iterations are required to obtain the necessary accuracy, we get:

Theorem 62 *Karmarkar's algorithm can be adapted to solve a general linear programming problem with m constraints and n variables in about $O(p^3L^2)$ arithmetical operations that can be carried out with $O(L)$ precision (here $p = m + n$).*

There is a modification of the algorithm with worst case complexness $O(p^{2,5}L^2)$ arithmetical operations [Kar 84]. Note that Dikin's method is monotonic: $(c, x(k)) > (c, x(k+1))$ for all k. Karmarkar's method guaranties only the monotonicity of potential functions, the sequence $\{(c, x_k)\}_{k=0}^{\infty}$ may be non-monotonic.

The Karmarkar's algorithm has been developed in many ways. In [Re 87b], [So 88] [BL 89] it was observed that Karmarkar's method can be interpreted as a projected Newton-type barrier method applied to logarithmic barrier function if we choose a special penalty parameter. This observation gave life to a class of "path following" methods (see [Gon 88], [KMY 89], [MA 87], [MA 90]). In a path following methods, the barrier and the objective functions are used to generate parametric family of penalty type functions, and, consequently, the trajectory (or path) of minimizers of these functions, which converges to optimal set of the problem when the penalty coefficient tends to zero. The main idea in path following algorithms is to try to approach the path of minimizers at each step.

This class of interior point methods is based on the notion of "analytic center", when the case of LP is considered.

If the polyhedron $P = \{x \in R^n : (a_i, x) \leq \beta_i, i = 1, \ldots, m\}$ has a nonempty interior Int$P$, and we define

$$f(x) = \ln \prod_{i=1}^{m} (\beta_i - (a_i, x)), \quad x \in IntP,$$

then ξ is called an "analytic center" of P if $f(\xi) \geq f(x)$ for all $x \in$ Int$P$. The notion of "analytic center" was explored in the works [So 88]. The idea of using the Newton method for approximate following of central path was first proposed by Renegar in [Re 87b]. His algorithm terminates in at most $O(\sqrt{n}L)$ iterations. Vaidya [V 90] obtained the modification of the algorithm with record (by order) complexity, requiring $O(n^3L)$ arithmetical operations.

The path-following algorithms can be easily extended to solve the convex quadratic problems ([NN 89],[NN 94]). For this aim Gonzaga [Gon 89] and Ye [Ye 92] proposed more effective algorithms based on projective and potential function methods.

All modifications of interior point methods for LP problems require the solution of a sequence of symmetric positive definite systems of linear equations with matrices that have large condition numbers. So the complex problems arise, due to instability of computations. Besides, for large LP problems it is interesting to use such computational methods that do not destruct the sparse structure of the initial LP problem. There are many works devoted to practical implementation of interior point methods that are based on the latest achievements in linear algebra computations (see, for example [KLSW 89], [KRR 91], [SM 85]).

In the book of Yu. Nesterov and Ar. Nemirovskii [NN 94] the generalized theory for interior point methods is given. The basic idea of this theory is the unconstrained successive smooth approximation of the problem under consideration by the so-called self-concordant functions and barriers.

Definition. *Let R^n be a n-dimensional real vector space, $Q \subseteq R^n$ be an open nonempty convex subset of R^n, $F : Q \to R$ be a function, $a > 0$. F is called self-concordant on Q with the parameter value a, if $F \in C^3$ is a convex function on Q, and, for all $x \in Q$ and $h \in R^n$ the following inequality holds:*

$$\frac{\partial^3 F(x)}{(\partial h)^3} \leq \frac{1}{2a}\left[\frac{\partial^2 F(x)}{(\partial h)^2}\right]^{3/2} \quad \textit{for all } h \in R^n \tag{4.37}$$

Function F, self-concordant on Q, is called strongly self-concordant if $F(x_i)$ tends to infinity along every sequence $\{x_i \in Q\}$ converging to a boundary point of Q.

Let $\lambda^2(F,x) = \frac{1}{a}([F''(x)]^{-1}F'(x), F'(x))$ where $F'(x)$,$(F''(x))$ are the gradient and the Hessian matrix of function F in the point x, a is the parameter in formula (4.37).

Definition. *Let G be a closed convex domain in R^n and let $\nu \geq 0$. A function F : int$G \to R$ is called a ν-self-concordant barrier for G if F is strongly self-*

concordant with $a = 1$ *on* $intG$ *and*

$$\nu(F) \equiv \sup\{\lambda^2(F, x) | x \in intG\} \leq \nu.$$

The value ν *is called the parameter of the barrier* F.

In [NN 94] it is shown that path-following methods known for LP, can be easily (in principle) extended for the case of general convex domains G. It is proven that number of iterations of those methods is proportional to $O(\sqrt{\nu})$, where ν is the parameter of the barrier. Note, that the usual logarithmic barrier for an m-facet polytope has $\nu = m$.

Every n-dimensional closed convex domain allows a self-concordant barrier with the parameter of order n, but in general this barrier is difficult to compute. But in the book [NN 94] a kind of calculas of "computable" self-concordant barriers is developed. It is shown how to construct the self- concordant barriers for rather complex domains by using the barriers for simple domains.

For example, we demonstrate one of the most important rules for constructing self-concording barriers that was proved in [NN 94] (so-called intersection rule).

Theorem 63 *Let* G_i *be closed convex domains in* R^n *and let* $F_i(x_i)$ *be* ν_i*-self-concordant barriers for* $G_i, 1 \leq i \leq m$. *Assume that the set* $G = \cap_{i=1}^m G_i$ *has a nonempty interior. Then the function*

$$F(x) = \sum_{i=1}^{m} F_i(x) : \ intG \rightarrow R$$

is a self-concordant barrier for the domain $G \subset R^n$ *with parameter*

$$\nu = \sum_{i=1}^{m} \nu_i$$

Let us apply this theorem to the quadratic-type problem:

$$\min(c, x), \ x \in R^n \tag{4.38}$$

subject to the constraints:

$$K_i(x) \leq 0, \ i = 1, \ldots, m, \tag{4.39}$$

where $K_i(x)$ are convex quadratic functions.

Introduce the notations:

$$\begin{aligned} G_i &= \{x \in R^n : K_i(x) \leq 0\},\ i = 1, \ldots, m, \\ G &= \cap_{i=1}^m G_i. \end{aligned}$$

Assume that $intG$ is nonempty. For any G_i one can use the logarithmic self-concordant barrier $s_i(x) = -\ln(-K_i(x))$. It is easy to calculate that parameter $\nu_i = 1,\ i = 1, \ldots, m$. So, the function

$$s(x) = -\sum_{i=1}^{n} \ln(-K_i(x))$$

is m-self-concordant barrier for domain G. Due to the general theory we can find the initial point $x_0 \in intG$ and form the penalty barrier function:

$$F_t(x) = t(c, x) + s(x)$$

Each iteration of the path-following interior-point method contains the calculation of Hessian matrix and gradient of $F_t(x)$ in current point, Newton-type procedure for the choice of the direction of move, special technique for calculating the step-size and changing penalty parameter t.

The arithmetic cost of a step does not exceed $O(1)(mn^2 + n^3)$ (it costs $O(mn^2)$ operations to form the Newton system and $O(n^3)$ operations to solve it). The number of iterations $N(\varepsilon)$ necessary for obtaining relative (in certain sense) accuracy ε with respect to objective function is of order $\sqrt{m} \ln \frac{2m}{\varepsilon}$. So, the total number of arithmetical operations for solution the problem (4.38)–(4.39) with relative accuracy ε does not exceed:

$$O(1)\sqrt{m}(mn^2 + n^3) \ln \frac{2m}{\varepsilon}.$$

The other well known barrier function is associated with problems of semidefinite programming.

Let S_n^+ be the cone of real symmetric positive-semidefinite matrices $n \times n$.

Theorem 64 *The function*

$$F(x) = -\ln \det x$$

is logarithmically homogeneous self-concordant barrier for the cone S_n^+ with parameter $\nu = n$, defined on the set $\tilde{S}_n^+ \subset S_n^+$ of positive definite matrices.

The standard-form of semidefinite programming is to find

$$\min(c, x) \text{ on } x \in S_n^+ \cap P, \tag{4.40}$$

where P is a polyhedral set of the form

$$P = \{x : (a_i, x) \leq b_i,\ i = 1, \ldots, m\}.$$

For this problem we can use the barrier function $F(x) = -\ln\det x + \sum_{i=1}^{m} s_i(x)$, where $s_i(x) = -\ln[b_i - (a_i, x)]$, with parameter $\nu = n + m$. So, if one uses the path-following interior-point method to solve the problem (4.40), $O(\sqrt{n+m})$ steps are enough for improving logarithm of accuracy by an absolute constant.

Due to this technique the polynomial-time interior point algorithms were developed for a number of classes of nonlinear convex programs, including convex problems with quadratic objective and constraint functions, minimization of the matrix norms on a convex parametric family of matrices, the problems of semidefinite programming and so on.

One can observe from this list of problems that the most of them are the subject of applications of the nonsmooth optimization methods described in our book. By the use of the ellipsoid method the polynomial-time algorithms were constructed for these problems. In many cases the worst case estimates for the number of arithmetical operations for interior point methods is better than for the ellipsoid method. But we know that among the subgradient-type methods with space transformations there are methods (for example, r-algorithm) which are practically more effective than the ellipsoid method by an order n, and may be in some cases practically more effective than interior point methods.

Fifteen years ago it was the era of ellipsoid method in optimization (let us remember the XI International symposium on Mathematical programming in Bonn, 1982). It seems that now is the time of blooming for interior point methods, and many mathematicians which earlier had elaborated the methods of nondifferentiable optimization now develop the interior-point methods. In our opinion there is a broad field of effective applications for each of the mentioned above classes of methods. Friendly competition among the two modern directions of mathematical programming (nondifferentiable optimization and interior-point methods) will be useful for progress in both of them.

4.4 REVIEW OF MAIN RESULTS ON COMPLEXITY THEORY OF POLYNOMIAL PROGRAMS

In the previous sections we have shown that LP problems with integer(rational) coefficients can be solved exactly by polynomial time algorithms. But what we can say about the complexity of polynomial programming problems on the whole?

One can find the full review of complexnity theory for polynomial problems in the works of L.G. Khachian [Kh 82] (see also [TK 80]). We will present here only main results.

In general the problems of polynomial programming have the following form: find

$$\inf f_0(x_1, \ldots, x_n) \tag{4.41}$$

subject to the constraints

$$f_i(x_1, \ldots, x_n) \leq 0, \ i = 1, \ldots, m, \tag{4.42}$$

where f_ν are polynomials of n real or integer variables, $\nu = 0, 1, \ldots, n$. Suppose in this section that all polynomials under consideration have integer coefficients that gives us possibility to determine the lenghth of input information for polynomial problems.

Degree d of a polynomial problem is the maximum of total degrees of related to the problem polynomials, **the height** h of a problem is the maximum of absolute values of coefficients determining the polynomials $f_\nu, \nu = 0, 1, \ldots, m$, **the length** L of a problem is the complete number of bits used for description of the list of nonzero polynomial coefficients. The main facts about the complexity of polynomial problems for integer and real variables are the following:

a) There is a certain family of Diophantine equations $f(a, x_1, \ldots, x_n) = 0$ such that the problem of their solving in integers for an arbitrary a is algorithmically nonsolvable [Mat 72]. As subsequent from this: there is no general algorithm for solution of systems of m quadratic inequalities of n integer variables (even for determining the existence of solution) for n and m large enough [Je 76].

b) For quadratic equation $a_1x_1^2 + a_2x_2^2 = a_3$ in two natural variables x_1, x_2 the problem of its solvability (input a_1, a_2, a_3 are natural numbers) is NP-complete.

c) The determination of feasibility of the system of linear Diophantine inequalities is NP-complete problem. Partially, all known algorithms of it solution demand exponential time in input length L. Note, that once n and m are fixed, then polynomial time algorithm exists [Len 81 65].

d) The simplest classes of nonlinear polynomial problems with boolean variables are NP-complete problems, because the booleaty of variables $x_1, \ldots, x_n$ may be expressed by inequalities

$$0 \leq x_i \leq 1,\ i = 1, \ldots, n;\ \sum_{i=1}^{n} x_i^2 \geq \sum_{i=1}^{n} x_i.$$

Because the determination of the feasibility of systems of linear boolean inequalities is NP-complete, the problem of minimization of quadratic nonconvex function subject to the system of linear inequalities is NP-hard. Besides if we could construct the polynomial time algorithm for the system of linear inequalities with one quadratic inequality with precision $\varepsilon \in (0, 1)$

$$f_i(\widehat{x}) \leq \varepsilon; i \in 1, \ldots, m; \widehat{x} \in R^n,$$

it would signify that $P \equiv NP$. So, in general, obtaining an approximate solution for systems of polynomial inequalities is similar by its complexity to the problem of obtaining precise solution.

As we know conditions of convexity in many cases facilitate considerably the solving of the corresponding problems. We see similar situation when we deal with polynomial problems. Let us introduce the notion of multidegree for system of constraints (4.42). Consider the inequalities in nonincreasing order of their degrees: $d_{i_1} \geq d_{i_2} \geq \ldots \geq d_{i_m}$, and let $\alpha = \min\{n, m\}$. The integer $D = d_{i_1} \ldots d_{i_2} \ldots d_{i_m}$ is a **multidegree** of the system (4.42). Tarasov and Khachian proved [TK 80]:

Theorem 65 *If the system with real and (or) integer variables of convex polynomial inequalities (4.42) has the degreee no more than d, $d \geq 2$, the multidegree D and heigth h, and it is compatible, then it has a solution in euclidean ball*

$$\|x\| \leq (hdn)^{Dd^{\frac{3d}{2}} n^{\frac{d}{2}}}.$$

If the extremal problem (4.41)–(4.42) is solvable, then it has a solution in the ball:

$$\|x\| \leq (hdn)^{D^2 d^{3d+3} n^d}.$$

The question arises if we can talk about the precise solution of the problem (4.41)–(4.42), since in most cases the solution can be expressed only by irrational algebraic numbers? But one can use the notion of precise solution in the sense of Tarsky's theory of the first order calcules for real numbers [Ta 51]. As a consequent of the previous theorem we obtain the following [Kh 82]

Theorem 66 *An algorithm exists for precise solution of arbitrary problems of convex polynomial programming with real and (or) integer variables which demands the twice exponential on length of input time, when the degree d of problem is fixed. This algorithm checks the solvability of the problem and if it is solvable, finds one of its algebraic solution $x^* = \{x_1^*, \dots, x_n^*\} \in R^n$, giving for each component x_j^* the equation $p_j(x_j^*) = 0$ with integer coefficients nonreduced in the field Q of rational numbers, and also rational intervals $\alpha_j \leq x_j^* \leq \beta_j$, where the polynomial $p_j(\cdot)$ doesn't have other real roots except x_j^*.*

Theorem 67 *The problem of recognizing the compatibility for system of convex Diophantine inequalities of fixed degree is NP-complete problem. Note that recognizing the existence of solution for one convex quadratic Diophantine equation is also NP-complete.*

Suppose that for the convex polynomial programming problem with integer coefficients of polynomials:

$$minimize f_0(x_1, \dots, x_n) \tag{4.43}$$

$$f_i(x_1, \dots, x_m) \leq 0, \; i = 1, \dots, m. \tag{4.44}$$

one knows radius ρ of a ball $S(0, \rho)$ containing the solution of the problem (4.43)–(4.44). If the problem (4.43)–(4.44) is solvable and has solution in $S(0, \rho)$ the modification of the ellipsoid method finds the solution with the precision ε, i.e. finds a vector $\overline{x} \in E^n$ such that

$$f_0(\overline{x}) \leq f^* + \varepsilon;$$

$$f_i(\overline{x}) \leq \varepsilon, \; i = 1, \dots, m.$$

If the problem (4.43)–(4.44) is nonsolvable in $S(0, \rho)$, the algorithm informs about it.

Theorem 68 *If d is the degree of the problem (4.43)–(4.44), n is the number of variables, h is the height and N is the number of nonzero coefficients,* *then for realization of ellipsoid method one has to fulfil no more than* $n^2(n^2 + d^2N)\log(d\rho^d n^d h/\varepsilon)$ *arithmetical operations with numbers which have no more than* $\log(d\rho^d n^d h/\varepsilon)$ *bits.*

In the partial case of linear programming problems¡ when $d = 1$, it is sufficient to make no more than

$$n^2[10.5n^2 + 3N]l(\lceil\frac{8\rho nh}{\varepsilon}\rceil)$$

arithmetical operations.

When $d = 2$, i.e. convex quadratic programming problem is given with quadratic constraints, it is sufficient no more than

$$n^2[13.5n^2 + 4.5N]l(\lceil\frac{32\rho^2n^2h}{\varepsilon}\rceil)$$

arithmetical operation with numbers that can be recorded by no more than

$$7l(\rho) + 3.5l(\lceil\frac{1}{\varepsilon}\rceil) + 3.5l(h) + 11.5l(n) + 35$$

bits (here by $l(a)$ the number of bits in the binary representation of number a is denoted.

Thus the approximate solution of convex polynomial programming problems of given degree in real variables can be obtained in time that is polynomial on input length L and $\log\frac{\rho}{\varepsilon}$, where ρ is the radius of solution localization, ε is the needed precision.

Remark. It is easy to construct for quadratic case simple examples where ρ is double exponential. For example, find

$$\min x_n$$

subject to constraints

$$x_1 \geq 4,\ x_{r+1} \geq x_r^2,\ 1 \leq r \geq n-1 \tag{4.45}$$

For this example $\rho \geq 2^{2^n}$, $l(\rho) \geq 2^n$. So if one uses usual binary system of number coding, in general, he cannot obtain polynomial algorithms for approximate convex polynomial problem when $d \geq 2$, because output may be exponential. But if one uses so-called additive-exponential representation of numbers, he can obtain a polynomial-time algorithm. This bright unexpected result (see [TK 80]) can be formulated as the following.

Let us consider for given $d \geq 2$ a system of convex polynomial inequalities of degree not exceeding d in real variables:

$$f_i(x) \leq 0; \; x \in R^n; \; i = 1, \dots, m \tag{4.46}$$

with integer coefficients.

The polynomial algorithm of regularization proposed in [TK 80] finds a subsystem of inequalities (4.46):

$$f_i(x) \leq 0; \; x \in R^n; \; i \in I \subseteq \{1, 2, \dots, m\}, \tag{4.47}$$

a family of integer vectors $k_1, \dots, k_r \in Z^n$ and natural numbers $\delta_1, \dots, \delta_r$ in binary coding ($r \leq \min(n, m) \cdot |I|$) such that

(i) the system (4.47) is feasible iff the system (4.46) is feasible;

(ii) if the system (4.47) is feasible, then there exists a solution x^0 in eucleadean ball

$$||x^0|| \leq (hdn)^{1/4} d^{3d/2} n^{d/2} \tag{4.48}$$

(iii) solution x^* of the original system (4.46) can be found if we know x^0 by formula:

$$x^* = k_r 2^{\delta_r} + \dots + k_1 2^{\delta_1} + x^0. \tag{4.49}$$

Note, that for fixed $d \geq 2$ $||x^0||$ is exponential relatively to the length of the problem (4.47).

Let the parametres x^0, k, δ be represented in usual binary codes. Then we shall call the presentation of x^* in the form (4.49) **additive-exponential.** Besides integer vectors $k_r, \dots, k_1$ and parametres $\delta_r, \dots, \delta_1$ possess the property, that the usual binary record of x^* can be reproduced from additive-exponential record in unique way. So using additive-exponential form of record x^* gives us possibility to condence the output information by excluding exponential number of zeroes in it binary representation in intervals amongst the fulled zones.

Note also that if x^0 satisfies the system (4.49) with accuracy ε then x^* satisfies (4.46) with the same precision. Exact or approximate value of x^0 can be obtained by modification of ellipsoid procedure with initial localization given by formula (4.48).

Theorem 69 *An ε-approximate solution of the system of convex polynomial inequalities of given degree in additive-exponential form can be received by polynomial (in L and $|\log\varepsilon|$) time.*

In the partial cases of linear programming, fractional-linear and convex quadratic programming the solution (if it exists) can be expressed in the form of rational numbers when the input is given in the form of list of integers or rational numbers. In this cases polynomial-time algorithms are constructed on the base of the ellipsoid method for obtaining explicit optimal solutions ([Kh 79], [KTK 79]).

Theorem 70 *For exact solution of linear (fractional-linear programming) with n variables and m constraints, and height h it is enough to make no more than $(n+m)\min^4(n,m)\log[h\ \min(n,m)]$ arithmetical operations with no more than $\min\{n,m\}\log[h\ \min(n,m)]$ - digit numbers.*

For exact solution of convex quadratic programming problem with n variables and m constraints of height h it is enough to make no more than $n^4(n+m)\log hn$ arithmetical operations with the numbers that have no more than $n\log hn$ digits.

Remark. Even a small deviation of convexity converts "good" problems into NP-hard ones. For example, if we substitute in quadratic programming problem the convexity of objective function by the condition that objective function is a quadratic form only with only one negative eigenvalue we obtain NP-hard problem ([NN 94]).

Remark. More effective polynomial-time algorithms are obtained for linear and convex quadratic programming by using of interior point methods (see [NN 94]).

In 1984 N.K. Karmarkar proposed his projective scaling algorithm for linear programming which possesses the property of polynomial time solvability but is rather more effective than the ellipsoid method. Note that similar method using so-called "affine scaling" was proposed in 1967 [Di 67] (see also similar algorithm in [Barn 86]) by I.Dikin (Dikin called his method as "interior point method").

Interior point methods were generalized for many important convex polynomial problems such as minimization of convex quadratic functions subject to convex quadratic constraints, solution of linear matrix inequalities (semidefinite programming), constructing optimal by volume inscribed and circumscribed ellipsoids [MA 87], [MA 90], [GLi 88], [J 87], [MSu 87], [MSu 88a], [MSu 88b] and so on. These methods developed by many researchers after Karmarkar's result gave possibility to obtain polynomial time algorithms for ε-approximate solution of the mentioned problems (see fundamental monography of Yu.Nesterov and A.Nemirovskii [NN 94]).

Now we shall recall some known results obtained by A. Nemirovskii and D. Yudin on information theory of complexity for mathematical programming [YN 76]. For many optimization problems it is hard to estimate their algorithmic complexity as a function of the input length especially if we receive information about functions, gradients and etc. by using outputs of some subroutines with unknown structure ("black boxes"). "We can think of an oracle as a subroutine R, which is called by the main program for some argument x and which returns some answer $R(x)$ about the hidden structure". [Ko 82]. We can consider $R(x)$ as a "portion of information" and use it in main algorithm in the form of corresponding proposition (for example, "the function value in the point $\bar{x}$ equal $\bar{y}$"). When we study the informational complexity of an algorithm we do not care in many cases about the time that oracle spends for finding answer, moreover, sometimes we are not interested in the time that the main program spends for processing of the answer. We assume that information complexity of the main algorithm is measured by the number of calls to the oracle (oracles). Below we shall reformulate the main results of A. Nemirovskii and D. Yudin on informational complexity of convex programming problems. Yudin and Nemirovskii considered the following convex optimization problems:

(C) Given a compact, convex subset $K \subseteq R^n$ and a continuous, convex function f on K, find a $x^* \in K$ such that

$$f(x^*) = \min\{f(x) : x \in K\}$$

and its ε-approximation:

(C_ε) for given tolerance $\varepsilon > 0$, find $x_\varepsilon^* \in R^n$ with $d(x_\varepsilon, K) \leq \varepsilon$ and

$$f(x_\varepsilon) \leq f(x^*) + \varepsilon$$

(here $d(x, K) = \min\{\|x - y\| : y \in K\}$)

Consider variance V of the function f on K: $V = \max\{f(x) : x \in K\} - \min\{f(x) : x \in K\}$ and separation oracle

$$R(x) = \begin{cases} c \in S_k(x) & , \quad \text{if } x \ni K \\ c \in G_f(x) & , \quad \text{if } x \in K, \end{cases}$$

where $S_K(x)$ is a set of the hyperplanes separating x from K, i.e. $c \in S_k(x)$ iff $(c, x) < (c, y)$ for all $y \in K$, $G_f(x)$ is the set of subgradients in the point $x \in K$.

Theorem 71 *Let $K \subset R^n$. Any separation oracle algorithm needs in the worst case at least $l(\varepsilon) = c_0 n \log |\frac{V}{\varepsilon}|$ oracle calls to solve (c_ε).*

For two known algorithm, namely, for Levin's method of central sections [Lev 65] (CS-method) and method of inscribed ellipsoids (IE-method) [KTE 88] the same order estimates are obtained.

Theorem 72 *CS-method needs at most $c_1 n \log |\frac{V}{\varepsilon}|$ calls to the separation oracle to solve C_ε. The similar result is true for IE-method.*

Thus, from the point of view of informational theory of complexity the CS-method and IE-method are almost equivalent by their efficiency. But if, for example, feasible region K is given by the "visible" system of linear inequalities and calculations of subgradient $g_f(x)$ are not too complex the practical efficiency of these methods are very differed, because finding by the main algorithm of the center of gravity of polytope in CS-method is a very hard problem even for small dimension, but finding of approximate center of inscribed in polytope ellipsoid optimal by volume in IE-method is polynomial-time solvable problem [KTE 88].

The ellipsoid method needs at most $c_2 n^2 \log \frac{V}{\varepsilon}$ calls of separation oracle, so from the point of view of informational complexity its efficiency is approximately n time worse in comparison with IE-method. But the ellipsoid method demands only $O(n^2)$ arithmetical operations at each step for processing new information.

For convex polynomial problems usually it is easy to estimate the computational complexity of one call of oracle. So by taking into account the processing time of the main algorithm at each step and its informational complexity one can obtain the relative efficiency of different algorithms for solving C_ε-problems.

5

DECOMPOSITION METHODS BASED ON NONSMOOTH OPTIMIZATION

Decomposition methods are used for solving large-scale linear and convex programming problems in order to save time by reducing the number of references to the external memory of a computer. Such methods convert the solution of the original problem into the solution of a series of problems of lower dimension (blocks). They are particularly efficient if the structure of each block permits the use of special, fast solution methods, or the structure of multiprocessor computer permits to make computations simultaneously for many blocks. We shall consider the use of the subgradient-type methods in iterative algorithms for solving linear, convex and some mixed programming problems with the aid of decomposition schemes.

In mathematical programming two general schemes of decomposition are mainly considered:

1. Partitioning of the set of constraints.
2. Partitioning of the set of variables.

Other, more complex decomposition schemes as a rule can be interpreted as a superposition of these main schemes (see [AW 63]. Let us begin from the first scheme in conformity to convex programming problems.

Consider a convex programming problem in the following form:

$$\text{to find } f^* = \inf f_0(x), \tag{5.1}$$

$$x \in X \subseteq E^n \tag{5.2}$$

$$f_i(x) \leq 0,\ i = 1, \ldots, m, \tag{5.3}$$

where X is a closed convex set, $f_\nu(x)$ are convex functions defined on some open convex set $Y \subseteq E^n$, including X, $\nu = 0, 1, \ldots, m$. One may form the Lagrange function $L(x, u)$ with vector $u = \{u_1, \ldots, u_m\}$ of uncertain Lagrange multipliers and consider the u-family of problems:

$$\text{to find } \Psi(u) = \inf_{x \in X} L(x, u), \tag{5.4}$$

where

$$L(x, u) = f_0(x) + \sum_{i=1}^{n} u_i f_i(x) \tag{5.5}$$

For arbitrary fixed $u \geq 0$ $L(x, u)$ is a convex function in x defined on $Y \supseteq X$, so for $u \geq 0$ the "inner" problem (5.4) is the convex programming problem (CPP). It is easy to show that if $\Psi(u)$ has nonempty domain ($dom\Psi$) in E^m, then $\Psi(u)$ is a concave function on $dom\Psi$. The value

$$\sup_{u \in dom\Psi} \Psi(u) = \Psi^* \tag{5.6}$$

is called the **dual lower bound** for f^* (if the problem (5.1)–(5.3) has no feasible solution, we set $f^* = +\infty$).

Due to the Kuhn–Tucker theorem, if the Slaiter condition holds, and $x^*$is the optimal point of CPP (5.1)–(5.3), then there exists $u^* = \{u_1^*, \ldots, u_m^*\} \geq 0$ such that the pair (x^*, u^*) forms saddle point of the Lagrange function $L(x, u)$:

$$L(x^*, u)_{u \geq 0} \leq L(x^*, u^*) \leq L(x, u^*)_{x \in X}$$

and $f^* = f(x^*) = \Psi^*$. In this case an arbitrary solution $\overline{u}^* \geq 0$ of the "outer" problem (5.6) forms the u-component of some saddle point $\{x(\overline{u}^*), \overline{u}^*\}$ for the Lagrange function L(x,u) (5.5). Denote by $X(u)$ the solution set for the u-problem (5.4). Suppose that u^* is the optimal Lagrange multiplier and $X(u^*)$ is unique. Then $x(u^*) \in X(u^*)$ is the solution of the problem (5.1)–(5.3). Otherwise only the part of $X(u^*)$ satisfies to all constraints (5.3). Note that the marginal concave function $\Psi(u)$ is differentiable almost everywhere in the $intdom\Psi$. The points of nondifferentiability correspond to nonunique $X(u)$.

Let $\overline{u} \in intdom\ \Psi$, and $\Psi(\overline{u}) = L(x(\overline{u}), \overline{u})$. Then the vector $g = \{f_i(x(\overline{u}))\}_{i=1}^m$ is the supergradient of the function Ψ in the point $\overline{u}$. If $x(\overline{u}) \in X$ is defined uniquely, then $\{f_i(x(\overline{u}))\}_{i=1}^m$ is the gradient of the function Ψ in the point $\overline{u}$.

Let u^* be the optimal vector of Lagrange multipliers, $u^* = \{u_1^*, \dots, u_m^*\}$. Let $I_0 = \{i \mid u_i^* = 0\}$. It means that the constraints $\{f_i(x) \leq 0,\ i \in I_0\}$ are unessential for the problem (5.1)–(5.3) and they may be removed from the model without changing the optimal solution. If the set $X(u^*)$ contains only one point $x(u^*)$, then $f_i(x(u^*)) = 0$ for $i \in I_1$, where $I_1 = \{1, \dots, m\} \setminus I_0$, $f_i(x(u^*)) \leq 0$ for $i \in I_0$, i.e. $x(u^*)$ is a feasible point. In this case $\Psi(u)$ is differentiable in the point u^*, and necessary and sufficient conditions of maximum $\Psi(u)$ on the set $u \geq 0$ are satisfied:

$$g_\Psi(u^*) = \{f_i(x(u^*))\}_{i=1}^m \begin{cases} = 0, & \text{for } i \in I_1 \\ \leq 0, & \text{for } i \in I_0 \end{cases}$$

In general case finding of feasible $x^* \in X(u^*)$ is a nontrivial problem. There are several ways for solution of this problem.

I. Finding u^* with good precision, then determination of the index set I_0 of zero's multipliers and removing of "unnecessary" constraints. After this one may solve reduced problem by some primal method. Such type procedures can be recommended for quasiblock convex programming models with large number of blocks with linear constraints.

II. "Smoothing" of the function $\Psi(u)$ by adding to the objective function (5.1) a strictly convex function with small parameter (for example, $\varepsilon \sum_{i=1}^n x_i^2$, $\varepsilon > 0$). Let u_ε^* be the optimal Lagrange multiplier for an ε-perturbed problem. Then $\tilde{x}(u_\varepsilon^*)$ (the unique solution of the corresponding "inner" problem) is a feasible approximation for optimal solution of the original problem (5.1)–(5.3).

It is shown in [Ge 81a] (see also [MTS 86]), that for quasiblock LP problems, if some natural regularity conditions are satisfied, there exists small enough $\varepsilon_0 > 0$ such that for $\varepsilon < \varepsilon_0$, $\tilde{x}(u_\varepsilon^*)$ is equal to an optimal solution of the original LP problem.

Consider the LP problem in the following form:

$$\text{to find } f^0 = \min(c, x) \tag{5.7}$$

$$(a_i, x) - b_i \leq 0,\ i = 1, \dots, N,\ x \in E^n \tag{5.8}$$

and the ε-perturbed problem:

$$\text{to find } f^\varepsilon = \min \left[(c, x) + \frac{\varepsilon}{2}(x, x)\right] \tag{5.9}$$

subject to the constraints (5.8). Let the optimal solution x^* of (5.7), (5.8) be unique and determined by intersection of n hyperplanes of the form:

$$P_i = \{x : (a_i, x) - b_i) = 0\},\ i \in \{1, \dots, N\}.$$

Without loss of generality, one may assume that

$$(a_i, x^*) - b_i = 0,\ i \in \overline{1, n},$$

$$(a_i, x^*) - b_i < 0,\ N \geq i > n,$$

and the vectors $a_i, \dots, a_n$ are linearly independent.

Consider the optimal dual vector for the problem (5.7), (5.8) $\overline{u} = \{\overline{u}_1, \dots, \overline{u}_n, \dots, \overline{u}_N\}$ E^n and note that

$$\overline{u}_i > 0, i = 1, \dots, n;\ \overline{u}_i = 0, i > n.$$

Consider the values:

$$D = \det \Gamma(a_1, a_2, \dots, a_n) > 0$$

$$D^{(i)} = \det \Gamma(a_1, \dots, a_{i-1}, a_{i+1}, \dots, a_n)$$

$$D_b^{(i)} = \det \begin{Vmatrix} \Gamma(a_1, \dots, a_{i-1}, a_{i+1}, \dots, a_n) & \begin{matrix} b_1 \\ \vdots \\ b_n \end{matrix} \\ (a_i, a_1)(a_i, a_2) \ \dots \ (a_i, a_n) & b_i \end{Vmatrix}$$

where $\Gamma(r_1, \dots, r_p)$ is the Gram matrix of vectors $(r_1, \dots, r_p)$.

Let

$$\varphi_\varepsilon(u) = \min_x \left[(c, x) + \frac{\varepsilon}{2}(x, x) + (u, Ax - b) \right] \tag{5.10}$$

Then

$$f^\varepsilon = \max_{u \geq 0} \varphi_\varepsilon(u) = \varphi_\varepsilon(u_\varepsilon^*).$$

Let x_ε^* be the optimal solution of (5.10) for $u = u_\varepsilon^*$. It is easy to understand that x_ε^* is the optimal solution of the ε-perturbed problem (5.7), (5.8).

In [Ge 81a] the following theorem is proved:

Theorem 73 *Let ε_0 be the* $\sup \varepsilon$, *for which the system of inequalities:*

$$\varepsilon D_b^i \leq \overline{u}_i D,\ i = \overline{1, n},$$

is satisfied. Then

a) $x_\varepsilon^* = x_0^* = x^*$ *for* $\varepsilon \in [0, \varepsilon_0)$

b) *if* $\varepsilon_0 < \infty$ *and* ε_1 *is the minimal value of* ε *for which there exists an index* i_0 *such that* $\varepsilon_0 D_b^{i_0} = \overline{u}_{i_0} D$*, and* i_0 *is unique then for* $\varepsilon \in [\varepsilon_0, \varepsilon_1]$

$$(c, x_\varepsilon^*) - (c, x^*) = \overline{u}_{i_0} D_b^{i_0} / D_0^{i_0} (1 - \frac{\varepsilon_0}{\varepsilon}).$$

Hence, if the LP-problem has a regular optimal solution, the ε-perturbed LP problem has the same solution for $\varepsilon \le \varepsilon_0$,where ε_0 is small enough. Thus, if we use the ε-perturbation in the decomposition scheme to obtain a feasible solution there is a probability that this prime solution will be optimal for the original problem. This probability increases when ε is decreased, but for small ε the high demands arise for precision of outer decomposition algorithm.

III. One may use the subgradient or ε-subgradient processes of minimizing $\Psi(u)$ with small steps in the neighborhood of u^* (null coordinates of u^* are fixed) with averaging $x(u_k)$ for generating by these processes sequences $\{u_k\}$. The basis of such procedures is given in [Zhu 77], [Shor 85].

5.1 COORDINATE PROBLEMS ARISING IN DECOMPOSITION ON CONSTRAINTS

Let $x_1, x_2, \ldots, x_N$ be vectors of dimensions $r_1, \ldots, r_N$ correspondigly, $z = (x_1, \ldots, x_N)$ be a vector of dimension $\sum_{k=1}^N r_k$. Consider the convex programming problem of the following form.

$$F_0(z) = \sum_{k=1}^{N} f_{0k}(x_k) \longrightarrow \text{minimize} . \tag{5.11}$$

$$F_\alpha(z) = \sum_{k=1}^{N} f_{\alpha k}(x_k) \le 0, \ \alpha \in A, \tag{5.12}$$

$$h_{ik}(x_k) \le 0, \ i = 1, \ldots, m_k, \ k = 1, \ldots, N, \tag{5.13}$$

where all functions f and h with indeces are convex. We name constraints of the type (5.13) for fixed k as k-th block, constraints of the type (5.12) as linking constraints and we say that the problem (5.11)–(5.13) has the **quasi-block structure** with linking constraints. Let us use for the solution of the problem (5.11)–(5.13) the decomposition scheme with partition of constraints into two parts: constraints of the type (5.12) and constraints of the type (5.13).

The Lagrange function has the following form:

$$L(x,u) = \sum_{k=1}^{N} \left[f_{0k}(x_k) + \sum_{\alpha \in A} u_\alpha f_{\alpha k}(x_k) \right],$$

where $u = \{u_\alpha\}_{\alpha \in A}$ is the vector of Lagrange multipliers.

The inner problem: to find $\Psi(u)$,

$$\text{to find } \Psi(u) = \inf_x L(x,u) \tag{5.14}$$

subject to the constraints

$$h_{ik}(x_k) \le 0,\ i = 1,\dots,m_k,\ k = 1,\dots,N \tag{5.15}$$

can be decomposed into N small subproblems:

$$\text{to find } \Psi_k(u) = f_{0k}(x_k) + \sum_{\alpha \in A} u_\alpha f_{\alpha k}(x_k) \longrightarrow \text{minimize} \tag{5.16}$$

$$h_{ik}(x_k) \le 0,\ i = 1,\dots,m_k,\ x_k \in E^k \tag{5.17}$$

for all $k = 1,\dots,N$, so

$$\Psi(u) = \sum_{k=1}^{N} \Psi_k(u),\ u \ge 0 \tag{5.18}$$

Let $\Psi(u)$ have a nonempty convex region of finite values $\overline{dom}\ \Psi$ in the subset $\{u : u \ge 0\} = Q^+$ and $\Psi^* = \sup_{u \in \overline{dom}\Psi} \Psi(u)$.

Let $\Psi^* = \Psi(u^*)$ be a finite number, and u^*_{ik} be optimal Lagrange multipliers associated with the corresponding constraints of the inner problems (5.17) when $u = u^* = \{u^*_\alpha\}_{\alpha \in A}$. Then the vector $\overline{u}^* = \{\{u^*_\alpha\}_{\alpha \in A},\ u^*_{ik},\ i = 1,\dots,m_k,\ k = 1,\dots,N\}$ is a vector of Lagrange multipliers for the original problem (5.11)–(5.13), and Ψ^* is equal to the optimal value of the original problem.

In comparison with the described above general scheme of decomposition on constraints, in the case of the quasiblock structure of the problem we have to use the partition of inner problems on block subproblems in accordance with this structure.

5.2 NONSMOOTH PROBLEMS LINKING WITH DECOMPOSITION ON VARIABLES

Consider a convex programming problem with variables grouped in two subsets. The problem is to find

$$\inf_{x,y} f_0(x,y), \tag{5.19}$$

subject to the constraints

$$f_i(x,y) \le 0, \; i = 1, \ldots, n, \tag{5.20}$$

where x and y are vectors of variables:

$$x = (x^{(1)}, \ldots, x^{(l)}), \; y = (y^{(1)}, \ldots, y^{(m)})$$

and f_0 and f_i, $i = 1, \ldots, n$, are convex functions of $(l+m)$-dimensional vector $z = (x,y) \in E^l \times E^m$. Let us fix $x = \overline{x}$ and consider the following problem: find

$$\inf_{y} f_0(\overline{x}, y) \tag{5.21}$$

subject to the constraints

$$f_i(\overline{x}, y) \le 0, \; i = 1, \ldots, n \tag{5.22}$$

It is easy to observe that the problem (5.21)–(5.22) is a convex programming problem. For the values of $\overline{x}$, for which a solution of the problem (5.21)–(5.22) exists, we define the function Φ as follows:

$$\Phi(\overline{x}) = \inf_{y \in D(\overline{x})} f_0(\overline{x}, y), \tag{5.23}$$

where $D(\overline{x})$ is the set of all y satisfying (5.22).

Theorem 74 *If $f_\alpha(z)$, $\alpha = 0, 1, \ldots, n$ are convex functions, and the problem (5.19)–(5.20) has an optimal solution $z^* = (x^*, y^*)$, then the function Φ defined by (5.23) is proper convex (finitely defined on some convex subset W of E^l). If for some $\overline{x} \in intW$ the Slater constraint qualification is fulfilled for (5.22), then subgradient of Φ at a point $x = \overline{x}$ can be calculated by the formula*

$$g_\Phi(\overline{x}) = g_{L^{x}_{u}}(\overline{x}, y(\overline{x})), \tag{5.24}$$

where

$$L_u(x,y) = f_0(x,y) + \sum_{i=1}^{n} u_i f_i(x,y),$$

$y(\bar{x})$ is one of the optimal values of y in the problem (5.21)–(5.22), $u = \{u_i\}$ are Lagrange multipliers of (5.21)–(5.22) obtained due to the Kuhn-Tucker theorem, and $g_{L_u^{\bar{x}}}(\bar{x}, y(\bar{x}))$ is the projection of a subgradient of the function $L_u(z) = L_u(x,y)$ on the subspace $E^l_{(x)}$, which has a null projection on the subspace $E^m_{(y)}$ (the subgradient is taken at the point $z = (\bar{x}, y(\bar{x}))$).

Proof. Consider two points $\bar{x}_1$ and $\bar{x}_2$ and their convex combination

$$\bar{x}_3 = \lambda_1\bar{x}_1 + \lambda_2\bar{x}_2 \ \ (\lambda_1 + \lambda_2 = 1;\ \lambda_1, \lambda_2 \geq 0).$$

Let $\bar{y}_1 = y(\bar{x}_1)$ and $\bar{y}_2 = y(\bar{x}_2)$ be two solutions of (5.21)–(5.22) for $\bar{x} = \bar{x}_1$ and $\bar{x} = \bar{x}_2$, respectively, and let $\bar{y}_3 = \lambda_1\bar{y}_1 + \lambda_2\bar{y}_2$. Then, by the convexity of $f_\alpha, (\alpha = 0, 1, \ldots, n)$, we have

$$f_0(\bar{x}_3, \bar{y}_3) \leq \lambda_1 f_0(\bar{x}_1, \bar{y}_1) + \lambda_2 f_0(\bar{x}_2, \bar{y}_2),$$

$$f_i(\bar{x}_3, \bar{y}_3) \leq \lambda_1 f_i(\bar{x}_1, \bar{y}_1) + \lambda_2 f_i(\bar{x}_2, \bar{y}_2) \leq 0,\ i = 1, \ldots, n.$$

Therefore $\bar{y}_3$ is feasible for the problem (5.21)–(5.22) for $\bar{x} = \bar{x}_3$. Consequently,

$$\Phi(\bar{x}_3) \leq f_0(\bar{x}_3, \bar{y}_3) \leq \lambda_1\Phi(\bar{x}_1) + \lambda_2\Phi(\bar{x}_2),$$

which proves the convexity of Φ.

We now pass to the proof of formula (5.24) for the subgradient of Φ. Due to the Kuhn-Tucker theorem, if $D(\bar{x})$ satisfies the Slater condition, then there exist multipliers $u(\bar{x}) = \{u_i(\bar{x})\}$, $i = 1, \ldots, n$ such that the following condition is fulfilled

$$\begin{aligned}\Phi(\bar{x}) &= \min_y \left[f_0(\bar{x}, y) + \sum_{i=1}^{n} u_i(\bar{x}) f_i(\bar{x}, y) \right] \\ &= \max_{u \geq 0} \min_y \left[f_0(\bar{x}, y) + \sum_{i=1}^{n} u_i f_i(\bar{x}, y) \right]\end{aligned}$$

If the Slater condition holds for $D(\bar{x})$, then it also holds for $D(x)$ if x belongs to some small enough neighborhood $S(\bar{x})$ of $\bar{x}$ (by the continuity of f_i for all *i*). For $x \in S(\overline{x})$ we have

$$\begin{aligned} \Phi(x) &= f_0(x, y(x)) + \sum_{i=1}^{n} u_i f_i(x, y(x)) \\ &\geq f_0(x, y(x)) + \sum_{i=1}^{n} \overline{u}_i f_i(x, y(x)) \\ &\geq L_{\overline{u}}(x, y), \end{aligned}$$

where $\overline{u} = \{\overline{u}_i\}$ are the Lagrange multipliers of the problem (5.21–(5.22) at $x = \overline{x}$. Let us choose a subgradient of L_u at $(x, y) = (\overline{x}, y(\overline{x}))$ such that its projection $g_L^y(\overline{x}, y(\overline{x}))$ on the subspace $E^m_{(y)}$ vanishes (it is always possible, since

$$L_u(\overline{x}, y(\overline{x})) = \min_y L(\overline{x}, y),$$

which means that the subdifferential of L_u at $(\overline{x}, y(\overline{x}))$ intersects the hyperplane $y = 0$). We obtain

$$\begin{aligned} \Phi(x) - \Phi(\overline{x}) &\geq L_{\overline{u}}(x, y(x)) - L_{\overline{u}}(\overline{x}, y(\overline{x})) \\ &\geq (x - \overline{x}, g_{L_u^x}(\overline{x}, y(\overline{x}))) + (y(x) - y(\overline{x}), g^y_{L_u}(\overline{x}, y(\overline{x}))) \\ &= (x - \overline{x}, g_{L_u^x}(\overline{x}, y(\overline{x}))), \end{aligned}$$

which yields (5.24) due to the definition of a subgradient.

Corollary 4 *If in the preceding theorem one assumes in addition that $f_\alpha(x, y)$, $\alpha = 0, 1, \ldots, n$, is continuously differentiable with respect to y, then formula (5.24) can be made more simple:*

$$g_\Phi(\overline{x}) = g^x_{f_0}(\overline{x}, y(\overline{x})) + \sum_{i=1}^{n} u_i(\overline{x}) g^x_{f_i}(\overline{x}, y(\overline{x})), \tag{5.25}$$

where $g^x_{f_\alpha}(\overline{x}, y(\overline{x}))$ are projections of arbitrary subgradients of f at $\overline{z} = (\overline{x}, y(\overline{x}))$ on $E^l_{(x)}$, $\alpha = 0, 1, \ldots, n$.

Proof. This result follows from the fact that in the case of continuous differentiability of f with respect to y, the projection $g^y_{L_u}(\overline{x}, y(\overline{x}))$ is uniquely determined as null. It remains to apply the formula representing a subgradient of the function L_u, which is a nonnegative combination of f_α, in terms of subgradients of the forming functions.

Formula (5.25) enables one to construct algorithms for solving convex programming problems (in particular, linear programs) by applying decomposition with respect to variables. To this end we have to assume that it is possible to solve the problem (5.21)–(5.22) in a finite number of steps, obtain Lagrange multipliers $u = \{u_i\}$, and compute "partial" subgradients $g^{\overline{x}}_{f_\alpha}$, $\alpha = 0, \ldots, n$. These assumption can be fulfilled, for instance, if (5.21)–(5.22) is a linear or quadratic programming problem. Additional complications may arise from the fact that the problem (5.21)–(5.22) may have no solutions for some $\overline{x}$. A standard way to overcome this difficulty is to use the method of penalty function, in which the problem (5.19)–(5.20) is replaced by the following one: find

$$\min_{x,y,v} f_0(x,y) + M \sum_{i=1}^{n} v_i \tag{5.26}$$

subject to the constraints

$$f_i(x,y) - v_i \leq 0, \; v_i \geq 0, \; i \in \{1, \ldots, n\} \tag{5.27}$$

where M is a sufficiently large positive number. The problem (5.21)–(5.22) is replaced by the problem: find

$$\min_{y,v} \left[f_0(\overline{x}, y) + M \sum_{i=1}^{n} v_i \right], \tag{5.28}$$

subject to the constraints

$$f_i(\overline{x}, y) - v_i \leq 0, \; v_i \geq 0, \; i = 1, \ldots, n. \tag{5.29}$$

It is clear that for any $\overline{x}$ the problem (5.28)–(5.29) has a nonempty feasible set, which satisfies the Slater condition. Thus we may suppose that the problem (5.19)–(5.20) is reduced to a form in which the function $\Phi(\overline{x})$ is defined for all $\overline{x}$ and one can compute a subgradient $g_\Phi(\overline{x})$ at any $\overline{x}$, i. e. one may use the subgradient algorithm for minimzing Φ. Consequently, we obtain the following schematic algorithm for solving the problem (5.19)–(5.20), whose **k**–th iteration consists of the following steps:

(a) for the previously calculated $\overline{x} = \overline{x}_k$ solve the problem (5.21)–(5.22) and find the Lagrange multipliers $u(\overline{x}_k) = \{u_i(\overline{x}_k)\}$,

(b) calculate the vector $g_\Phi(\overline{x}_k)$ by (5.24), or, in favourable cases, by (5.25),

(c) compute $\overline{x}_{k+1}$ by realization an iteration accordingly to simple subgradient process or to other subgradient-type method (for example, r-algorithm).

The stepsizes h_k are selected according to the analysis and recommendations of [Shor 85].

For the special case of (5.19)–(5.20) being a linear programming problem, we shall describe the above algorithm in more detail. Suppose one wants to find

$$\max[(c,x)+(d,y)] \tag{5.30}$$

subject to the constraints

$$Ax + By \leq e, \tag{5.31}$$

$$x \geq 0,\ y \geq 0, \tag{5.32}$$

where x and y are vector variables of dimension l and m, respectively, A is an $n \times l$-matrix, B is an $n \times m$-matrix; c, d and e are fixed vectors of dimension l, m and n, respectively. Let us fix $x = \overline{x}$ and consider the problem: to find

$$\max_y [(c,\overline{x})+(d,y)], \tag{5.33}$$

subject to the constraints

$$By \leq e - A\overline{x}, \tag{5.34}$$

$$y \geq 0, \tag{5.35}$$

and the dual problem: to find

$$\min_v (e - A\overline{x}, v), \tag{5.36}$$

subject to the constraints

$$B^* v \geq d, \tag{5.37}$$

$$v \geq 0, \tag{5.38}$$

where v is a variable vector of dimension n, B^* is the transpose of the matrix B.

Let $\overline{v}(\overline{x})$ denote a solution of (5.36)–(5.38) and α_j denote the j-th row of the matrix $A (j = 1, \ldots, l)$. Define the j-th component of the l-vector $g(\overline{x})$ as follows:

$$g_j(\overline{x}) = (-\alpha_j, \overline{v}(\overline{x})) + c_j,\ \ g(\overline{x}) = (g_1(\overline{x}), \ldots, g_l(\overline{x})), \tag{5.39}$$

where c_j is the j-th component of the vector c. Then the iterative algorithm for solving problem (5.30)-(5.32) according to the above decomposition scheme and its realization by subgradient method can be described as follows. We choose an initial point $\overline{x}_0$, and after k iterations obtain a point $\overline{x}_k$.

On the $(k+1)$-st iteration we proceed as follows:

(1) solve the problem (5.36)–(5.38) at $\overline{x} = \overline{x}_k$ and find the vector $v(\overline{x}_k)$,

(2) compute the subgradient $g(\overline{x}_k)$ by formula (5.39) with $\overline{x} = \overline{x}_k$,

(3) calculate $\overline{x}_{k+1}$ by the formula

$$\overline{x}_{k+1} = \max\{0, \overline{x}_k + h_{k+1} g(\overline{x}_k)\}, \tag{5.40}$$

where h_{k+1} is the stepsize and "max" denotes coordinatewise maximization.

The described above decomposition method can be used for solving various optimization problems, particularly, for analysing large transportation networks (see [MTS 86], [AV 73]).

5.3 EXAMPLES OF SOLVING LARGE-SCALE PROBLEMS BY USING DECOMPOSITION SCHEMES IN COMBINATION WITH R-ALGORITHM ON HIGH-SPEED PARALLEL COMPUTERS

Example 1. The quasiblock LP problems with linking constraints ([Shor 67a]). Consider the LP problem of the following structure:

$$(c, z) = \sum_{k=1}^{N} (c_k, x_k) \longrightarrow \text{minimize} \tag{5.41}$$

$$\sum_{k=1}^{N} (a_k^{(\alpha)}, x_k) - b_\alpha \leq 0, \ \alpha \in A \tag{5.42}$$

$$(h_{ik}, x_k) - b_{ik} \leq 0, \ i = 1, \ldots, m_k, \ k = 1, \ldots, N \tag{5.43}$$

The Lagrange function has the form:

$$L(x,u) = \sum_{k=1}^{N} \left(c_k + \sum_{\alpha \in A} u_\alpha a_k^{(\alpha)}, x_k \right) - \sum_{\alpha \in A} u_\alpha b_\alpha \tag{5.44}$$

and the inner problems decompose into N "small" LP subproblems $SP_k(u)$: to find

$$\Psi_{ik}(u) = \min_{x_k} \left(c_k + \sum_{\alpha \in A} u_\alpha a_k^{(\alpha)}, x_k \right), \ x_k \in E^{r_k} \tag{5.45}$$

subject to the constraints

$$(h_{ik}, x_k) - b_{ik} \leq 0, \ i = 1, \ldots, m_k \tag{5.46}$$

Suppose that for every k polytopes $\Pi_k = \{x_k : x_k \text{ satisfies } (5.46)\}$ are nonempty and bounded. Then for arbitrary u and every $k = 1, \ldots, N$ the optimal solution set $X_k(u)$ is nonempty, and we can choose for every k $x_k(u) \in X_k(u)$. Then

$$\begin{aligned} \Psi(u) &= \sum_{k=1}^{N} \left(c_k + \sum_{\alpha \in A} u_\alpha a_k^{(\alpha)}, x_k(u) \right) - \sum_{\alpha \in A} u_\alpha b_\alpha = \\ &= \sum_{\alpha \in A} u_\alpha \left[\sum_{k=1}^{N} (a_k^{(\alpha)}, x_k(u)) - b_\alpha \right] + \sum_{k=1}^{N} (c_k, x_k(u)) \end{aligned}$$

The vector $g_\Psi(u) = \{\sum_{k=1}^{N} (a_k^{(\alpha)}, x_k(u)) - b_\alpha\}_{\alpha \in A}$ is the supergradient of Ψ in the point u. Thus, for solving the outer coordinating problem: to find

$$\max_{u \geq 0} \Psi(u) = \Psi^* \tag{5.47}$$

one may use subgradient methods for minimizing the convex function $-\Psi(u)$. Note that when we change u it affects only on coefficients of objective functions in local problems (5.45)–(5.46). Therefore for solution of these problems the using of primal simplex procedures is appropriate. As a partial case consider the so-called **generalized distribution LP problem** that arises in optimal production-transportation planning.

Example 2. Generalized distribution LP problems. The generalized distribution LP problems, are in our terminology, problems of the following form:

$$\text{to find } \min \sum_{(i,j) \in P} c_{ij} x_{ij} \tag{5.48}$$

subject to

$$\sum_{(i,j)\in P} p_{ij}^{(\alpha)} x_{ij} - T^{(\alpha)} \leq 0,\ \alpha \in A \tag{5.49}$$

$$\sum_{\{j/(i,j)\in P\}} x_{ij} = d_i,\ i \in I = \{1, \ldots, N\} \tag{5.50}$$

$$x_{ij} \geq 0\ \forall (i,j) \in P \tag{5.51}$$

Here index i is the index of "product" , index j is the index of "technology", the set P is the set of pairs (i,j) such that "product" i can be produced by "technology" j. The set I will be called the set of demands, the set A will be called the set of technological constraints (with technological coefficients $p_{ij}^{(\alpha)}$ which may be interpreted as input of resource $\alpha \in A$ necessary for production of unit volume of product i by technology j). We can consider each demand constraint (5.50) with corresponding part of constraint (5.51) as a block associated with vector $x_i = \{x_{ij}/(i,j) \in P\}$, the technological constraints (5.49) are linking constraints. Due to the mentioned above scheme of decomposition we form the Lagrange function:

$$L(x,u) = \sum_{(i,j)\in P} (c_{ij} + p_{ij}^{(\alpha)} u_\alpha) x_{ij} - u_\alpha T^{(\alpha)} \tag{5.52}$$

and obtain the inner problem: to find

$$\Psi(u) = \sum_{i=1}^{N} \psi_i(u) - \sum_{\alpha} u_\alpha T^{(\alpha)} \tag{5.53}$$

where for all $i \in I$

$$\psi_i(u) = \min \sum_{\{j/(i,j)\in P\}} (c_{ij} + p_{ij}^{(\alpha)} u_\alpha) x_{ij} \tag{5.54}$$

subject to

$$\sum_{\{j/(i,j)\in P\}} x_{ij} = d_i,\ x_{ij} \geq 0\ \forall j, (i,j) \in P. \tag{5.55}$$

Let $x_i(u)$ be the solution of the inner porblem (5.54), (5.55). One can easy obtain $x_i(u) = \{x_{ij}(u)\}$ by following formulas: $x_{ij}(u) = d_{ij}$ if $j = j_i^*(u)$, where $j_i^*(u)$ is the smallest index that minimize $c_{ij}(u) = c_{ij} + \sum_\alpha u_\alpha p_{ij}^{(\alpha)}$ on $j, (i,j) \in P$;

$$x_{ij}(u) = 0 \text{ for all } j \neq j_i^*(u).$$

Thus,

$$\psi_i(u) = d_i \min_{\{j,(i,j)\in P\}} \left[c_{ij} + \sum_{\alpha\in A} u_\alpha p_{ij}^{(\alpha)} \right], \; i = 1, \ldots, N$$

and

$$\Psi(u) = \sum_{i=1}^{N} d_i \min_{\{j,(i,j)\in P\}} \left[c_{ij} + \sum_{\alpha\in A} u_\alpha p_{ij}^{(\alpha)} \right] - \sum_{\alpha\in A} u_\alpha T^\alpha \text{ for } u \geq 0$$

$$\begin{aligned} g_\Psi(u) &= \{g_\Psi^{(\alpha)}(u)\}_{\alpha\in A} = \left\{ \sum_{(i,j)\in P} x_{ij}(u) p_{ij}^{(\alpha)} - T^{(\alpha)} \right\}_{\alpha\in A} \\ &= \left\{ \sum_{i\in I} d_i p_{ij_i^*(u)}^{(\alpha)} - T^{(\alpha)} \right\}_{\alpha\in A} \end{aligned}$$

Problems of the type (5.48)–(5.51) arise very frequently in practice, especially in the solution of industrial planning problems. The research on subgradient-type methods for solving distribution problems stems from the necessity to solve real-life, large scale problems, partially large scale problems connected with optimum planning of turning lathes of the former USSR and distributing the production among customers. The size of the problems and the necessity to use relatively small capacity computers dictated certain requirements on the algorithm. The realization of the described above scheme of decomposition by subgradient-type methods with space dilation in the direction of difference of two successive subgradients (r-algorithm) turned out to be applicable to the solution of such problems.

Distribution problems of particularly large scale were solved in planning loads of turning lathes and in planning of tube production (see [MTS 86]). In those problems the number of demands reached 10.000, the range of products comprised about 1000 items, the number $|A|$ of linking technological constraints reached 300, the technological information included about 200.000 of numbers. The $|A|$-dimensional problems of maximization concave function $\Psi(u)$ on $u \geq 0$ were solved by about 1000 iterations of r-algorithm with relative error in objective function error of order $10^{-6} - 10^{-7}$. We used the ε-quadratic perturbations of objective function [Ge 81a] (see [MTS 86]) for finding the prime solution of relaxed problems. Similar decomposition schemes were used with success for solution a dynamic transportation problem [BZS 78], for the static and dynamic problems of allocating passenger aircrafts to airlines [ZPSY 76],for the problem of distribution of coking coals among coking plants [Ge 77] and etc. (see [SR 67], [SI 69], [SGo 67], [SGM 78], [MSGZ 77]).

Example 3. Concentrator location problem and parallel implementation of r-algorithm [DMi 95] Consider the **capacitated simple plant location problem** (SPLT) model that describes the location of concentrators in centralized computer networks. This model was studied with the aim of testing possibilities of the massive parallel machine (the Connection Machine CM 200) to solve large scale mixed programming problems by using decomposition scheme and subgradient-type method with space dilation. The results of this research were described in [DMi 95].

The decomposition scheme is applied to the following mixed-variable problem (see [Ben 62]): A number of known terminals must be connected, via concentrators, to a central processing unit (CPU). Let $T = \{t_1, \ldots, t_n\}$ be the set of terminals, and let $C = \{c_1, \ldots, c_m\}$ be the set of possible concentrator location from which a subset can be chosen. Each terminal may be assigned one to one of the located concentrators. Concentrator c_i has a fixed cost f_i consisting of the hardware cost and the cost of the line connecting c_i to the CPU. Each concentrator can have a maximum of k terminals assigned to it. Let c_{ij} be the cost for the line t_{ij} from t_j to c_i and let the $x_{ij} = 1$, if t_j is connected to c_i; and 0 otherwise; $y_i = 1$, if c_i is actually located, and 0 otherwise. The **concentrator location problem** (CLP) can be formulated as follows:

$$\min \sum_{i=1}^{m} f_i y_i + \sum_{i=1}^{m} \sum_{j=1}^{n} c_{ij} x_{ij} \tag{5.56}$$

subject to

$$\sum_{i=1}^{m} x_{ij} = 1, \; j = 1, \ldots, n \tag{5.57}$$

$$\sum_{j=1}^{n} x_{ij} \le k y_i, \; i = 1, \ldots, m \tag{5.58}$$

$$x_{ij} \in \{0, 1\}, \; y_i \in \{0, 1\}, \; \forall i, j. \tag{5.59}$$

From (5.57) and (5.58) one can deduce that

$$\sum_{i=1}^{m} y_i \ge \lceil (n/k) \rceil \tag{5.60}$$

We can add the constraint (5.60) to the model (5.56)–(5.59) without changing the problem.

Let us cut the constraints into two parts: (i) the constraints (5.58), (ii), (5.57),(5.59),(5.60), and formulate Lagrangian dual problem [LD] with multiplier vector $u = (u_1, \ldots, u_m)$ associated with the constraints (5.58):

$$[\text{LD}]\ \Psi(u) = \min \sum_{i=1}^{m} f_i y_i + \sum_{i=1}^{m}\sum_{j=1}^{n} c_{ij} x_{ij} + \sum_{i=1}^{m} u_i \left(\sum_{j=1}^{n} x_{ij} - k y_i \right)$$

subject to

$$\sum x_{ij} = 1,\ \forall j = 1, \ldots, n$$

$$\sum_{i=1}^{m} y_i \geq \lceil (n/k) \rceil$$

$$x_{ij} \in \{0,1\},\ y_i \in \{0,1\},\ \forall (i,j).$$

This problem falls into two easily solvable subproblems:

$$(\text{SUB-C})\ \min \sum_{i=1}^{m} (f_i - k u_i) y_i$$

subject to

$$\sum_{i=1}^{m} y_i \geq \lceil (n/k) \rceil$$

$$y_i \in \{0,1\},\ \forall i$$

and

$$(\text{SUB-SA})\ \sum_{i=1}^{m}\sum_{j=1}^{n} (c_{ij} + u_i) x_{ij}$$

$$\sum_{i=1}^{m} x_{ij} = 1,\ \forall j = 1, \ldots, n$$

$$x_{ij} \in \{0,1\},\ \forall i, j.$$

The first problem can be easy solved by known greedy procedure: i) rank the costs $f_i - k u_i$ in ascending order and ii) let $y_i(u) = 1$ if $(f_i - k u_i \leq 0)$ or (inclusive) the rank number of $(f_i - k u_i)$ no more than $\lceil (n/k) \rceil$, otherwise, $y_i(u) = 0$.

The second problem is a greedily solvable semi-assignment problem and the optimal solution $x(u)$ can be computed by finding the least cost row $i_j(u)$ (with minimal index if such rows is nonunique):

$$x_{ij}(u) = \begin{cases} 1, & \text{where } i = i_j(u) \\ 0, & \text{otherwise} \end{cases}$$

From the general theory of Lagrangian dual estimates we know that for $u \geq 0$

$$\Psi(u) = \sum_{i=1}^{m} (f_i - ku_i)\, y_i(u) + \sum_{i=1}^{m}\sum_{j=1}^{n} (c_{ij} + u_i)\, x_{ij}(u)$$

provides a lower bound on the objective of (CLP). $\Psi(u)$ is a concave piecewise linear function and its subgradient at the point $\overline{u} \geq 0$ can be obtained from the subproblem solutions as $g_\Psi(\overline{u}) = \{\sum_{j=1}^{n} x_{ij}(\overline{u}) - ky(\overline{u})\}_{i=1}^{m}$.

Let u^* be an optimal solution of Lagrangian dual problem. As a rule we have a non-unique set of optimal boolean solutions $B(u^*)$ of the form $\{y_i(u^*), i = 1, \ldots, m,\ x_{ij}(u), \forall(i,j)\}$.

If among these solutions there are feasible (that satisfy (5.59)) they represent the optimal solution for the original problem (5.56)–(5.59). But the original problem is NP-hard, therefore we have to expect that in many such problems $B(u^*)$ does not contain feasible solutions. Therefore we need a special algorithm for obtaining a feasible solution of [CLP]. It is proposed in [DMi 95] to use for this aim "many-to-one" modification of assignment problem. Let $\overline{C}$ denote the index set of located concentrators obtained from the solution of SUB-C for u which lie in some neigborhood of u^*. Let $\overline{n} = k|\overline{C}| - n$. If $k|\overline{C}| > n$ we introduce "dummy" terminals with zero cost to obtain a balanced assignment-type problem [MA]

$$\min \sum_{i \in \overline{C}} \sum_{j=1}^{n} c_{ij} x_{ij}$$

subject to

$$\sum_{j \in \overline{C}} x_{ij} = 1,\ j = 1, \ldots, n + \overline{n}$$

$$\sum_{j=1}^{n} x_{ij} = k,\ i \in \overline{C}$$

$$x_{ij} \in \{0, 1\},\ \forall i, j.$$

It is the partial case of transportation problem, so it has integer optimal solutions if we change constraint (5.60) on linear one $x_{ij} \in [0,1]$. Thus, [MA]-problem can be solved for polynomial-time.

Since [LD] is a nondifferentiable problem some subgradient-type algorithms were proposed [DMi 95].

1. Ordinary subgradient algorithm with relaxation step length formula

$$h_k = \rho^{(k)} \frac{\overline{\Psi}_k - \Psi(u_k)}{\|g_\Psi(u_k)\|^2}, \tag{5.61}$$

where $\overline{\Psi}_k$ is the upper bound received by solution of MA; $\rho^{(k)} \in (0,2)$ is relaxation parameter (see [Po 69]);

$$u_{k+1} = \max\{0, u_k + h_k g_\Psi(u_k)\} \tag{5.62}$$

2. As we know the condition number of the function to minimize is an important factor in the efficiency of the subgradient algorithm. In order to get better convergence one can use space transformations. In [DMi 95] the algorithm for solution of [LD] problem is described that is a modification of the space dilation algorithm with dilation along the difference of two successive subgradients. This procedure differs from our version of r-algorithm by line search: the relaxation step length rule is proposed (see formula (5.61)).

One can see that the computations in the mentioned above algorithm for solution [LD] may be easy decomposed on many parallel branches. For Lagrangian multiplier updating procedure, it is easy to see that the sequentual code has a worst case time complexity $O(m^2)$ per iteration since the procedure involves some matrix-vector multiplications and one rank-one update of the dilation matrix of size m by m. The solutions of supproblem [SUB-C] can be computed in $O(m \log_2(m))$ time, since it is dominated by the ranking stage; all other operations are clearly done in $O(m)$ time. Finally the time complexity for the [SUB-SA] procedure is $O(mn)$ (the calculation of the cheapest concentrator for each terminal). The total worst case complexity for one iteration of the sequential space dilation algorithm could therefore be estimated with $O(m^2) + O(m \log^2(m)) + O(mn) = O(mn)$, since $n \geq m$. Turned the attention to the parallel algorithm, a few details must be noted:

1. For simplicity, we assume in our analysis that the required number of processors is available.

2. The dilation matrix B is mapped elementwise to a $m \times m$ grid of processors.

For the cost matrix $\{c_{ij}\}$ the column j is mapped to processor j. All vectors (of length m and n) are distributed elementwise across the processors. It is shown that the total worst case complexity for one iteration of the parallel space dilation algorithm could be estimated with $O(n)+O(m)$. Hence this gives some hints that a parallel implementation is worthwhile for the algorithm. The numerical experiments support this notion.

The numerical experiments were performed by O. Damberg and A. Migdalas on a massively parallel Connection Machine CM200 with 8196 one-bit processing elements and 256 64-bit floating point chips and a total of 1GByte of memory. For comparison the sequential machine used was a Sun SPARC station IPX with 32 MByte of memory. 24 randomly generated instances of varying size $n = 128 \times 2^k, k \in \{0, 1, 2, 3, 4, 5\}, m(n) \le \frac{n2^r}{8}(r = 0, 1, 2, 3,)$. The maximum size problem that was observed on IPX was $m \times n$ 1024×1024 while by CM 200 the problem 4096×4096 was computed. The numerical experiments demonstrated very good properties of subgradient-type algorithm with space dilation relatively to parallel implementation. The computational time for each iteration increases almost quadratically with the problem size in the sequential enviroment, while it is close to linear for the parallel implementation where the number of processors is enough. The large total memory give the possibility to solve with good relative precision very large combinatorial problems by using nondifferentiable optimization algorithms with space dilation for obtaining lower bounds, and special polynomial time algorithms to get feasible solutions and upper bounds (see also [MPS 97], [MPV 97]).

Example 4. A minimal cost network flow problem. Consider a **directed network** or a **digraph** G, consisting of a finite set of **nodes** (vertices) $N = (1, 2, \ldots, m)$ and a set of **directed arcs** (links, branches, edges, or lines) A, each element of A is the ordered pair (i, j) of nodes. Arc (i, j) is said to be **incident** at nodes i and j and is directed from node i to node j. With each node i in G we associate a number b_i. Nodes with $b_i > 0$ are called **sources** and nodes with $b_i < 0$ called **sinks**. If $b_i = 0$, node i is called an intermediate (or **transshipment**) node. For $(i, j) \in A$ let x_{ij} be the amount of flow on the arc (i, j) and c_{ij} be the unit shipping cost along the arc. We shall assume that the total supply equals the total demand within the network i.e. $\sum_{i=1}^{m} b_i = 0$. Vector $x_A = \{x_{ij}\}_{(i,j) \in A}$ is called **flow** if so-called Kirchhoff equations (or nodal balance)

$$\sum_{j \in S_i^-} x_{ij} - \sum_{j \in S_i^+} x_{ij} = b_i, \; i = 1, \ldots, m \tag{5.63}$$

are satisfied. Here S_i^- denotes the set of nodes j such that $(i,j) \in A$ and S_i^+ denotes the set of nodes j such that $(j,i) \in A$.

The minimal cost network flow problem may be stated as follows: minimize

$$f(x) = \sum_{(i,j) \in A} c_{ij} x_{ij} \tag{5.64}$$

subect to

$$x_{ij} \geq 0, \; \forall (i,j) \in A. \tag{5.65}$$

When digraf G is acyclic then no more than one feasible solution exists. Let G have directed cycles. Sequentualy removing from set A the arcs which enter in some dicycle we shall obtain an acyclic partial digraph $\overline{G}$ with the arc set $\overline{A} = A \backslash Y$, where Y is the set of removed arcs. Let $\overline{x} = \{x_{ij}\}_{(i,j) \in A}$ be a feasible solution of the system (5.63). Using the Kirchhoff equation we can uniquely represent $x_{ij}, \forall (i,j) \in \overline{A}$ as affine functions of $\{x_{ij}\}_{(i,j) \in Y}$.

Let $y_{ij} = x_{ij}$, if $(i,j) \in Y$, $y = \{y_{ij}\}_{(i,j) \in Y}$. Then $x_{ij} = l_{ij}(y)$, and the minimal cost network problem (5.64), (5.63), (5.65) reduces to the following problem in variables y : to find

$$\min \sum_{(i,j) \in Y} c_{ij} y_{ij} + \sum_{(i,j) \in \overline{A}} c_{ij} l_{ij}(y)$$

subject to:

$$y_{ij} \geq 0, \; (i,j) \in Y; \; l_{ij}(y) \geq 0, \; (i,j) \in \overline{A}.$$

Using the nonsmooth penalty function method we obtain the problem of minimization of a piecewise linear function $\Psi_s(y)$, where $\Psi_s(y)$ has the form:

$$\begin{aligned} \Psi_s(y) \; &= \; \sum_{(i,j) \in Y} c_{ij} y_{ij} + \sum_{(i,j) \in \overline{A}} c_{ij} l_{ij}(y) - \\ &- \; \sum_{(i,j) \in Y} s_{ij} \min(0, x_{ij}) - \sum_{(i,j) \in \overline{A}} s_{ij} \min(0, l_{ij}(y)), \end{aligned}$$

where $\{s_{ij}\}_{(i,j) \in A}$ is a vector of penalty multipliers. Thus, we can reduce the problem (5.64), (5.63), (5.65) to a simple problem of unconstrained nondiffirentiable optimization using the variable set decomposition.

One can see that it is possible to use the same technique for convex nonlinear cost functions:

$$f(x) = \sum_{(i,j)\in A} f_{ij}(x_{ij})$$

and for two-side constraints on arc flows:

$$a_{ij} \leq x_{ij} \leq b_{ij}, \forall (i,j) \in A.$$

Similar algorithms can be used for optimization of flows in nondirected connected nets.

Example 5. The multicommodity minimal cost flow problem. The multicommodity minimal cost flow problems arise when we want to find optimal plan of reconstruction of transportation net by changing capacities of its nodes and edges. As a rule, we have to consider the transportation of different kinds of loads at this net. Let $G = (V, E)$ be the connected graph.The set of paths (ways) connecting two nodes i and j from V we denote by Π_{ij} N kinds of products are transported in the net. The sets P^k and S^k are correspondingly the set of sources and of sinks for product k, $k = \overline{1,N}$. For each node $i \in P^k$ ($j \in S^k$) we know the volume of the k-th product $a_i^k (b_j^k)$ that is produced at a point i (that is consumed at a point j). We assume that transportation problems are balanced:

$$\sum_{i\in P^k} a_i^k = \sum_{j\in S^k} b_j^k, \; k = \overline{1,N}.$$

Each edge $(i,j) \in E$ has a length (cost) d_{ij} and capacity ρ_{ij}. Let $x_{ij}^{(k)}(\pi)$ be the volume of transportation of the k-th product from the i-th point of production to the j-th point of consumption along the path $\pi \in \Pi_{ij}$. Then the problem of optimal transportation of N products in capaciated net has the following form: to find $x = \{x_{ij}^{(k)}(\pi)\}$ that minimize

$$f(x) = \sum_{k=1}^{N} \sum_{i\in P^{(k)}} \sum_{j\in S^{(k)}} \sum_{\pi\in\Pi_{ij}} x_{ij}^{(k)}(\pi) d(\pi) \tag{5.66}$$

subject to the constraints

$$\sum_{j\in S^{(k)}} \sum_{\pi\in\Pi_{ij}} x_{ij}^{(k)}(\pi) = a_i^{(k)}, \; i \in P^k, \; k = \overline{1,N} \tag{5.67}$$

$$\sum_{i\in P^{(k)}} \sum_{\pi\in\Pi_{ij}} x_{ij}^{(k)}(\pi) = b_j^{(k)}, \; j \in S^k, \; k = \overline{1,N} \tag{5.68}$$

$$x_{ij}^{(k)}(\pi) \geq 0,\ k = \overline{1, N};\ i \in P^k,\ j \in S^k,\ \pi \in \Pi_{ij} \tag{5.69}$$

$$\sum_{k=1}^{N} \sum_{i \in P^{(k)}} \sum_{j \in S^{(k)}} \sum_{\pi \in A(\alpha,\beta)} x_{ij}^{(k)}(\pi) \leq \rho_{\alpha\beta},\ (\alpha, \beta) \in E \tag{5.70}$$

(here $d(\pi) = \sum_{(\alpha,\beta)\in\pi} d_{\alpha\beta}$, $A(\alpha, \beta)$ is the set of paths that contain the edge $(\alpha, \beta) \in E$.) For solution of the above problem we can propose some decomposition schemes.

I. The first way is based on decomposition on constraints. The constraints (5.70) with unknown Lagrange multipliers $u = \{u_{\alpha\beta}\}_{(\alpha,beta)\in E}$ are passed to the objective function (5.66). Let $L(x, u)$ be the corresponding Lagrange function. If we fix $\overline{u} > 0$, then the problem of minimizing $L(x, \overline{u})$ under the conditions (5.67)-(5.69) is decomposed into N transportation problems with length of (α, β)-th edge $\overline{d}_{\alpha,\beta}(u) = d_{\alpha,\beta} + \overline{u}_{\alpha\beta}$. Assume that we have efficient enough procedure for the solution of transportation problems in net form (see [MTS 86]). Solving N transportation problems for given $\overline{u}$, we obtain total Lagrange objective function value $\Psi(\overline{u})$ and total flows along each edge (α, β):

$$\Psi(\overline{u}) = \sum_{k=1}^{N} \Psi_k(\overline{u}),$$

$$y^*_{\alpha\beta}(\overline{u}) = \sum_{k=1}^{N} |y^{*k}_{\alpha\beta}(\overline{u})|,$$

where $\Psi_k(\overline{u})$ is optimal value of objective function and $y^{*k}_{\alpha\beta}$ is the optimal flow along (α, β) of the k-th transportation problem. Vector $\{y^*_{\alpha\beta}(\overline{u}) - \rho_{\alpha\beta}\}_{(\alpha,\beta)\in E}$ determines generalized gradient $g_\Psi(\overline{u})$ of concave dual function $\Psi(u)$ in the point $\overline{u}$. Now we are ready to use one of the known efficient methods of nonsmooth optimization to maximize $\Psi(u)$ on $u \geq 0$. . If $\sup_{u\geq 0} \Psi(u) = +\infty$, then the problem (5.66)– (5.70) has no solution. Otherwise we obtain dual variables u^* and have to use some of the methods of obtaining primal solution.

II. The second way also based on partition of the constraints, but in this case the constraints (5.67), (5.70) with unknown Lagrange multipliers $w = \{w_i^k, w_{\alpha\beta}\}$ are passed to the objective function. After that the inner problem is splitted on $\sum_{k=1}^{N} |S^k|$ simple blocks which are solved by procedure of finding the shortest paths in network (see [MTS 86]). Let $\varphi(\overline{w})$ be the total minimal value of the Lagrange function for given $\overline{w} \geq 0$. Discrepancies in constraints (5.67),

(5.70) give us the components of generalized gradient $\varphi(w)$ at the point $\overline{w}$. Maximization of $\varphi(w)$ gives us the optimal values for dual variables w.

III. The third way of solving the problem (5.66)- (5.70) is the decomposition on variables by use of "circuit" variables (similar to technique that we used for solution of maximal flow net problem). Separate in the graph $G(V, E)$ a spanning tree with the set of edges T, and let $Y = E \backslash T$. Then $x_{\alpha\beta}^{(k)}$ for $(\alpha\beta) \in T$ will be the linear function $l_{\alpha\beta}^{(k)}(y)$, where $y^{(k)} = \{y_{\alpha\beta}^{(k)}\}$, $y_{\alpha\beta}^{(k)} = x_{\alpha\beta}^{(k)}$ for $(\alpha\beta) \in E \backslash T$. In terms of $\{y^{(k)}\}_{k=1}^{N}$ the problem (5.66)–(5.70) reduces to the following problem:

$$\min \sum_{k=1}^{N} \sum_{(\alpha,\beta) \in R} d_{\alpha\beta} |l_{\alpha\beta}^{k}(y^{(k)})| \tag{5.71}$$

subject to

$$\sum_{k=1}^{N} |l_{\alpha\beta}^{k}(y^{(k)})| \leq \rho_{\alpha\beta}, \ (\alpha\beta) \in E \tag{5.72}$$

Note, that $l_{\alpha\beta}(y^{(k)}) = y_{\alpha\beta}^{(k)}$ for $(\alpha\beta) \in Y$. We obtain the problem of convex programming in variables $y = \{y^{k}\}_{k=1}^{N}$. For solving this problem one can use the nonsmooth penalty function method, by which one reduces the problem to the minimization of the piecewise linear function

$$f(y) = \sum_{k=1}^{N} d_{\alpha\beta} |l_{\alpha\beta}^{k}(y^{(k)})| + \sum_{(\alpha,\beta)} s_{\alpha\beta} \max \left\{ 0, \sum_{k=1}^{N} |l_{\alpha\beta}^{k}(y^{(k)})| - \rho_{\alpha\beta} \right\}, \tag{5.73}$$

where $s_{\alpha\beta} > 0$ are penalty multipliers. The number of variables is $N\lambda(G)$, where $\lambda(G)$ is cyclomatik number of graph G. The advantage of the separation of spanning tree variables consists in possibility to solve more complex, for example, nonlinear or dynamical flow problems in framework of such technique.

The Institute of Cybernetics of Ukrainian National Academy of Sciences in cooperation with leading scientific industrial branch institutes have developed the methods of solution of nonlinear large-scale network transportation problems. Such problems are reduced to minimization of nonsmooth functions of the form:

$$\overline{f}(y) = \sum_{(\alpha,\beta)} \varphi_{\alpha\beta}(l_{\alpha\beta}(y)) + \max \left\{ 0, \sum_{(\alpha,\beta)} l_{\alpha\beta}(y) - \rho_{\alpha\beta} \right\}, \tag{5.74}$$

where $\varphi_{\alpha\beta}$ are given objective functions, $l_{\alpha\beta}$ are affine functions that were obtained by the same technique as we used above for receiving formula (5.73). If all functions φ_{ij} are convex we can use some of the known subgradient-type procedures directly. If some of φ_{ij} are nonconvex we used two-stage procedure for finding a local minimum. At the first stage we obtain the lower convex approximations of φ_{ij} and solve the corresponding convex problem of the type (5.73). Then we use the obtained optimal point $\overline{y}^*$ as an initial point for local optimization of $\overline{f}(y)$ (5.74) with original φ_{ij}, at the second stage of algorithm. Such approach with good success was used in research of optimal distribution of gase flows in Unit Gas Supplying System of former USSR as for certain year period so with the account of dynamic of development. The more detail description of this work one can find in the book [SGM 78]. Other instances were linked with the problems of design of communication networks, with optimal production of oil-products and supplying them by complex transportation network including railroads, river and sea transportation means, etc. (see [MTS 86]).

Example 6. Solution of large-scale nonlinear problems of special structure. The "might" of nondifferentiable optimization was demonstrated in the work [Mai 91] for solving large-scale convex nonlinear problems of special structure. In that case structure implies only that strictly linear variables can be treated separately and they form the major part of the total set of variables. Therefore it is natural to use the decomposition scheme with partition of variables on "linear" and "nonlinear" part in such situation.

Consider a convex programming problem: to minimize

$$f(y,w) = \{f(y) + (b,w)\} \tag{5.75}$$

subject to the constraints

$$A^T w + c(y) \leq 0, \; c(y) = \{c_i(y)\}_{i=1}^m \tag{5.76}$$

where $y \in E^s$, $w \in E^m$, $A \in R^{m \times n}$, f and c_i are convex functions defined in E^s. For simplicity assume that an inner problem $\text{InP}(\overline{y})$: to find

$$\Psi(\overline{y}) = \inf_w [(b,w) + f(\overline{y})] \tag{5.77}$$

subject to

$$A^T w + c(\overline{y}) \leq 0 \tag{5.78}$$

has a nonempty optimal solution set $X(\overline{y})$ for arbitrary $\overline{y} \in E^s$. Let $u \in E^n$ be the vector of Lagrange multipliers for the problem (5.75), (5.76) Form the Lagrange function:

$$L(w,y,u) = (b,w) + f(y) + (u, A^T w) + (u, c(y))$$

and define

$$L_{\overline{y}}(w,u) = L(w,\overline{y},u)$$

for fixed $y = \overline{y}$ as the Lagrange function of the problem InP($\overline{y}$). Let $\{w(\overline{y}) \in X(\overline{y}),\ u(\overline{y}) \in U(\overline{y})\}$ be a pair of optimal primal and dual variables forming Kuhn-Tucker point for the LP problem (5.77), (5.78). Then

$$\Psi(\overline{y}) = (b, w(\overline{y})) + f(\overline{y});$$

$\Psi(\overline{y})$ is a convex function defined on E^s. A subgradient of Ψ at the point $\overline{y}$ can be obtained by the formula:

$$g_\Psi(\overline{y}) = g_f(\overline{y}) + (u(\overline{y}), g_c(\overline{y})),$$

where $g_c(y) = \{g_{c_1}(y), \ldots, g_{c_m}(y)\}$. Function Ψ is differentiable almost everywhere excepting the points where f or some $c_i(y)$ are nondifferentiable, and the points $\overline{y}$ where $u(\overline{y})$ is nonunique.For finding optimal y one can use the subgradient-type methods of nondifferentiable optimization. The computational results for large-scale, structured problems, where A has block-diagonal form and each block is sparce, are given in the work [Mai 91]. The number of blocks k changes from 1 to 64, every block contains 40 linear variables and 100 constraints, the number of nonlinear variables is 20. A class of model problems was devised to allow experimental testing of various algorithms. $f(y)$ was chosen linear with pseudo- random coefficients, $c_i(y)$ were convex quadratic functions with coefficients generated also by pseudorandom procedure.Three procedures of optimization were used for comparison:

(i) procedure MINOS 5.3 (see [MSa 83]) which is believed to be the best available code for smooth problems when the number of nonlinear variables is small compared to the number of linear variables and when the problems are characterized by relatively mild nonlinearties. MINOS employs a projected augmented Lagrangian algorithm. This algorithm uses the reduced gradient method to solve a sequence of constrained subproblems, involving constraint linearization.

(ii) the Madsen algorithm [Mad 75], which applies the trust region concept to sequential linear programming, giving a globally convergent method. This algorithm was improved by Fletcher and Sainz de la Mara which used a Newton-like method on the second stage of the algorithm, when the active constraints are determined [FSM 89]. This method makes sure the local quadratic rate of convergence when the usual properties of regularity of a problem are satiafied

(iii) the subgradient-type method with space dilation in the direction of two successive subgradients in the form of r-algorithm with changes in the procedure of directional search (but this changes, in our opinion, only made worse the convergence of the algorithm).

In spite of that r-algorithm turned out to be the most powerful mean for solving structured testing large scale problems, as it is shown in [Mai 91].

6

ALGORITHMS FOR CONSTRUCTING OPTIMAL ON VOLUME ELLIPSOIDS AND SEMIDEFINITE PROGRAMMING

6.1 INTRODUCTION

The solution of problems of constructing optimal with respect to the volume circumscribed and inscribed ellipsoids is of great interest due to the broad range of their applications. Primarily these problems can be used to define the "upper" and "lower" approximations of convex sets, for example, the regions of attainability in the theory of control and differential games; to refine regions of localization of possible values of dynamic system parameters, using **a priori** information and results of measurements [Ch 88]; then in statistics, in planning experiments to determine parameters of regression models [Fed 71], in numerical mathematics to describe sets of possible solutions of linear and nonlinear equations with perturbed coefficients, [FF 60], etc.

In recent years interest in the problems of constructing circumscribed and inscribed ellipsoids has been increased considerably due to their possible application in mathematical programming for approximating regions of localization of optimal points in extremal problems. Thus, the widespread "ellipsoid method" and its modifications are based on constructing ellipsoids of minimal volume, circumscribed around regions of optimum localization; the centers of optimal inscribed ellipsoids were used in [KTE 88] to construct fast converging cutting plane algorithms for finding the minimum of a convex function on a polyhedron.

Consider the problems of constructing optimal in a certain sense inscribed and circumscribed ellipsoids . Let W_n be a set of convex compact bodies of n - dimensional Euclidean space R^n including interior points. For each $M \in W_n$ let $O_1(M)$ be a set of ellipsoids circumscribed around M (including M) and $O_2(M)$ be a set of n - dimensional ellipsoids inscribed in M (included in M);

$V(M)$ be n - dimensional volume of compact $M \subset R^n$. Let $E^*(M) \in O_1(M)$ be a circumscribed ellipsoid of minimal volume and $E_*(M) \in O_2(M)$, be the inscribed ellipsoid of maximal volume for a convex body M.

The existence and uniquness of $E^*(M)$ and $E_*(M)$ were first justified by Lewner in an oral presentation, so these ellipsoids are sometimes called "Lewner's" ellipsoids (see [DLL 57]). Behrend [Beh 37] for $n = 2$ and John [Jo 48] for arbitrary n substantiated the existence and uniquness of the optimal circumscribed and inscribed ellipsoids for any $M \in W_n$. John proved also that for each $M \in W_n$ the pair of homotetic ellipsoids $E^+(M)$ (circumscribed) and $E_+(M)$ (inscribed) with a common center and coefficient of homotety $1/n$ exists. Examples of such pairs can be:

(a) ellipsoid $E^*(M)$ and ellipsoid $E^*_+(M)$ homotetic to it with respect to the center of $E^*(M)$ with the coefficient of homotety $1/n$;

(b) ellipsoid $E_*(M)$ and ellipsoid $E^+_*(M)$ homotetic to it with the coefficient of homotety n with respect to the center of $E_*(M)$.

If a set M is a central-symmetric body, then the constant n in John's theorem can be replaced by $n^{1/2}$ [Jo 48].

It should be noted that the problems connected with inscribed and circumscribed ellipsoids are closely linked and are in a sense equivalent. This follows from the following properties of the so-called polar transformation $poM = \{y : (x, y) \leq 1, x \in M\}$ introduced by Minkovski (see [GLS 88] and Chapter 1) for a set $M \in W_n, 0 \in intM$:

$po(poM) = M$;

$po(\cap_{i=1}^m M_i) = conv(\cup_{i=1}^m poM_i)$;

if $M_1 \subset M_2$, then $poM_2 \subset poM_1$;

if M is an ellipsoid, then poM is an ellipsoid too;

if $M = (K, 0) = \{y : (Ky, y) \leq 1, K \succ 0\}$ is an ellipsoid with center at the point 0, then $poM = (K^{-1}, 0)$ and $V(M) * V(poM) = w_n^2$, where w_n^2 is the volume of the sphere $\{x : \|x\| \leq 1\}$;

if E^* is an ellipsoid of minimal volume, circumscribed around M (the ellpsoid center is fixed at the origin), then poE^* is the ellipsoid of maximal volume inscribed in poM (with the same center).

The transition from the algorithms for solving problems of constructing ellipsoids with fixed ellipsoid centers to solving problems with any centers is based on the next lemma.

Lemma 3 *Let a $(n+1)$-dimensional ellipsoid E_{n+1} be intersected by a hyperplane P and this hyperplane does not contain the center of E_{n+1}. Let E_n be a n-dimensional ellipsoid defined by $E_n = E_{n+1} \cap P$. Then the ratio of volumes*

$$\frac{V(E_n)}{V(E_{n+1})} = \frac{w_n}{w_{n+1}} \frac{(1 - h^2/h_*^2)^{n/2}}{h_*},$$

where h_* is the distance from the center of E_{n+1} to the hyperplane tangent to the ellipsoid E_{n+1} and parallel to P; and h is the distance from the center of E_{n+1} to P (see [SB 92a], [Ti 75]).

Proof. Consider a $(n+1)$ - dimensional ellipsoid E_{n+1}:

$$E_{n+1} = \{x \in E^{n+1} : (Qx, x) \leq 1; x = \{x_1, \dots, x_{n+1}\}\},$$

where Q is $(n+1) \times (n+1)$ symmetric positive definite matrix ($Q \succ 0$). Let $\overline{x}_{n+1} = \max x_{n+1}$, $x \in E_{n+1}$, $\overline{x} = \arg\max x_{n+1}$, $x \in E_{n+1}$. For an arbitrary $z \in (-x_{n+1}, x_{n+1})$ we can construct cutting plane $P(z) = \{x : x_{n+1} = z\}$. The intersections $E_{n,z} = P(z) \cap E_{n+1}$ are the similar ellipsoids of dimension n with centers lying on the intersection of $P(z)$ and straight line $x = t\overline{x}$, $t \in R$.

It is easy to understand that linear sizes of these ellipsoids are proportional to $\sqrt{\overline{x}_{n+1}^2 - z^2}$, so their n - dimensional volume equals:

$$V(E_n, z) = V(E_n, 0)\sqrt{\left(1 - \frac{z^2}{\overline{x}_{n+1}^2}\right)^n} \tag{6.1}$$

The $(n+1)$ - dimensional volume of E_{n+1}

$$\begin{aligned} V(E_{n+1}) &= V(E_n, 0)\int_{-\overline{x}_{n+1}}^{\overline{x}_{n+1}} \sqrt{(1 - \frac{z^2}{\overline{x}_{n+1}^2})^n}\, dz & (6.2)\\ &= V(E_n, 0)\overline{x}_{n+1} \int_{-1}^{1} (1 - y^2)^n dy = V(E_n, 0)\overline{x}_{n+1}\frac{\omega_{n+1}}{\omega_n} & (6.3)\end{aligned}$$

where ω_k is the volume of k - dimensional ball of radius equal to one. From (6.1) and (6.2) we obtain the following result:

$$V(E_n, z) = V(E_{n+1})\frac{\omega_n}{\omega_{n+1}\overline{x}_{n+1}}\sqrt{\left(1 - \frac{z^2}{\overline{x}_{n+1}^2}\right)^n} \tag{6.4}$$

for all $z \in (-\overline{x}_{n+1}, \overline{x}_{n+1})$.

Lemma is proved.

Let $S \subset E^n$ be a finite set of points $\{a_i\}_{i=1}^m$. Transit this set from the plane $P(0) = \{y : y_{n+1} = 0\}$ to the plane $P(1) = \{y : y_{n+1} = 1\}$ in $(n+1)$ - dimensional space. Thus, we obtain the set S' in $(n+1)$ - dimensional space:

$$S' = \{a_i'\}_{i=1}^m, a_i' = (a_i^{(1)}, \dots, a_i^{(n)}, 1),$$

where $a_i = (a_i^{(j)})_{j=1}^n, i = 1, 2, \dots, m$.

Let E_{n+1} be an ellipsoid with center in 0_{n+1} of minimal $(n+1)$ -dimensional volume, containing S'. Then n - dimensional ellipsoid $E_{n,1} = E_{n+1} \bigcap P(1)$ contains S'. If there is another ellipsoid $\overline{E}_{n,1}$ with n-dimensional volume less than $E_{n,1}$, containing S' , then there is the corresponding $(n+1)$ - dimensional ellipsoid $\overline{E}_{n+1}$, containing S' with the same hight and with smaller volume. This contradicts to the fact that E_{n+1} has the minimal $(n+1)$ - dimensional volume. So, $E_{n,1}$ has the minimal n - dimensional volume among the set of ellipsoids, containing S'. If Q^* is such that

$$E_{n+1} = \{y \in E^{n+1} : (Q^* y, y) \le 1\} \tag{6.5}$$

then we can obtain the defining inequality for the n - dimensional ellipsoid of minimal volume, containing S, by setting in (6.5) $y_{n+1} = 1$.

So, the problem of constructing minimal on volume circumscribed ellipsoid with variable center in n - dimensional space can be fully reduced to the similar problem with fixed center in $(n+1)$ - dimensional space.

Now we know the polynomial-time algorithms for constructing an approximation with given accuracy of the minimal on volume circumscribed ellipsoid, which containes m given points in n-dimensional space E^n, and the maximal on volume inscribed ellipsoid, which is contained in given polytope M, described by m linear inequalities. Essential contribution to complexity theory of such algorithms was made by the works of Nesterov and Nemirovskii [NN 89] and Khachian and Todd [KT 90]. The proposed by the author of this book algorithms based on nonsmooth optimization are simpler than those algorithms and can be competitive for large n and m when the high accuracy of defining the parameters of the extremal ellipsoids is not required [SB 92a].

The problem of constructing an ellipsoid of minimal volume containing a finite set of points

We can limit ourselves by the problem with the fixed center.

Let $\{a_i\}_{i\in I}$ be a finite family of points $a_i = (a_i^{(1)}, \ldots, a_i^{(n)}) \in E^n$. Consider the problem of finding n - dimensional ellipsiod of minimal volume $E(0, Q)$ with fixed in origin center containing a set of points $\{a_i\}_{i\in I}$.

Since the n - dimensional volume of $E(0, Q)$ can be obtained by formula:

$$v_n(E(0, Q)) = w_n \sqrt{\det Q^{-1}},$$

where w_n is the volume of an unit ball in E^n, this problem can be reduced to the following problem of convex programming: to find f^*:

$$f^* = \inf(-\ln \det Q), Q \succ 0 \tag{6.6}$$

subject to linear constraints

$$(Qa_i, a_i) \leq 1, i \in I. \tag{6.7}$$

Consider the polytope

$$P = \text{conv}\{(\{a_i\}, i \in I) \cup 0_n\}$$

If P is n - dimensional, then for arbitrary $Q \succ 0$ satisfying (6.7), $v_n E(0, Q) > v_n(P) > 0$, because $Q \supset P$, and there is unique $Q^* \succ 0$ such that $f^* = -\ln \det Q^*$. Otherwise $f^* = -\infty$. Further we suppose that P has full dimension, so $|I| \geq n$. Consider the problem dual to (6.6), (6.7). Let $U = \{u_i\}_{i\in I}$ be the vector of Lagrange multipliers. Consider the Lagrange function :

$$L(Q, u) = -\ln \det Q + \sum_{i\in I} u_i[(Qa_i, a_i) - 1]$$

In the region $\Omega^+ = \{Q : Q \succ 0\}$ and $\overline{u} \geq 0$, $L(Q, u)$ is a strictly convex smooth function of Q. The value of the marginal function

$$\psi(\overline{u}) = \min_{Q\succ 0} L(Q, \overline{u}) = L(Q(\overline{u}), \overline{u}), \tag{6.8}$$

may be found by solution of nonlinear system of equations:

$$\frac{\partial L}{\partial q_{ij}}(Q, \overline{u}) = 0, (i, j) \in I \times I. \tag{6.9}$$

Denote by $R = \{r_{ij}\}_{i,j=1}^{|I|}$ the symmetric matrix Q^{-1}. Since the gradient of the function ($\ln \det Q$) equals to Q^{-1} , we obtain the following system of equations, equivalent to (6.9):

$$-r_{ij}(\overline{u}) + \sum_{k\in I} \overline{u}_k a_k^{(i)} a_k^{(j)} = 0, \forall (i, j) \in I \times I$$

or in matrix form:

$$R(\overline{u}) = \sum_{k \in I} \overline{u}_k a_k a_k^t \qquad (6.10)$$

If all $\overline{u}_k$ are nonnegative and the family $\{a_k\}_{i \in I}$ has a subset of n linearly independent vectors a_k with the corresponding positive $\overline{u}_k$, then $R(\overline{u})$ is the nonsingular positive definite matrix and $Q(\overline{u}) = [R(\overline{u})]^{-1}$.

Let $m = |I| \geq n$. Then

$$\begin{aligned} \psi(\overline{u}) &= \ln\det R(\overline{u}) + \sum_{k=1}^{m} u_k[(Q(\overline{u})a_k, a_k) - 1] \qquad (6.11) \\ &= \ln\det R(\overline{u}) + \sum_{k=1}^{m} \overline{u}_k[(R(\overline{u})^{-1}a_k, a_k) - 1] \end{aligned}$$

$\psi(\overline{u})$ is a concave function of $\overline{u} = \{\overline{u}_k\}_{k \in I}$ defined on the subset $U^+ = \{\overline{u} : \overline{u}_k \geq 0\}$.

Theorem 75 *In accordance with the Kuhn–Tucker theorem if Q^* is the optimal solution of the problem (6.6),(6.7), then there exists $u^* \geq 0$ such that $R(u^*) = (Q^*)^{-1}, u^*$ is a solution of the problem*

$$\max_{\overline{u} \geq 0} \psi(\overline{u}),$$

where $\psi(\overline{u})$ is defined by (6.11).

Thus, we obtain the dual problem of constructing the optimal circumscribed ellipsoid with the center in origin:

$$\text{maximize } [\ln\det R(\overline{u}) + \sum_{k=1}^{m} \overline{u}_k[(R(\overline{u})^{-1}a_k, a_k) - 1)] \qquad (6.12)$$

subject to the constraints:

$$\overline{u} \geq 0 \qquad (6.13)$$

Let

$$f^*(\varepsilon) = \min(-\ln\det Q), Q \succ 0 \qquad (6.14)$$

subject to the constraints

$$(Qa_i, a_i) \leq 1 + \varepsilon, i \in I, \varepsilon \geq 0 \qquad (6.15)$$

or

$$(\frac{Q}{1+\varepsilon}a_i, a_i) \le 1, i \in I. \tag{6.16}$$

Let us denote $\overline{Q}_\varepsilon = \frac{Q}{1+\varepsilon}$. Then the problem (6.14)-(6.15) is equivalent to the problem:

$$\min\,[-\ln\det\overline{Q}_\varepsilon + n\ln(1+\varepsilon)].$$

subject to the constraints

$$(\overline{Q}_\varepsilon a_i, a_i) \le 1, i \in I, \overline{Q}_\varepsilon \succ 0.$$

So $f_\varepsilon^* = f_0^* - n\ln(1+\varepsilon)$, and $\lim_{\varepsilon\to 0}\frac{f_\varepsilon^* - f_0^*}{\varepsilon} = n$, i.e. $\frac{df_\varepsilon^*}{d\varepsilon}(0) = n$.

One can see that

$$R(u) = \sum_{i\in I} u_k a_k a_k^{(t)} = AUA^{(t)},$$

where $U = \text{diag}\{u_1, \ldots, u_m\}$, A is $n \times m$ matrix with a_j, as the j-th column, $j = 1, \ldots, m$. Since $\frac{df_\varepsilon^*}{d\varepsilon}(0) = \sum_{i=1}^m u_i^* = n$, the dual problem can be represented in the following form:

$$\max\ \ln\det(AUA^t) \tag{6.17}$$

$$\sum_{j=1}^m u_j = n;\ u_i \ge 0,\ i = 1, \ldots, m. \tag{6.18}$$

Such a form has adequate interpretation in some problems of design of statistical experiments.

An example of application of optimal circumscribed ellipsoids to design of statistical experiments. Let our statistical experiments be described by the following model:

$$y_{ij} = (\theta, f(x_i)) + \xi_{ij},\ i = 1, \ldots, m;\ j = 1, \ldots, r_i,\ \sum_{i=1}^m r_i = N \tag{6.19}$$

where $x_i \in X$ are controlled variables, $\theta \in \Omega \subset E^n$ is the vector of unknown parameters, $y_{ij} \in Y$ are observations, ξ_{ij} are the errors of these observations. N is a given total number of experiments, r_i are numbers of experiments in the given points $\{x_i\}_{i=1}^m$, which may be chosen by experimenter. The equation (6.19) is usually called a regression equation, and the function $(\theta, f(x_i))$ is called a regression function.

Suppose that:

(a) functions $f(x_i) = \{f_1(x_i), \ldots, f_n(x_i)\}$ are known for all $i \in \{1, \ldots, m\}$;

(b) we have to estimate parameters $\theta \in E^n$;

(c) errors are additive with null average and with given variance and they are non-correlated for different experiments;

(d) there is a possibility to control x, i.e. to choose $x_i \in X$ according to experimenter judgement. The pair of vectors $\xi_N = \{x, p\}$, where $x = \{x_1, \ldots, x_m\}$, is **the set of based points** of design, $p = \{\frac{r_i}{N}\}_{i=1}^m$ are the measure of this points, which gives us **the design of experiments.**

The best linear unbiased estimate for θ in the model (6.19) can be received by the method of the least squares. The variance matrix is fully determined by design of experiments

$$D[\hat{\theta}_N] == \frac{1}{N} D(\xi_N) = \frac{1}{N} M^{-1}(\xi_N)$$

where $M(\xi_N)$ is the Fisher normed information matrix, $D(\xi_N) = M^{-1}(\xi_N)$ [Fed 71]. Due to the supposition about non-correlation of errors ε_{ij} , $M(\xi_N)$ has the following form: $M(\xi_N) = \sum_{i=1}^m p_i \mu(x_i)$, where $\mu(x_i)$ is the update of information matrix, caused by one experiment at the point x_i, $\sum_{i=1}^m p_i = 1$, $p_i \geq 0, i = 1, \ldots, m$. In our case matrices $M(\xi_N)$ and $D(\xi_N)$ are independent of estimated parameter $\hat{\theta}$. Therefore we can have the optimal a priori design of experiments that minimizes a given function φ of information matrix:

$$\xi_N^* = \arg\inf_{\xi_N} \varphi[NM(\xi_N)].$$

One of the most useful criteria of design optimization is the so-colled D - optimality, when the volume of ellipsoid of concentration for estimates $\hat{\theta}$, i.e. the volume of the set $\{\theta : E[(\theta - \hat{\theta})^T D[\theta_N](\theta - \hat{\theta})] = const\}$ is minimized.

Due to this criterion, we must find

$$\xi^* = \arg\inf \det M^{-1}(\xi). \tag{6.20}$$

If observations y_{ij} are scalar (one-dimensional), the matrix $\mu(x_i)$ has the range equal to 1, namely

$$\mu(x_i) = \sigma^{-2} r_i [f(x_i) f(x_i)^T], i = 1, \ldots, m \tag{6.21}$$

where $\sigma^2 = E(\varepsilon_{ij}^2)$.

Let

$$f(x_i) = \{f_1(x_i), \ldots, f_n(x_i)\},$$

$$a_i = \frac{f(x_i)}{\sigma_i}, \quad a_i = \{a_i^{(1)}, \ldots, a_i^{(n)}\}, \quad i = 1, \ldots, m \tag{6.22}$$

Denote by A a matrix $n \times m$ with columns $\{a_j\}_{j=1}^m$, $P = \mathrm{diag}(p), p = (p_1, \ldots, p_m)$. Then $M(\xi_N)$ can be represented in the following form:

$$M(\xi_N) = APA^T, \tag{6.23}$$

and the problem (6.20) is equivalent to

$$\max_p (\ln \det(APA^T)) \tag{6.24}$$

$$\sum_{i=1}^n p_i = 1; \;\; p_i \geq 0, \;\; i = 1, \ldots, m. \tag{6.25}$$

One can easily see that the problem (6.24)–(6.25) is equivalent to the dual problem of construction a minimal on volume ellipsoid with the center at the origin containing all points a_i, if we set $u = np$ in (6.17)–(6.18).

Let $u^* = \{u_k^*\}_{k=1}^m$ be the optimal vector of the dual problem (6.12), (6.13), I^* be the subset of I, containing all indeces $k \in I$ with positive u_k^*. Then all points $a_i, i \in I^*$, are on the surface $([R(u^*)]^{-1}x, x) = 1$ bounding the optimal ellipsoid. Family $a_i, i \in I^*$ contains a system of n linearly independent vectors, although $|I^*|$ may be much greater than n. If u^* is unique, then removing of an arbitrary point from I^* implies that u^* becomes nonoptimal. Let B^* be an $n \times n$ matrix such that $B^*(B^*)^t = Q^* = (R(u^*))^{-1}$. Multiplying the left and the right sides of the equality (6.10) by $B^*(B^*)^t$ and seting $\overline{u} = u^*$, we obtain:

$$I_n = \sum_{k \in I} u_k^*(B^* a_k)(B^* a_k)^t \tag{6.26}$$

Let A be a matrix $n \times m$ containing the vectors $a_k, k \in I$, as columns. Then we can interprete the equality (6.26) as the following:

Theorem 76 *The rows of the matrix $B^* A\sqrt{diagu^*}$ form an orthonormed system.*

Note that B^* is the transformation of E^n which transforms the optimal ellipsoid Q^* in a unit ball. This fact gives us an idea of obtaining B^* as a limit matrix of a sequence of phase space transformations converting the minimizing sequence of circumscribed ellipsoids in a sequence of balls.

If $m << \frac{n(n+1)}{2}$ it is convenient to solve the dual problem (6.12), (6.13) by using r-algorithm. But for the case of large m we propose a simple algorithm for solving direct problem (6.6)–(6.7) that is based on the subgradient method with transformation of phase space.

6.2 SIMPLE ALGORITHMS FOR CONSTRUCTING OPTIMAL ON VOLUME CIRCUMSCRIBED ELLIPSOIDS

To begin with, reduce the problem (6.6)–(6.7) to minimizing the penalty function F in the form of maximum of residuals:

$$F_p(Q) = -\ln\det Q + p\max_{i\in I}[\max\{0, (Qa_i, a_i) - 1\}] \tag{6.27}$$

Using Theorem 27 about the penalty functions in the form of maximum we obtain:

Theorem 77 *If penalty coefficient p in (6.27) is not less than $n (p \geq n)$, then* $\min_{Q\succ 0} F_p(Q) = f^*$ *and* $\arg\min F(Q)$ *equal to* Q^*. *[SB 92a]*

Let us try to apply the simple subgradient process to minimizing $F_p(Q)$ on $Q \succ 0$. Subgradient $g_{F_p}(Q)$ can be calculated by the following formula:

$$g_{F_p}(Q) = Q^{-1} + p\begin{cases} 0 & \text{if } Q \text{ is feasible} \\ a_{i(Q)} \cdot a_{i(Q)}^{(t)} & \text{otherwise,} \end{cases}$$

where $i(Q)$ is one of the indeces from I at which the maximum in (6.27) is obtained.

The inversion of Q demands $O(n^3)$ arithmetical operations and finding $i(Q)$ demands $O(mn^2)$ operations, so the total number of arithmetical operations for calculating $g_{F_p}(Q)$ is $O(mn^2)$ (take into account that $m \geq n$). But if Q is unit matrix I_n then only $O(mn)$ operations for calculating $g_{F_p}(I_n)$ are needed. Basing on this we proposed the following algorithm of minimizing $F_p(Q)$ (see [SSt 90], [SB 92a]).

Denote the set of points $\{a_i\}_{i \in I}$ by M_0. Sphere S_0 of minimal radius r_0 including M_0 is chosen as the initial approximation of circumscribed ellipsoids, with its center at the point 0. Then we use the linear transformation of the space R^n that "comresses" the space in the direction of the maximal norm point contained in M_0 (this maximal norm equals to r_0). In the transformed space denote the image of M_0 by M_1, and next we choose in the transformed space the sphere S_1 of minimal radius containing M_1, with center in origin, etc. From the sequence of circumscribed spheres $S_0, S_1, \ldots, S_N, \ldots$ in the transformed spaces one can generate the corresponding set of ellipsoids $E_0, E_1, \ldots, E_N, \ldots$ containing M_0 in the original space. Under certain conditions the sequence E_N converges to the optimal circumscribed ellipsoid.

Formally the algorithm can be written as follows:

0 - **step.** $B_0 = I_n$, $|I| = m$; $\{a_i^{(0)}\}_{i=1}^m = \{a_i\}_{i=1}^m$,

$$r_0 = \max_{i \in I} \|a_i\| = \|a_{i(0)}^{(0)}\|;$$

After N iterations we have the resulting matrix B_N of the space transformation, $\{a_i^{(N)}\}_{i \in I} = \{B_N a_i\}_{i \in I}$ and $r_N = \max_i \|a_i^{(N)}\| = \|a_{i(N)}^{(N)}\|$.

Calculate:

$$\xi_{N+1} = \frac{a_{i(N)}^{(N)}}{\|a_{i(N)}^{(N)}\|}; \tag{6.28}$$

$$a_i^{(N+1)} = R_{1-\beta_N}(\xi_{N+1}) a_i^{(N)} \text{ for all } i \in I; \tag{6.29}$$

$$B_{N+1} = R_{1-\beta_N}(\xi_{N+1}) B_N, \tag{6.30}$$

$$r_{N+1} = \max_{i \in I} \|a_i^{(N+1)}\| = \|a_{i(N+1)}^{(N+1)}\|. \tag{6.31}$$

It is proved that the algorithm (6.28)–(6.31) can be interpreted as a subgradient method with variable metric applied to the penalty function $F_p(Q)$, given by (6.27). The changes in metric satisfy the certain conditions of regularity.

Theorem 78 *Let $Q_N = B_N^t B_N / r_N^2$ and $conv\{0, \{a_i\}_{i\in I}\}$ be a convex body of dimension n. If:*

$$0 < \beta_N < 1, \quad \textit{for all } N \tag{6.32}$$

$$\sum_{N=1}^{\infty} \beta_N = \rightarrow \infty; \quad \beta_N \rightarrow 0 \quad when N \rightarrow \infty, \tag{6.33}$$

then $\lim Q_N = Q^*$ where Q^* is the matrix of the unique optimal ellipsoid corresponding to the minimum of the penalty function $F_p(Q)$ (6.27), $p \geq n$.

Proof. Let $Q_0 = r_0 I_n$, $r_0 > 0$. Carry out one step of a subgradient descent from the point Q_0 for the function $F_p(Q)$

$$\begin{aligned} \overline{Q}_1 &= Q_0 - h_0 g_{F_p}(x_0) = r_0 I_n - h_0(-\frac{I_n}{r_0} + p a_{i(0)} a_{i(0)}^t) \\ &= \left(r_0 + \frac{h_0}{r_0}\right)\left[I_n - p h_0 \frac{\|a_{i(0)}\|^2}{r_0 + \frac{h_0}{r_0}} \xi_1 \xi_1^t\right] = d_1 R_{1-\overline{\beta}_0}(\xi_1) \end{aligned} \tag{6.34}$$

where $\|a_{i(0)}\| = \max_i \|a_i\|$,

$$d_1 = \frac{d_0 + h_0}{d_0}, \quad d_0 = r_0; \quad \overline{\beta}_0 = \frac{p h_0 \|a_{i(0)}\|^2}{d_1} \tag{6.35}$$

$$\xi_1 = \frac{a_{i(0)}}{\|a_{i(0)}\|}; \tag{6.36}$$

(here $R_\alpha(\xi) = I_n + (\alpha - 1)\xi\xi^t$, $\|\xi\| = 1$, $\alpha \geq 0$ is the matrix of space dilation in the direction ξ with the dilation coefficient α).

Denote $1 - \sqrt{1 - \overline{\beta}_0}$ by β_0 , i.e.

$$(1 - \beta_0)^2 = 1 - \overline{\beta}_0.$$

Let $B_1 = R_{1-\beta_0}(\xi_1)$. Use of the formulas $B_1 a_i = a_i^{(1)}$, $\forall i \in I$, results in a new family of points $\{a_i^{(1)}\}_{i\in I}$.

Substitute $a_i = R^{-1}_{1-\beta_0}(\xi_1)a_i^{(1)}, i \in I$ into (6.27). Then the problem (6.27) reduces to the problem of minimization of the function

$$F_p^1(Q^{(1)}) = -\ln\det Q^{(1)} + 2\ln(1-\beta_0) + p\max\{0, \max_{i\in I}[(Q^{(1)}a_i^1, a_i^1) - 1]\},$$

where $Q^{(1)} = B_1^t Q B_1$.

The approximation $\overline{Q}_1 = d_1 R_{1-\overline{\beta}_0}(\xi_1)$ in the original space is equivalent to $Q_1^{(1)} = d_1 I_n$ for a new family of points $M^{(1)} = \{a_i^{(1)}\}_{i\in I_n}$ in transformed space. To find the feasible ball for $M^{(1)}$ we must calculate

$$r_1 = \max_{i\in I}\|a_i^{(1)}\|$$

In the original space the ellipsoid

$$E_1 = \{x : (B_1^t Q B_1 \leq r_1^2)\} = \left\{x : \left(\frac{B_1^t B_1}{r_1^2}x, x\right) \leq 1\right\}$$

is the next feasible approximation of Q^* . Thus, $Q_1 = \frac{B_1^t B_1}{r_1^2}$ can be interpreted as a result of applying one subgradient-type step to initial point $Q_0 = r_0 I_n$.

After the first step we are in the same position as before the null-step, but with a new family of points $\{a_i^{(1)}\}_{i\in I_n}$, so we may repeat the process of the type (6.28)–(6.31) as many times as necessary. From formula (6.35) one can see that the value of β_0 is proportional to h_0. For convergence of the usual subgradient method we use the conditions: $h_i > 0, \forall i; \sum_{i=1}^{\infty} h_i = +\infty, \lim_{i\to\infty} h_i = 0$. Therefore it is natural to demand the analogous conditions for $\{\beta_i\}$ and to hope get a prove of our theorem by using the same technique as for the mentioned method. First, consider the lemma proving the boundedness of the sequence $\{Q_N\}_{N=0}^{\infty}$.

Lemma 4 *The sequence $\{c_N\}_{N=0}^{\infty}$ of the ratios of the largest axis lengths to the smallest axis lengths for the ellipsoids*

$$E_N = \{x \in E^n : \|B_N x\| \leq 1\}$$

is limited in the case when polytope $P = conv\{0 \cup (\cup_{i=1}^m a_i)\}$ has dimension n.

Proof. Let $W_n = conv\{\cup_{i=1}^m (a_i \cup (-a_i))\}$ has full dimension n. Without loss of generality, assume that $r_0 = \max_{i\in I}\|a_i\| = 1$ and $\|x\| \geq a_{\min} > 0$, for $x \in$ boundary W_n.

Consider $a_i^N = B_N a_i$; $K_N = B_N^{(t)} \cdot B_N$, $(\lambda_1^{(N)})^2 \geq (\lambda_2^{(N)})^2 \geq \dots \geq (\lambda_n^{(N)})^2$ are eigenvalues of K_N and $\{v_1^{(N)}, \dots, v_n^{(N)}\}$ is the corresponding orthonormal system of eigenvectors. B_N can be represented as $K^{\frac{1}{2}} \cdot O_N$, where O_N is an orthogonal operator. Let $u_i^{(N)} = O_N v_i^{(N)}$; $i = 1, \dots, n$. Let us represent $a_i, i \in I$ in the following form:

$$a_i = \sum_{j=1}^{n} a_{ij} \cdot u_j^{(N)},\ i \in I$$

Then $a_i^{(N)} = \sum_{j=1}^{n} \lambda_j^{(N)} a_{ij} u_j^{(N)}$, where $\{u_j^{(N)}\}_{j=1}^{N}$ is an orthonormal system of vectors. It is clear that

$$r_N^2 = \max_i \|a_i^{(N)}\|^2 = \|a_{i^*}^{(N)}\|^2 \geq a_{\min}^2 \cdot (\lambda_1^{(N)})^2,\ a_{i^*}^{(N)} = \sum_{i=1}^{n} \lambda_i^{(N)} a_{i^*j} u_j^{(N)} = \sum_{i=1}^{n} \beta_j u_j^{(N)} \tag{6.37}$$

where $\beta_j = \lambda_j^{(N)} a_{i^*j}$, $i = 1, \dots, n$. Note that $\sum_{j=1}^{n} a_{i^*j}^2 \leq 1$, so

$$\sum_{j=r}^{n} \beta_j^2 \leq (\lambda_r^{(N)})^2 \tag{6.38}$$

Set:

$$\xi_{N+1} = \frac{a_{i^*}^{(N)}}{\|a_{i^*}^{(N)}\|} = \left\{ \frac{\beta_j}{\|a_{i^*}^{(N)}\|} \right\}_{j=1}^{n} = \left\{ \xi_{N+1}^{(j)} \right\}_{j=1}^{N} \tag{6.39}$$

$$B_{N+1} = R_{1-\beta_N}(\xi_{N+1}) \cdot B_N$$

and let $v_n^{(N+1)} \in R^n (\|v_n^{(N+1)}\| = 1)$ be a vector such that $\|B_{N+1} v_n^{(N+1)}\| = \lambda_n^{(N+1)}$.

Let $\overline{z}_N = B_N v_n^{(N+1)}$, and let $\overline{x} = \frac{\overline{z}_N}{\|\overline{z}_N\|}$. $x = \overline{x}$ satisfies the following constraints:

$$\sum_{j=1}^{n} x_j^2 \quad = \quad 1 \tag{6.40}$$

$$\sum_{j=1}^{n} \left(\lambda_j^{(N)}\right)^2 x_j^2 \quad \leq \quad (1+\delta)[\lambda_n^{(N)}]^2, \tag{6.41}$$

where $1 + \delta = \frac{1}{(1-\beta_N)^2}$. Consider c, $1 < c < \frac{\lambda_1^{(N)}}{\lambda_n^{(N)}}$ and replace (6.41) by the weaker inequality:

$$c^2 \sum_{j \in I_1(c)} x_j^2 + \sum_{j \in I_2(c)} x_j^2 \leq 1 + \delta,$$

where $I_1(c) = \{j : c^2 \left(\lambda_n^{(N)}\right)^2 < \left(\lambda_j^{(N)}\right)^2\}$; $I_2(c) = \{j : c^2 \left(\lambda_n^{(N)}\right)^2 \geq \lambda_j^{(N)}\}$. Set $c = \frac{\lambda_1^{(N)}}{\lambda_n^{(N)}}$. Then from (6.38) and (6.39) we obtain:

$$\sum_{j \in I_2(c)} \left(\xi_{N+1}^{(j)}\right)^2 \leq \frac{\lambda_1^{(N)} \lambda_n^{(N)}}{\|a_{i*}^{(N)}\|} \leq \frac{\lambda_n^{(N)}}{\lambda_1^{(N)} a_{\min}^2} \tag{6.42}$$

Consider $\max(\xi_{N+1}, x)$ subject to the constraints (6.40), (6.41). Let

$$\rho_1^2(c) = \sum_{j \in I_1(c)} x_i^2; \; \rho_2^2(c) = \sum_{i \in I_2(c)} x_i^2$$

We obtain:

$$\rho_1^2(c) \leq \frac{\delta}{c^2 - 1} = \frac{\delta_N}{\left(\frac{\lambda_1^{(N)}}{\lambda_n^{(N)}}\right)^2 - 1}$$

Let $\mu_N = \frac{\lambda_1^{(N)}}{\lambda_2^{(N)}}$. Then

$$(\xi_{N+1}, x) \leq \sqrt{\sum_{j \in I_1(\mu)} \left(\xi_{N+1}^{(j)}\right)^2 \cdot \sum_{j \in I_1(\mu)} x_j^2} +$$
$$+ \sqrt{\sum_{j \in I_2(\mu)} \left(\xi_{N+1}^{(j)}\right)^2 \cdot \sum_{j \in I_2(\mu)} x_j^2} \leq \sqrt{\frac{\delta_N}{\mu_N^2 - 1}} +$$
$$+ \sqrt{\frac{1}{\mu_N a_{\min}^2}}$$

This inequality is true for all x, satisfying (6.40), (6.41). For an arbitrary $\varepsilon > 0$ one can find $\mu(\varepsilon)$ such that if $\mu_N > \mu(\xi)$, $(\xi_{N+1}, x) < \varepsilon$ for any x, $\|x\| = 1$.

Consider

$$\|R_{1-\beta_N}(\xi_{N+1})x\|, \; \|x\| = 1.$$

It is easy to obtain, that

$$\|R_{1-\beta_N}(\xi_{N+1})x\| = \sqrt{1 - (2\beta_N - \beta_N^2)(\xi_{N+1}, x)^2}.$$

So we can find $\overline{\mu}$ such that if $\mu_N > \overline{\mu}$, $|R_{1-\beta_N}(\xi_{N+1})\overline{z}_N| > \sqrt[n]{1 - \beta_N} \cdot |\overline{z}_N|$ and

$$\lambda_n^{(N+1)} > \lambda_n^{(N)} \sqrt{1 - \beta_N} \tag{6.43}$$

Note that $\{\mu_N\} = \left\{ \frac{\lambda_1^{(N)}}{\lambda_n^{(N)}} \right\}_{N=1}^{\infty}$ is unbounded iff

$$\{\rho_N\}_{N=1}^{\infty} = \left\{ \frac{\sqrt[n-1]{\prod_{i=2}^{n} \lambda_i^{(N)}}}{\lambda_1^{(N)}} \right\}_{N=1}^{\infty}$$

is unbounded.

But when $\mu_N > \overline{\mu}$, (6.43) is satisfied and $\rho_{N+1} < \rho_N$. Hence, $\{\mu_N\}$ cannot be unbounded. The lemma is proved.□

Algorithm (6.28)–(6.31) can be interpreted as a special version of the subgradient method with variable metric. Below we shall prove the general theorem about convergence of such methods.

Theorem 79 *Let $f(x)$ be a convex function defined on E^n, and let x^* be unique minimum point. Consider the subgradient-type method with variable metric:*

$$x_{k+1} = x_k - h_k B_k \frac{B_k^t g_f(x_k)}{\|B_k g_f(x_k)\|} \tag{6.44}$$

$$h_k > 0, \forall k; \lim_{k \to \infty} h_k = 0, \sum_{k=0}^{\infty} h_k = +\infty. \tag{6.45}$$

If sequences $\{\|B_k\|\}_{k=0}^{\infty}$ and $\{\|B_k^{-1}\|\}_{k=0}^{\infty}$ are limited, then

$$\lim_{k \to \infty} x_k = x^*$$

(x_k is generated by (6.44) starting from an arbitrary initial point x_0).

Proof. Let us fix $k = \overline{k}$. Consider the transformation $x = B_{\overline{k}} y$, and let $\varphi_k(y) = f(B_{\overline{k}} y)$. Using for both parts of (6.44) the operator $(B_k)^{-1}$ we obtain

$$\overline{y}_{\overline{k}+1} = (B_k)^{-1} x_{k+1} = y_{\overline{k}} - h_{\overline{k}} \frac{g_{\varphi_{\overline{k}}}(y_{\overline{k}})}{\|g_{\varphi_{\overline{k}}}(y_{\overline{k}})\|}$$

So one step of the subgradient descent for $\varphi_k(y)$ can be interpreted as variable metric step (6.44) for $f(x)$ from the corresponding point.

Note that

$$(\varphi_{\overline{k}}(y), y - y^*) = (\frac{B_{\overline{k}}^t g_f(x_{\overline{k}})}{\|B_{\overline{k}}^t g_f(x_{\overline{k}})\|}, y - y^*) = (\frac{B_{\overline{k}}^t g_f(x_k)}{\|B_{\overline{k}}^t g_f(x_k)\|}, x - x^*) \geq a_{\varphi_{\overline{k}}}(y_{\overline{k}})$$

where $a_{\varphi_{\overline{k}}}(y_{\overline{k}})$ is the minimal distance from $B_{\overline{k}}^{-1}x^*$ to the set $S_{\overline{k}}$ of equal values of function $\varphi_{\overline{k}}(y)$:

$$S_{\overline{k}} = \{y : \varphi_{\overline{k}}(y) = \varphi(y_{\overline{k}})\}.$$

Since $\|B_k\|_{k=1}^{\infty}$, $\{\|B_k^{-1}\|\}_{k=1}^{\infty}$ are bounded, there is a number $c > 1$ such that $\frac{\|y\|}{c} \leq \|B_k y\| \leq c\|y\|$. Hence

$$\frac{1}{c} a_{\varphi_{\overline{k}}}(y_{\overline{k}}) \leq a_f(x_k) \leq c a_{\varphi_{\overline{k}}}(\overline{y}_k) \tag{6.46}$$

where $a_f(x_k)$ is the minimal distance from x^* to the equal value set:

$$S_{x_k}^f = \{x : f(x) = f(x_k)\}$$

From (6.44) we can obtain:

$$\begin{aligned} \|x_{k+1} - x^*\|^2 &= \|x_k - x^*\|^2 - \\ & 2h_k(\frac{B_k B_k^t g_f(x_k)}{\|B_k g_f(x_k)\|}, x_k - x^*) + h_k^2 \frac{\|B_k B_k^t g_f(x_k)\|^2}{\|B_k g_f(x_k)\|^2} \\ &\leq \|x_k - x^*\|^2 - 2h_k a_{\varphi_k}(y_k) + ch_k^2 \\ &\leq \|x_k - x^*\|^2 - \frac{2h_k}{c}(a_f(x_k) - \frac{c^2 h_k}{2}) \end{aligned} \tag{6.47}$$

Because $h_k \to 0$ and $\sum h_k = +\infty$, the inequality (6.46) implies the existence of subsequence $\{x_{k_r}\}_{r=1}^{\infty}$ that converges to x^* (cf. with Theorem 31) and further we can easy obtain the result about convergence $\|x_k - x^*\|$ to zero.

Let us continue the proof of Theorem 78. We use the algorithm (6.28)–(6.31) for minimization of the convex function $F_p(Q)$ given by (6.27). At the N-th step of the algorithm we use a new variable vector $\overline{Q}$, where Q and $\overline{Q}$ are linked by the linear transformation:

$$Q = (B_N^t \overline{Q} B_N) r_N^{-2},$$

or

$$\overline{Q} = r_N^2 (B_N^{(t)})^{-1} Q (B_N^{(t)})^{-1}.$$

Further,

$$\max_{\|x\|=1} \|B_N x\| \leq \frac{\lambda_1^{(N)}}{\lambda_n^{(N)}} \cdot \frac{r_0}{\|a_{i\bullet}\|} \|B_N a_{i\bullet}^{(N)}\|,$$

where $\|B_N a_{i\bullet}^{(N)}\| = r_N$. So $\frac{\|B_N\|}{r_N} \leq \frac{\lambda_1^{(N)}}{\lambda_n^{(N)}} \cdot \frac{r_0}{\|a_{i\bullet}\|}$. The sequence $\left\{\frac{\|B_N\|}{r_N}\right\}_{N=1}^{\infty}$ is bounded due to Lemma 4.

Similary,

$$\max_{\|x\|\leq 1} \|B_N^{-1} x\| \leq \frac{\|a_{i\bullet}\|}{r_N} \cdot \frac{\lambda_1^{(N)}}{\lambda_n^{(N)}},$$

so the sequence $\left\{\varepsilon_N \|B_N^{-1}\|\right\}_{N=1}^{\infty}$ is also bounded.

Note that the conditions $\beta_N \to 0$ and $\sum \beta_N = \infty$ where $1 - \beta_N$ is the coefficient of space compressing in Theorem 78 are equivalent to the conditions $h_N \to 0$ and $\sum_{N=1}^{\infty} h_N = \infty$, (see (6.29)) where h_N is the corresponding step multiplier in "subgradient-step" interpretation of the algorithm (6.28)–(6.31) (see formulas (6.33)–(6.35)).

Thus, all conditions of the previous theorem are fulfilled in algorithm (6.28)–(6.31). Therefore, $\frac{Q_N}{r_N^2}$ tends to the optimal solution Q^*. Theorem 78 is proved.□

To test the algorithm similar to (6.28)-(6.31), the problems were generated by using special stochastic procedures. The algorithms parameters were regulated in the following way: in k_0 initial iterations the space was compressed with a fixed coefficient $\alpha = 1 - \beta$, $\beta = 0.2$ and, starting from the $k_0 + 1$ iteration, α was chosen as $\alpha_{k_0+k} = 1 - \beta_{k_0+k}$, $\beta_{k_0+k} = \frac{1}{(20+k)}$, $k = 1, 2, \ldots$ Some results of these computational tests are given in Table 6.1, where n is the dimension of the space, m is the number of points, ϵ is the relative precision on the criterion function, and k is the number of iterations (in more detail these experiments were described in [SSt 90]).

Table 6.1

n	2	3	4	5	6	7	8	9	10
m	30	50	70	70	90	100	110	120	130
ε	0.001	0.001	0.001	0.01	0.01	0.01	0.01	0.01	0.01
k	200	300	300	400	400	500	600	700	700

6.3 ALGORITHMS FOR CONSTRUCTING INSCRIBED ELLIPSOIDS OF MAXIMUM VOLUME

Let a compact n - dimensional polytope M be described by the system of linear inequalities:

$$M = \{x : (a_i, x) \leq 1, a_i \in E^n, i = 1, \dots, m\}, m \geq n + 1 \qquad (6.48)$$

$E = \{x : (K(x - c), x - c) \leq 1\}$ be an arbitrary ellipsoid defined by positive definite matrix K and center $c \in E^n$. Let

$$P_i = \{x : (a_i, x) \leq 1\} \qquad (6.49)$$

$$\overline{P_i} = \{x : (a_i, x) = 1\}, \; i = 1, \dots, m. \qquad (6.50)$$

Condition $E \subset M$ is equivalent to the system of inequalities:

$$f(a_i, c) = \min_{x \in \overline{P}_i} (K(x_i - c), x_i - c) \geq 1, \; i = 1, \dots, m. \qquad (6.51)$$

By using the Lagrange technique to the problem (6.51) one can easily obtain:

$$f(a_i, c) = \frac{[1 - (a_i, c)]^2}{(K^{-1} a_i, a_i)}, \; i = 1, \dots, m. \qquad (6.52)$$

So, from $E \subset M$ it follows that

$$\frac{[1 - (a_i, c)]^2}{(K^{-1} a_i, a_i)} \geq 1, \; i = 1, \dots, m \qquad (6.53)$$

Let P be a matrix that satisfies the equality:

$$P^t P = K^{-1}.$$

Then due to (6.53)

$$\frac{[1-(a_i,c)]^2}{||Pa_i||^2} \geq 1,$$

and for $c \in M$

$$1-(a_i,c) \geq ||Pa_i||, \; i=1,\ldots,m.$$

Thus, the problem of finding maximum volume ellipsoid inscribed in M can be reduced to the following problem of convex programming: find

$$\min[-\ln|\det P|] \tag{6.54}$$

subject to the constraints

$$||Pa_i||+(a_i,c)-1 \leq 0, \; i=1,\ldots,m \tag{6.55}$$

Note that if $c=0$ and $0 \in M$, then the corresponding problem has the form:

$$\min[-\ln \det Q] \tag{6.56}$$

$$(Qa_i,a_i) \leq 1, \; i=1,\ldots,m, \text{ where } Q=K^{-1} \tag{6.57}$$

The last problem has the same form as (6.48)-(6.49) for circumscribed ellipsoids. This result can be simply achieved by using polar transformations. Indeed, let a compact n - dimensional polyhedron M be described by a system of linear inequalities:

$$M=\{x:(a_i,x) \leq 1, a_i \neq 0, i=1,\ldots,m\} \tag{6.58}$$

and $0 \in intM$. Let $\overline{E}$ be an arbitrary inscribed in M ellipsoid with center in origin.

$$M=\cap_{i=1}^m P_i, \quad P_i=\{x:(a_i,x) \leq 1\}.$$

Using the properties of polar transformations (see p.170) we have

$$\text{po}M=\text{conv}\{\cup_{i=1}^m a_i\}.$$

Let

$$\overline{E}=\{x:(Kx,x) \leq 1, K \succ 0\}. \tag{6.59}$$

There is a one to one correspondence of inscribed in M ellipsoids $\bar{E}$ to circumscribed ellipsoids pol$\bar{E}$, containing po M (it is equivalent to containing $S=\{a_i\}_{i=1}^m$). It is clear that the polar circumscribed ellipsoid E^* of minimum volume corresponds to the inscribed in M ellipsoid of maximum volume $\overline{E}^*$, because vol$\bar{E}$vol(po)$\bar{E}=w_n^2$.

Thus, if one wants to obtain an ellipsoid of maximal volume with the center in origin inscribed in M, he may formulate the "polar" problem of finding circumscribed ellipsoid of minimal volume with the same center containing $\{a_i\}_{i=1}^m$ and use, for instance, the "simple" algopithm (6.28)–(6.31). Suppose, that after N steps one finds a feasible solution K_N that is good enough approximation for the "polar" problem. Then $Q_N = K_N^{-1}$ gives us a feasible matrix for the original problem of the maximum volume inscribed in M ellipsoid with the same relative error on the volume.

In general case of the variable center c we can replace the problem (6.56)-(6.57) by the penalty function minimization problem:

Theorem 80 *If $N \geq 2n$ then the solution of the problem (6.56)-(6.57) is equivalent to the problem of finding minimum of convex penalty function $\Phi_N(P, c)$, in the domain $P \succ 0$; where*

$$\Phi_N(P, c) = -\ln\det(P^{(t)}P) + N\max\{0, \max_i(\|Pa_i\| + (a_i, c) - 1)\} \qquad (6.60)$$

Proof of this theorem is very similar to that of Theorem 77. (see [SB 89])

Now we try to construct a "simple" algorithm for finding optimal inscribed in M ellipsoid with variable centre.

Let us use the subgradient method for minimization of penalty function $\Phi_N(P, c)$. Similar to the case of circumscribed ellipsoids the calculation of subgradient $g_{\Phi_N}(P, c)$ demands $O(mn^2)$ arithmetical operations. But if $P = \rho I_n$ the number of arithmetical operations for subgradient calculations is $O(mn)$.

The main idea of the proposed below algorithm is to make in the process of subgradient descent such transformations of the space of variable matrices P which convert the current value of P into unit matrix I_n.

Let the $(\overline{\rho} I_n, \overline{c})$ be the initial point of subgradient process of minimization of $\Phi_N(P, c)$

$$g_{\Phi_N}(\overline{\rho} I_n, \overline{c}) = \begin{cases} -\frac{I_n}{\rho^2} + \frac{\overline{\rho} N a_{i^*} \cdot a_{i^*}^{(t)}}{2\|a_{i^*}\|} \\ a_{i^*} \end{cases}, \qquad (6.61)$$

where i^* is the index at which the maximum in formula (6.60) is obtained, the upper part of formula (6.61) gives us the "partial" subgradient on P, and the lower one – on c. Let us make one step of the subgradient process with the step multiplier $\overline{h}$.

$$\begin{aligned} P_1 &= \overline{\rho} I_n - \overline{h}\left(-\frac{I_n}{\overline{\rho}^2} + \frac{\overline{\rho} N a_{i\bullet} \cdot a_{i\bullet}^{(t)}}{2\|a_{i\bullet}\|}\right) \\ &= \left(\overline{\rho} + \frac{\overline{h}}{\overline{\rho}^2}\right) I_n - \overline{h} \frac{\overline{\rho} N}{2\|a_{i\bullet}\|} a_{i\bullet} \cdot a_{i\bullet}^{(t)} \\ &= \left(\overline{\rho} + \frac{\overline{h}}{\overline{\rho}^2}\right)[I_n - \overline{\beta} \xi_{i\bullet} \cdot \xi_{i\bullet}^t], \end{aligned}$$

where

$$\overline{\beta} = \frac{\overline{h}\overline{\rho} N \|a_{i\bullet}\|}{2\rho_1}, \quad \rho_1 = \overline{\rho} + \frac{\overline{h}}{\overline{\rho}^2}, \quad \xi_{i\bullet} = \frac{a_{i\bullet}}{\|a_{i\bullet}\|} \tag{6.62}$$

Thus,

$$P_1 = \rho_1 R_{1-\overline{\beta}}(\xi_{i\bullet}),$$

where $R_\alpha(\xi)$ denotes the matrix of space dilation in direction ξ with coefficient α_i. Note that if

$$\overline{h} < \frac{2\rho_1}{\overline{\rho} N \|a_{i\bullet}\|}, \tag{6.63}$$

then $\overline{\beta} < 1$.

Let us use the transformation of the phase space:

$$a_i' = R_{1-\overline{\beta}}(\xi_{i\bullet}) a_i, \tag{6.64}$$

$$\|P a_i\| = \|P R_{1-\overline{\beta}}^{-1}(\xi_{i\bullet}) a_i'\|, \tag{6.65}$$

$$(c, a_i) = (R_{1-\overline{\beta}}^{-1}(\xi_{i\bullet}) c, a_i'), \tag{6.66}$$

$$\ln\det P = \ln\det(P R_{1-\overline{\beta}}^{-1}(\xi_{i\bullet})) + \ln\det R_{1-\overline{\beta}}(\xi_{i\bullet}) \tag{6.67}$$

$$= \ln\det(P R_{1-\overline{\beta}}^{-1}(\xi_{i\bullet})) + \ln\det(1-\overline{\beta}). \tag{6.68}$$

Set

$$P R_{1-\overline{\beta}}^{-1}(\xi_{i\bullet}) = P'; \tag{6.69}$$

$$R_{1-\overline{\beta}}^{-1}(\xi_{i\bullet}) c = c.' \tag{6.70}$$

We can see that the problem (6.60) is equivalent to the problem of the same type with respect to $\{a_i'\}$, P' and c': find

$$\min[-\ln\det P'(P')^{(t)}] + 2\ln(1-\overline{\beta}) \tag{6.71}$$

subject to the constraints:

$$\|P'c'\| + (a_i', c') - 1 \le 0, \; i = 1, \ldots, m \tag{6.72}$$

$P_1' = \rho_1 I_n$ corresponds to $P_1 = \rho_1 R_{1-\overline{\beta}}(\xi_{i^*})$. The described above one step of the procedure may be iterated infinitely. Thus, we may consider the following algorithm for solving the problem (6.55)-(6.56).

0 - step. Set

$$c_0 = 0;\ \rho_0 = \min(\|a_i\|^{-1}) = \|a_{i^*}\|^{-1};\quad \overline{P}_0 = \rho_0 I_n.$$

We have the feasible approximation for inscribed ellipsoid:
$E_0 = \{y : \frac{(y,y)}{\rho_0^2} \leq 1\}$.

(k+1) -st step. After k steps we have a matrix P_k, vectors $\{a_i^{(k)}\}$, central point $c^{(k)}$ and a feasible approximation of optimal inscribed ellipsoid:

$$E_k = \{y : (K_k(y - P_k c^{(k)}), y - P_k c^{(k)})\},$$

where $K_k = (P_k^t P_k)^{-1}$. Calculate:

a)

$$\rho_k = \min_i \frac{1 - (a_i^{(k)}, c^{(k)})}{\|a_i^{(k)}\|} = \frac{1 - (a_{i(k)}^{(k)}, c^{(k)})}{\|a_i^{(k)}\|} \tag{6.73}$$

($i(k)$ is the index on which the minimum in (6.73) is obtained, ρ_k is the radius of the ball in transformed space with center at the point $c^{(k)}$ inscribed in the corresponding image of M).

b) $\xi_k = \frac{a_{i(k)}^{(k)}}{\|a_{i(k)}^{(k)}\|}$;

c) for given $\beta_k, 0 < \beta_k < 1$, set:

d) $P_{k+1} = \rho_k P_k R_{\frac{1}{1-\beta_k}} xi_k)$

e) $c^{k+1} = R_{\frac{1}{1+\beta_k}}(\xi_k)(c^k - h_k \xi_k)\rho_k$, where

$$h_k = \frac{2\beta_k}{N\|a_{i(k)}^{(k)}\| - \beta_k};$$

f) $a_i^{k+1} = \frac{R_{1-\beta_k}(\xi_k) a_i^{(k)}}{\rho_k}$, $i = 1, \ldots, m$.

g) $E_{k+1} = \{y \in E^n : ((P_k^t P_k)^{-1}(y - P_k c^{k+1}), y - P_k c^{k+1}) \leq 1\}$

Theorem 81 *Let a compact convex n - dimensional polyhedron M be given by a system of linear inequalities:*

$$M = \{x : (a_i, x) \leq 1, i = 1, \ldots, m, m \geq n + 1\}$$

and $\{\beta_k\}_{k=0}^{\infty}$ *be a sequence of positive numbers* $\beta_k < 1$ *for all k and*

$$\lim_{k \to \infty} \beta_k \to 0; \quad \sum_{k=0}^{\infty} \beta_k = \infty \tag{6.74}$$

Then the sequence of matrices $K_k = (P_k^t P_k)^{-1}$, *generated by the described above algorithm, converges to matrix* K^*, *the sequence of vectors* $\{P_k c_k\}_{k=0}^{\infty}$ *converges to* c^*. K^* *and* c^* *are the matrix and central point of inscribed in M maximal volume ellipsoid* E^* *respectively:*

$$E^* = \{y : (K^*(y - c^*), y - c^*) \leq 1\} \tag{6.75}$$

The proof of this theorem is analogous to that of theorem 78 and bulky enough. Therefore we shall not cite it here. Note only, that the main part of the proof contains the proposition that the minimal axes of sequence of feasible inscribed ellipsoids

$$E_k = \left\{(Q_k^{-T}(x - c^{(k)}), x - c^{(k)}) \leq 1\right\}$$

are bounded from below by some positive number. Further one may use the theorem 79 about convergence of subgradient method with variable metric.

Note that conditions (6.74) in theorem 81 may be replaced by almost equivalent conditions:

$$\lim_{k \to \infty} \beta_k(h_k) = 0; \quad \sum_{i=1}^{\infty} \beta_k(h_k) = +\infty \tag{6.76}$$

Practical effectiveness of the described above algorithm was tested by using it as a subroutine in central cutting plane method for convex function minimization – so called "inscribed ellipsoid method" (see [KTE 88]). This method is based on sequential cutting of polyhedron that contains the optimal point of minimized convex function $f(x)$, by constructing a hyperplane, orthogonal to subgradient calculated in the point that is an approximated center of maximal inscribed in the polyhedron ellipsoid. Note that in inscribed ellipsoid method we need not to determine the optimal parameters of inscribed ellipsoids with high accuracy, it is sufficient to have the relative error on ellipsoid volume of the order 0.01-0.05, and such deviation does not virtually influence the efficiency of the algorithm as a whole.The numerical experiments were made by Dr. O.A.Beresovsky who conducted the research and obtained good practical results. (see [SB 92a])

6.4 INTERIOR POINT METHODS FOR DETERMINATION OF OPTIMAL CIRCUMSCRIBED AND INSCRIBED ELLIPSOIDS

Essential contribution in development of polynomial-time algorithms for constructing γ-optimal inscribed in polytope (circumscribed around polytope) ellipsoids was made by Nesterov and Nemirovskii [NN 89] and Khachian and Todd [KT 90]. They considered the following problems:

Problem A. Given a polytope M:

$$M = \{x \in R^n : (a_i, x) \leq 1, \; i = 1, \dots, m\}$$

and a relative accuracy $\gamma \in (0,1)$ by the volume, find a γ-maximal ellipsoid E, contained in the polytope M, such that

$$\frac{\text{vol} E}{\text{vol} E^*} \geq \gamma,$$

where E^* is the ellipsoid of maximum volume inscribed in M.

Problem B. Given a polytope $\bar{M}$:

$$\bar{M} = \text{conv}\left\{\cup_{i=1}^{m} a_i; \; a_i \in R^n\right\}, \text{ and } \gamma \in (0,1),$$

find a γ-minimal ellipsoid $\bar{E}$, containing $\bar{M}$(or set of points $\cup_{i=1}^{m} a_i$) such that

$$\frac{\text{vol} \bar{E}_*}{\text{vol} \bar{E}} \geq \gamma,$$

where $\bar{E}_*$ is the ellipsoid of minimum volume, containing $\bar{M}$. In some algorithms for solving problems A and B a priori known parameter R is used, that is ratio of the radii of two Euclidean balls, the first of which is circumscribed about M, and the second inscribed in M.

The existence of a polynomial time bound for the problem of finding γ-maximal ellipsoid inscribed in M can be deduced from application of the ellipsoid method to the following problem of convex programming that we use for constructing maximal inscribed ellipsoids:

$$f(P) = -\ln \det P \rightarrow \min \tag{6.77}$$

(P-is a symmetric matrix $n \times n$)

$$\|Pa_i\| + (a_i, c) \le 1, \ i = 1, \ldots, m \ P \succeq 0 \tag{6.78}$$

with unknown P and c. The ellipsoid method yields a polynomial bound:

$$O[n^6(n^2 + m)\ln[\frac{Rn}{\ln\frac{1}{\gamma}}]] \tag{6.79}$$

on the number of arithmetical operations sufficient to solve the problem (6.77)-(6.78)

This problem is equivalent to the problem:

$$f(Q) = -\ln\det Q \to \min \tag{6.80}$$

$$(Qa_i, a_i) \le (1 - (c, a_i))^2, \quad i = 1, \ldots, m, \quad Q \succ 0. \tag{6.81}$$

Using one of the interior point methods namely, the path-following Newtons method for minimizing the function

$$F_t = -\ln\det Q - t\sum_{i=1}^{m} \ln[(1 - (c, a_i)^2) - (Qa_i, a_i)]$$

with penalty parameter $t \to 0$, Nesterov and Nemyrovsky in [NN 89] reduced the number of arithmetical operations sufficient for computing a γ - maximal ellipsoid to the bound

$$O(m^{4.5}\ln[Rm[\ln(\frac{1}{\gamma})]]).$$

When $c = 0$ is fixed, they obtained better bound:

$$O(m^{3.5}\ln[Rm[\ln(\frac{1}{\gamma})]]).$$

Since by polar transformation the problem (A) with fixed center $c = 0$ is converted to the problem (B) with fixed center, this bound can also be used to estimate the complexity of finding a γ - minimal circumscribed ellipsoid containing m given points in E^n (see [KTE 88]).

In the paper [KT 90] authors reduces the problem (6.80)–(6.81) to a small number of subproblems $P(b)$:

$$f(Q) = -\ln\det Q \to \min \tag{6.82}$$

$$(Qa_i, a_i) \leq (1 - (c, a_i))(1 - (a_i, b)), \quad i = 1, \dots, m, \quad Q \succ 0 \qquad (6.83)$$

with fixed value of $b \in R^n$. (After the substitutions $c \to b$, $P^2 \to Q$, we obtain a problem, similar to the problem (6.82)–(6.83)).

So, each $P(b)$ is a problem with variables Q and c subject to linear constraints with parameter b. Khachiyan and Todd [KT 90] proposed the algorithm that was named ElliP for computing γ-maximal ellipsoid for polytope M:

Step 0. Set

$$\delta := \frac{1}{3}\ln(\frac{1}{\gamma}),$$
$$k := 0,$$
$$b_k := b_0 := \text{ an arbitrary interior point in } Q;$$

Step 1. Find an approximate solution $c_k = c(b_k)$ and $Q_k = Q(b_k)$ to problem $P(b_k)$, (see (6.82)–(6.83)), with absolute error δ in the functional.

Step 2. Update

$$b_{k+1} := \frac{1}{2}(b_k + c_k),$$
$$k := k + 1.$$

Go to Step 1 and start a new iteration.

Let $\gamma(b) = volE^*(b)/volE^*$ be the ratio of the volumes of the b approximate and maximal ellipsoids for M. The main theorem is the following:

Theorem *Algorithm ElliP convergs in at most $(k+1)$ iterations, where*

$$k = \log\left[\ln(\frac{1}{\gamma(b_0)})/\ln(\frac{1}{\gamma})\right] + 1,$$

with the ellipsoid

$$E_k = \{x \in E^n : x = b_{k+1} + Q_k^{1/2}z, \|z\| \leq 1\},$$

γ - maximal for M.

Selecting $b_0 = 0$ as the starting point, we obtain

$$k \leq \log[\frac{2n\ln R}{\ln(\frac{1}{\gamma})}].$$

The resulting estimate

$$O\left(m^{3.5}\ln\left[\ln\left[\frac{mR}{\ln\left(\frac{1}{\gamma}\right)}\right]\right]\ln\left[\frac{nR}{\ln(\frac{1}{\gamma})}\right]\right) \tag{6.84}$$

for the number of arithmetical operations sufficient for solving Problem A by algorithm ElliP is obtained in [KT 90]. This estimate is consequence of the estimate

$$O\left(m^{3.5}\ln\left[\frac{m\bar{R}}{r(b)\ln(\frac{1}{\gamma})}\right]\right)$$

for the number of arithmetical operations sufficient for solving the subproblem (6.82)–(6.83) by one of the versions of the interior point algorithm. Here: $\bar{R}$ is the radius of Euclidean ball that includes M;

$$\frac{1}{r(b)} = \max_{i=1,\dots,m} \frac{|a_i|}{1-(a_i,b)}.$$

6.5 METHOD OF INSCRIBED ELLIPSOIDS FOR MINIMIZATION OF CONVEX FUNCTIONS ON POLYTOPES

One of the most interesting applications of inscribed ellipsoids is linked with a special modification of the central cutting plane method for solving convex programming problems, that is called "inscribed ellipsoid method" [KTE 88].

The method of inscribed ellipsoids [KTE 88] is based on the following bright result. Let $K \subset E^n$ be a n-dimensional compact convex body. There is unique ellipsoid of maximal volume inscribed in K [Jo 48] that will be denoted by $E^*(K)$. Center x^* of this ellipsoid will be called center of K and be denoted by $a(K)$. Draw through the center of K a hyperplane $P_g = (g, x - a(K))$, $0 \neq g \in E^*$, and let $K_g = \{x | x \in K, (g, x - x^*) \leq 0\}$ be one of two bodies, which are formed from K by cutting hyperplane P_g. Let $\mu(\overline{K})$ be the volume of $E^*(\overline{K})$ for an arbitrary convex body $\overline{K} \subset E^n$. The main theorem that was proved in [KTE 88] is

Theorem 82

$$\mu(K_g) \leq 0.843\mu(K) \tag{6.85}$$

Proof. An ellipsoid E in E^n can be described in the form

$$E = \{x : (A^{-2}(x-a),\ x-a) \leq 1\}$$

where A is a symmetric positive definite matrix. Denote E by the symbol: $E \sim (a, A)$. This ellipsoid can be described in the form:

$$E = \{x/x = a + Az, ||z|| \leq 1\},$$

vol$E = w_n \det A$, where w_n is the volume of the unit ball in E^n.

Let $E^*(K)$ and $E^*(K_g)$ be maximal ellipsoids, inscribed in K and K_g respectively. One can find an affine transformation of coordinates such that ellipsoids $E^*(K)$ and $E^*(K_g)$ will have the following description:

$$E^*(K) \sim (a, B),\ E^*(K_g) \sim (-a, B^{-1}),\ \text{where } B = \text{diag}(b_1, \ldots, b_n) > 0.$$

Let $K = \{x/(c,x) \leq 1; c \in C\}$, where $C \subset E^n$ be a compact set (note that origin $O_n \in K$). Since $K \supseteq E^*(K) \supset E^*(K_g)$ we obtain:

$$||cB|| \leq 1 - (c,a); ||cB^{-1}|| \leq 1 + (c,a) \qquad \text{for } c \in C \tag{6.86}$$

Since $E^*(K_g) = \{x : \sum_{i=1}^n (x_i + a_i)^2 b_i^2 \leq 1\}$ and $a \notin intE^*(K_g)$, $4||Ba_i||^2 \geq 1$, i.e.

$$||Ba|| \geq \frac{1}{2}. \tag{6.87}$$

Note that

$$\text{vol}(E^*(K))\text{vol}(E^*(K_g)) = w_n^2 \tag{6.88}$$

From (6.86) one can receive:

$$\frac{1}{2}||c(B + B^{-1})|| \leq \frac{1}{2}\left(||cB|| + ||cB^{-1}||\right) \leq 1,$$

and deduce that K has inscribed ellipsoid $E^+(K)$ with the center in origin and half-axes $(b_i + b_i^{-1})/2 \geq 1$, $i = 1, \ldots, n$. Let $b = \max_{i \in \overline{1,n}} b_i$. Then

$$\text{vol}E^+ \geq w_n(b + b^{-1})/2.$$

Since $(c_i, c_i) = (Bc_i, B^{-1}c_i) \leq ||Bc_i|| ||B^{-1}c_i|| \leq 1 - (c,a)^2$ (see (6.86)), all $c_i \in C$ are located in ellipsoid

$$E_C = \{x : (x,x) + [(x,a)]^2 \leq 1\}$$

which can be obtained by contraction of unit ball in the direction a in $\sqrt{1 + ||a||^2}$ times. Thus, the body K contains polE_C, and

$$\text{vol}(\text{pol}E_C) = w_n \sqrt{1 + ||a||^2}.$$

Using (6.87), we obtain:

$$b^2(a,a) \geq ||Ba||^2 \geq \frac{1}{4}, \text{ or } (a,a) \leq \tfrac{1}{4b^2}.$$

Thus, one can inscribe ellipsoid E^0 in the body K such that

$$\begin{aligned} \text{vol}E^0 &\geq \max\{\text{vol}E^+, \text{vol pol}E_C\} \\ &\geq w_n \min_{0<b<\infty} \max\left\{\frac{b+b^{-1}}{2}, \sqrt{1+\frac{1}{4b^2}}\right\} = w_n \frac{3}{2\sqrt{2}} \end{aligned} \quad (6.89)$$

Since vol$E^*(K) \geq$ volE^0, one can obtain from (6.88) that

$$\frac{\mu(K_g)}{\mu(K)} = \frac{\text{vol}(E^*(K_g))}{\text{vol}(E^*(K))} = \frac{w_n^2}{[\text{vol}E^*(K)]^2} \leq \frac{8}{9} = 0.8889.$$

If the described above transformation $E^*(K), E^*(K_g)) \rightarrow E^{(0)}$ is used recursively $(E^*(K), E^{(0)}) \rightarrow E^{(1)}, (E^*(K), E^{(1)}) \rightarrow E^{(2)}, \ldots$ as it is shown in [KTE 88] one can obtain more precise estimate:

$$\frac{\mu(K_g)}{\mu(K)} \leq \left[\frac{2\sqrt{2}}{3} / \sum_{m=1}^{\infty} \left(1 + 2^{-4-2m+2^{2-m}} 3^{-2^{1-m}}\right)\right]^2 \leq 0.8429\ldots \quad (6.90)$$

The theorem is true. □

Note that if $K_n = \text{conv}\left\{0 \cup \left\{x : x_n = \sqrt{\sum_{i=1}^{n-1} x_i^2} = 1\right\}\right\}$ and $g = (0, i = 1, \ldots, n-1; 1)$, then $\frac{\mu(K_g)}{\mu(K_n)} \rightarrow \frac{\sqrt{e}}{2} \approx 0.8243\ldots$, so the estimate (6.90) is precise enough.

Let $a_\gamma(k)$ be a center of γ-maximal ellipsoid, and let

$$K_g^\gamma = \{x | x \in K, (g, x - a_\gamma(K)) \leq 0\}.$$

Then Theorem 82 can be generalized:

Theorem 83 *[KTE 88]*

$$\mu(K_g^\gamma) \leq \gamma^{-2} 0,843 \mu(K), \qquad \gamma < 1.$$

Consider the convex programming problem:

$$\text{minimize} f(x), \; x \in K \subset E^n, \tag{6.91}$$

where $f(x)$ is a convex function defined on E^n, K is a compact convex body of full dimension n. Let $\text{var}_K f$ be the variation of f on K: $\text{var}_K f = \max_{x \in K} f(x) - \min_{x \in K} f(x)$, $X^* \subset K$ the set of optimal points. The method of inscribed ellipsoids for solution of convex programming problems of the type (6.91) is a cutting plane method in which cutting hyperplanes, defined by subgradient direction, go through the center of γ-maximal inscribed in L ellipsoid, where L is the convex body containing X^* at the current iteration.

Let $L_0 = K$ be the initial set of localization of X^*, X^*_γ is the center of γ-maximal inscribed in K ellipsoid. One can calculate subgradient $g_f(X^*_\gamma(K))$ and consider half-space

$$H_0 = H\left(g_f(X^*_\gamma(K)),\; X^*_\gamma(K)\right) = \left\{x : (g_f(x^*_\gamma(K)),\; x - X^*_\gamma(K)) \leq 0\right\}$$

if $g_f(X^*_\gamma) \neq 0$. (Otherwise the point $X^*_\gamma(K)$ is the optimal point for the problem (6.91).

Set $L_1 = L_0 \cap H_0$ It is clear that $X^* \subset L_1$. After k steps we obtain convex compact set L_k containing X^*.

$k+1$-st step. Define $X^*_\gamma(L_k)$ and subgradient $g_f(X^*_\gamma)(L_k)$. If $g_f(X^*_\gamma(L_k)) = 0$, then X^*_γ is the optimal point. Otherwise, consider half-space

$$H_k = \left\{x : (g_f(X^*_\gamma(L_k)),\; x - X^*_\gamma(L_k))\right\}$$

and a new set of localization of X^*:

$$L_{k+1} = L_k \cap H_k.$$

Due to Theorem 83, $\text{vol} L_{k+1} \leq \frac{0.843}{\gamma^2} \text{vol} L_k$

If we set $\gamma = 0.99$,

$$\text{vol} L_{k+1} \leq 0.86 \text{vol} L_k.$$

Thus, using the algorithm of inscribed ellipsoids one can build the sequence of sets $\{L_k\}_{k=1}^{\infty}$, containing X^*,

$$K,\; L_1, \cdots, L_k, \cdots$$

Moreover, volL_k < 0.86^kvol$K$ Taking into account that γ-maximal ellipsoid $E^*_\gamma(G)$ is an approximated ellipsoid of John for any compact convex body $G \subset E^n$ we can use the following relation (see [KTE 88]):

$$E^*_\gamma \subset G \subset \left(\frac{1+3\sqrt{1+\gamma}}{\gamma} n E^*_\gamma(G)\right)$$

6.6 SEMIDEFINITE PROGRAMMING PROBLEMS

6.6.1 Nonsmooth matrix functions

Many combinatorial optimization problems can be formulated as boolean LP problems, and the corresponding dual bounds may be obtained by LP relaxations of such models. But in some cases the nonlinear quadratic-type formulation of a combinatorial problem is more convenient and may give more exact dual bounds. In these cases, as a rule, the problem of obtaining dual bounds reduces to the convex programming problems with nonsmooth matrix function (or to the equivalent problems of semidefinite programming). The most typical examples of such functions are maximal and minimal eigenvalues of symmetric matrices and sums of k largest eigenvalues.

Let S_n be the class of real $n \times n$ symmetric matrices. Any matrix $A \in S_n$ has n real eigenvalues (with account of their multiplicity) and a pair of eigenvalues associated with two different eigenvectors are orthogonal. Let $A \in S_n$, $A = \{a_{ij}\}^n_{i,j=1}$,

$$\lambda_1(A) \geq \lambda_2(A) \geq \ldots \geq \lambda_n(A)$$

be the eigenvalues of A, ordered in nonincreasing way.

The Rayleigh–Ritz formula is known for maximal eigenvalue $\lambda_1(A)$:

$$\lambda_1(A) = \max_{\|y\|=1}(Ay, y) = \max_{\|y\|=1} \sum_{i,j=1}^{n} a_{ij} y_i y_j, \tag{6.92}$$

where $y = (y_1, \ldots, y_n) \in E^n$. $\lambda_1(A)$ is a convex function defined on S_n since formula (6.92) gives representation of this function as maximum of a family of linear functions in entries $\{a_{ij}\}^n_{i,j=1}$.

Denote by $Y^*(A)$ a set of normed eigenvectors of A associated with $\lambda_1(A)$, i.e. $\lambda_1(A) = (Ay^*, y^*)$ for all $y^* \in Y^*(A)$. From (6.92), one may obtain the subgradient set $G_{\lambda_1}(\overline{A})$ of function $\lambda_1(\cdot)$ in the point $\overline{A}$:

$$G_{\lambda_1}(\overline{A}) = conv\left\{\cup_{y\in Y^*(\overline{A})} yy^{(t)}\right\}.$$

The calculation of a subgradient $g_{\lambda_1}(\overline{A}) \in G_{\lambda_1}(\overline{A})$ may be reduced to finding arbitrary $y^*(\overline{A}) \in Y^*(\overline{A})$ and applying next formula:

$$g_{\lambda_1}(\overline{A}) = y^*(\overline{A})[y^*(\overline{A})]^{(t)}.$$

(Note that $yy^{(t)}, y \in E^n$, is the symmetric matrix of range 1 with entries $\{y_i y_j\}_{i,j=1}^n$). If $\lambda_1(\overline{A})$ has multiplicity 1, then $g_{\lambda_1}(\overline{A})$ is unique and function $\lambda_1(A)$ is differentiable at the point $\overline{A}$. When multiplicity of $\lambda_1(\overline{A})$ is more than 1, the function $\lambda_1(A)$ is nondifferentiable in $\overline{A}$.

Now we introduce very interesting class of convex matrix functions defined on symmetric matrices $A \in S_n$, namely, the sums of the k largest eigenvalues:

$$S_{n,k}(A) = \sum_{r=1}^{k} \lambda_r(A),\ 1 \le k \le n.$$

where $\lambda_1(A) \ge \lambda_2(A) \ge \ldots \ge \lambda_n(A)$ are eigenvalues of A.

Famous mathematician Fan Ky gave in 1949 variational description of $S_{n,k}(A)$ that is far going generalization of Rayleigh-Ritz formula [Fan 49]. Let M_n^k be the class of rectangular matrices Y of size $n \times k$ which columns y_i, $i = 1,\ldots,k$, $k \le n$, form the orthonormed system of n-dimensional vectors, i.e. $Y^tY = I_k$ (I_k is identity matrix $k \times k$).

Theorem 84 (Fan Ky) .

$$S_{n,k}(A) = \max_{Y\in M_n^k}\{tr(AYY^t)\},\ \forall A \in S_n. \tag{6.93}$$

The maximum in formula (6.93) is obtained on orthonormal system of eigenvectors $y_1^*,\ldots,y_k^*$ corresponding to eigenvalues $\lambda_1(A),\ldots,\lambda_k(A)$. Indeed,

$$\begin{aligned} Ay_i^* &= \lambda_i(A)y_i^*,\ i = 1,\ldots,k, \\ \mathrm{tr}(AYY^t) = (A, YY^t) &= \textstyle\sum_{i=1}^k(\lambda_i(A)y_i^*, y_i^*) = \\ &= \textstyle\sum_{i=1}^k \lambda_i(A) = S_{n,k}(A). \end{aligned}$$

When $k = 1$, formula (6.93) is reduced to the expression (6.92).

Consider the class of $n \times n$ matrices

$$C_n^k = \operatorname{conv}\{YY^t : Y \in M_n^k\}$$

Theorem 85 *C_n^k coincides with the class of all positive semidefinite matrices C with $\lambda_1(C) \leq 1$ and with trace, equal to k [Fan 49].*

Proof. Note that $\mathrm{tr}YY^t = \mathrm{tr}Y^tY = \sum_{i,j=1}^k (y_i, y_j) = k$ for an arbitrary $Y \in M_n^k$. Since "trace X" is a linear matrix function of any convex combination of the matrices $Z_\alpha = Y_\alpha Y_\alpha^{(t)}$, $Y_\alpha \in M_n^k$ for all α, has the trace equal k. Thus,

$$trZ = k \text{ for all } Z \in C_n^k.$$

Since $(YY^tx, x) = (Y^tx, Y^tx) \geq 0$ for all $x \in R^n$, YY^t are positive semidefinite, so all $Z \in C_n^k$ are positive semidefinite. Denote by $x(Y)$ the projection of vector $x \in R^n$ into subspace generated by column vectors Y_i, $i = 1, \ldots, k$. It is clear that

$$\|Y^tx\| = \|x(Y)\| \leq \|x\|, \text{ so } (YY^tx, x) \leq (x, x).$$

It means that $\lambda_1(YY^t) \leq 1$ for all $Y \in M_n^k$. Since $\lambda_1(\cdot)$ is a convex matrix function, $\lambda_1(Z) \leq 1$ for all $Z \in C_n^k$.

Let Q be the orthogonal transformation in R^n. Then if $Y \in M_n^k$ and $(QY)(QY)^t = QYY^tQ^t \in C_n^k$, $QY \in M_n^k$ also. So the set C_n^k is invariant relatively the group of orthogonal transformations in R^n, i.e. if $Z \in C_n^k$, then $QZQ^t \in C_n^k$ also.

Particularly, for any $Z \in C_n^k$ one can find orthogonal transformation $\overline{Q}$ such that $\overline{Z} = \overline{Q}Z\overline{Q}^t \in C_n^k$ and $\overline{Z}$ is a diagonal matrix.

So, without loss of generality, a consider diagonal matrix $\overline{Z} \in C_n^k$, $\overline{Z} = diag\{d_1, \ldots, d_n\}$. In addition,

$$\text{(i) } \sum_{j=1}^n d_j = k;$$

$$\text{(ii) } 0 \leq d_i \leq 1.$$

Consider the convex polyhedral set $D \subset R^n$ determined by the conditions (i) and (ii). Extreme points δ_α of the set D have the following structure: all entries of δ_α are 0 or 1, moreover, the number of ones equal k, the number of zero entries equal $(n-k)$. Let $\delta_\alpha^{(i)}$ be the i-th entry of δ_α, so

$$\delta_\alpha^{(i)} = \begin{cases} 1, & \text{for } j_1, \dots, j_k \\ 0, & \text{otherwise,} \end{cases}$$

where $j_1, j_2, \dots, j_k$ is the subset of indeces $1, 2, \dots, n$.

It is clear that δ_α can be represented as $Y^{(\alpha)}(Y^{(\alpha)})^t$, where $Y^{(\alpha)} \in M_n^k$ is gnerated by columns vectors: y_i, $i = 1, \dots, k$, where

$$y_i = \{y_i^{(p)}\}_{p=1}^n, \quad y_i^{(p)} = \begin{cases} 1, & \text{when } p = j_i \\ 0, & \text{otherwise} \end{cases} \quad i = 1, \dots, k; \; p = 1, \dots, n.$$

So, $\delta_\alpha \in M_n^k$.

Each element of polyhedral set D can be represented as a convex combination of its extremal points. Hence, each semidefinite matrix with maximal eigenvalue not exceeding 1 and with trace k belongs to C_n^k. This proposition can be expanded as we showed above on all symmetric positive semidefinite matrices with maximal eigenvalue not exceeding one and trace k. **The theorem is proved.**

Due to this theorem, we obtain a new variational formula for $S_{n,k}(A)$:

$$S_{n,k}(A) = \max_{C \in C_n^k} (A, C), \tag{6.94}$$

where $(A, C) = \sum_{i,j=1}^n a_{ij} c_{ij}$ $(A = \{a_{ij}\}_{i,j=1}^n,\ C = \{c_{ij}\}_{i,j=1}^n)$.

Formulas (6.93) and (6.94) give us the representation of the function $S_{n,k}(A)$ as a pointwise maximum function on infinite family of linear (in matrix variable $A \in S_n$) functions. So, $S_{n,k}(A)$ is a convex function for any $n, k \leq n$. The structure of subgradient set $G_{S_{n,k}}(A)$ is determined by (6.94):

$$G_{S_{n,k}}(A) = \{C^* \in C_n^k : S_{n,k}(A) = (A, C^*)\}$$

For almost all A, C^* is unique and gives us the gradient $g_{S_{n,k}}(A)$ of $S_{n,k}$ in the point A.

For finding an approximate value of C^* one may use for some $\varepsilon > 0$) the barrier function method: to find

$$\varphi_\varepsilon^* = \max_{\{C \in S_n, \operatorname{tr} C = k\}} \{(A, C) + \varepsilon[\log\det C + \log\det(I_n - C)]\} \tag{6.95}$$

Function $\varphi_\varepsilon(C)$, which we have to maximize in (6.95), is defined on interior points of the set C_n^k and tends to $-\infty$, when C moves from interior of C_n^k to its boundary. So, the problem (6.95) can be solved by the interior point method, but it is not convenient to use this technique for obtaining subgradients of $S_{n,k}$.

More strict way for obtaining a subgradient $g_{S_{n,k}}(A)$ is the following:

(i) solve the eigenvalue problem for matrix A and find the eigenvalues $\lambda_1(A), \ldots, \lambda_k(A)$ and the corresponding orthonormal system of eigenvectors $Y_j(A) = \{y_i^j\}_{i=1}^n$, $j = 1, \ldots, k$;

(ii) construct the $n \times k$ matrix $Y = \{y_i^j\}_{i=1,\ldots,n}^{j=1,\ldots,k}$

(iii) set $\overline{g}_{S_{n,k}}(A) = YY^t$.

In general, when not all k largest eigenvalues of A have multiplicity 1, the subgradient $g_{S_{n,k}}$ is nonunique, because the system of eigenvectors $\{Y_i(A)\}_{i=1}^k$ is determined nonuniquely in this case. But if one is interested in calculating any subgradient from $G_{S_{n,k}}(A)$ he can use the described above procedure for arbitrary orthonormal system $Y(A)$ of eigenvectors associated with the k largest eigenvalues $\lambda_1(A) \geq \lambda_2(A) \geq \ldots \geq \lambda_k(A)$.

Let A be a diagonal matrix and $a_{11} \geq a_{22} \geq \cdots \geq a_{kk} \geq \cdots \geq a_{nn}$, $1, k, n$. Consider two cases:

(I) $a_{kk} > a_{(k+1)(k+1)}$. In this case in formula:

$$S_{n,k}(A) = \max_{C \in C_n^k} (A, C) = (A, C^*).$$

C^* is determined uniquely (C^* is the diagonal matrix with $c_{ii} = 1$ for $i \leq k$ and $c_{ii} = 0$ for $i > k$). The subgradient of the function $S_{n,k}$ in the point A, $g_{S_{n,k}}(A)$ equal C^*, and the function $S_{n,k}$ is differentiable at the point A;

(II) $a_{kk} = a_{(k+1)(k+1)}$. In this case the subgradient set $G_{S_{n,k}}(A)$ contains more than one extreme points. For example, if $a_{ii} = a_{kk}$ for all i, $k - s \leq i \leq k + p$ ($s \geq 0$, $p \geq 1$), then an arbitrary diagonal matrix A with properties:

(1) $a_{ii} = 1$ for $i < k - s$;

(2) $a_{ii} = 0$ for $i > k + p$;

(3) the set of values $\{a_{ii}\}$, $k-s \le i \le k+p$ contains exactly s ones and p zeroes;

is an extreme point of $G_{S_{n,k}}(A)$. So, in the case (II) the function $S_{n,k}$ is nondifferentiable at the point A.

In general, if for a symmetric matrix A, $\lambda_k(A) > \lambda_{k+1}(A)$ then $S_{n,k}$ is differentiable at the point A; otherwise (i.e. $\lambda_k(A) = \lambda_{k+1}(A)$) the function $S_{n,k}$ is nondifferentiable at the point A.

In many applications we meet with weighted sum of k largest eigenvalues:

$$S_{n,k}(A,w) = \sum_{i=1}^{k} w_i \lambda_i(A), \text{ where } w = (w_1, \ldots, w_k) \ge 0 \tag{6.96}$$

Lemma 5 *If $w_1 \ge w_2 \ge \ldots \ge w_k$, then $S_{n,k}(A,w)$ is a convex function defined on S_n.*

Proof. For $k=1$, $S_{n,1}(A,w) = w_1\lambda_1(A)$ is a convex function. If $k>1$,

$$S_{n,k}(A,w) = w_k S_{n,k}(A) + S_{n,k-1}(A, w^{(k-1)}), \tag{6.97}$$

where $w^{(k-1)} = (w_1 - w_k, \ldots, w_{k-1} - w_k) \ge 0$. We can use the decomposition (6.97) $(k-1)$ times and as a result obtain:

$$S_{n,k}(A,w) = w_k S_{n,k}(A) + (w_{k-1} - w_k) S_{n,k-1}(A) + \ldots + (w_1 - w_2) S_{n,1}(A).$$

So, $S_{n,k}(A,w)$ can be represented as sum of convex functions. **The lemma is proved.**

The weighted sum of the largest eigenvalues $S_{n,k}(A,w)$ can be represented also by the variational formula similar to formula (6.93):

$$S_{n,k}(A,w) = \max_{Z \in M_n^k(w)} \{\mathrm{tr}(AZZ^t)\}, \tag{6.98}$$

where $M_n^k(w)$ is the class of rectangular matrices Z of size $n \times k$ which columns Z_i form orthogonal system of n-dimensional vectors, and $\|Z_i\|^2 = w_i$, $i = 1, \ldots, k$, $w_1 \ge w_2 \ge \ldots \ge w_k$.

Due to the expression (6.93) and Theorem 16 the subgradient set for the function $S_{n,k}(\cdot)$ in the point X is given by the following expression:

$$G_{S_{n,k}}(X) = conv(\sum_{i=1}^{k} y_i y_i^T),$$

where $y_i, \quad i = 1, \ldots, k$, form an arbitrery orthonormal system of vectors, associated with eigenvalues $\lambda_1(X), \ldots, \lambda_k(X)$. If multiplicity of all eigenvalues $\lambda_1(X), \ldots, \lambda_k(X)$ is equal to one, then the matrix $\sum_{i=1}^{k} y_i y_i^T$ is determined uniquely and coinsides with gradient $S_{n,k}(X)$ in the point X.

The functions $\lambda_m(X), \; 1 < m < n$ are quasi-differentiable functions (in the sence of Demjanov and Rubinov [DR 80]. They may be considered as difference of two convex functions

$$\lambda_m(X) = S_{n,m-1}(X) - S_{n,m}(X).$$

If $w = \{w_1, \ldots, w_n\} \geq 0$ and $w_k \geq w_{k+1}$ for $k = 1, \ldots, n-1$, then subgradient set of the convex function $S_n^w(X) = \sum_{i=1}^{n} w_i \lambda_i(X)$ can be represented by the following expression:

$$G_{S_n^w}(X) = conv\{\sum_{i=1}^{n} w_i y_i y_i^T\},$$

where $\{y_i\}_{i=1}^n$ is any orthonormal system of eigenvectors of matrix X (each y_i is associated with $\lambda_i(X)$). If all eigenvectors $y_i(X)$ with $w_i > 0$ have multiplicity 1, then S_n^w is differentiable at X.

The properties of directional derivatives of the first and the second orders of the nondifferensiable functions $\lambda_m(X), \; m = 1, \ldots, n, \; S_{n,k}(X), \; k = 1, \ldots, n, \; S_n^w(X)$ were studied in many works (see, for example [HY 91], [HY 92], [OW 92], [OW 93], [Se 90], [DR 80], [Chu 90], [HUL 93]). The optimization of the mentioned above matrix functions may be reduced to the so-called semidefinite programming in many practically important cases.

6.6.2 Models of semidefinite programming in automatic control

Below we shall use the symbols $\succ$ ($\succeq$) in the following sense: $A \succeq B$ ($A \succ B$) means that $A - B$ is positive semidefinite (positive definite) for $A, B \in S_n$.

Hence, $A \succeq 0$ ($A \succ 0$) is equivalent to $\lambda_n(A) \geq 0$ ($\lambda_n(A) > 0$).

Many mathematical and real-life problems are reduced to the so-called semidefinite programming i.e. to the mathematical programming problems with symmetric matrices, which entries are functions in one or several variables and there are constraints, which demand for some classes of such matrices to be positive semidefinite. As a rule the problems of semidefinite programming can be reduced to minimization of nonsmooth matrix functions.

Remark. In some books and articles term "semidefinite programming" is used in more restricted sense (as analogy to linear programming). Let A and B be two $n \times n$ matrices. (A, B) denotes inner (scalar) product of these matrices:

$$(A, B) := tr(A \cdot B^t) = \sum_{i,j}^{n} a_{ij} b_{ij}.$$

Let S_n be the class of $n \times n$ symmetric matrices.

The following problem:

$$\sup(c, x), \tag{6.99}$$

$$(A_i, X) \leq b_i,\ i = 1, \ldots, m, \tag{6.100}$$

$$X \succeq 0, \tag{6.101}$$

where b_i are real numbers, matrices C, A_i, $i = 1, \ldots, m$, and variable matrix X belong to S_n, is called **the semidefinite programming problem** (SDP). It is interesting that duality theory for (linear) semidefinite programming is very similar to the LP duality. The problem DSDP dual to SDP problem (6.99)–(6.101) is the following:

$$\inf \sum_{i=1}^{n} b_i y_i \tag{6.102}$$

$$\sum_{i=1}^{m} Y_i A_i - C \succeq O \tag{6.103}$$

$$y = \{y_i\}_{i=1}^{m} \geq 0. \tag{6.104}$$

It is shown (see e.g. [BF 63], [Al 92b]) that the following result is true:

Theorem 86 *If both problems (SDP and DSDP) have a feasible solution, then the strong duality holds (i.e. the supremum attained in (SDP) coincides with the infinum attained in DSDP). If, in addition, there exist a feasible $\overline{X} \succ 0$ in SDP and a feasible $\overline{y}$ satisfying*

$$\sum_{i=1}^{m} \overline{y_i} A_i - C \succ 0$$

in DSDP, then optimal values in both problems are attained.

Let A and X be a pair of $n \times n$ symmetric matrices, and $X \succeq 0$. The following results were proven:

Lemma 6 *[PK 97] For any $R > 0$*

$$\min\{(A \cdot X),\ \ X \succeq 0,\ trX = R\} = R\lambda_n(A), \tag{6.105}$$

$$\min\{(A \cdot X)\ \ X \succeq 0,\ trX \leq 0\} = \min\{0, R\lambda_n(A)\}. \tag{6.106}$$

Proof. To show the identity (6.105), observe that $\min\{A \cdot X | X \succeq 0, trX = R\} = R\lambda_n(A) + \min\{(A - \lambda_n(A)I_n) \cdot X|\ X \succeq 0, trX = R\} = R\lambda_n(A)$ because minimum of the expression in brackets is attained on one-rank matrix $X^* = yy^T$, (where y is an eigenvector of A corresponding to $\lambda_n(A)$), and equal zero. To prove the identity (6.106), note that if $A \succeq 0$, then the minimum equal zero. Otherwise $\lambda_n(A) < 0$, so the minimum on the left hand side of (6.106) is negative and hence, it is attained at a matrix X such that $trX = R$. Then (6.106) becomes a consequence of (6.105).

Recently Porkolab and Khachiyan [PK 97] obtained important results on the complexity of semidefinite programs, which generalized the complexity theory of solving systems of linear inequalities with integer coefficients.

Consider general semidefinite feasibility problem (F). Given integral $n \times n$ symmetric matrices $A_1, \ldots, A_m$ and integers $b_1, \ldots, b_m$. Determine whether there exists a real $n \times n$ symmetric matrix X such that

$$(A_i, X) \leq b_i,\ \ i = 1, \ldots, m;\ \ X \succeq 0. \tag{6.107}$$

We also consider the following polynomially equivalent problem (G).

Given integral $n \times n$ symmetric matrices $Q_0, Q_1, \dots, Q_m$, determine whether there are real numbers $x_1, \dots, x_m$ such that

$$Q_0 + \sum_{i=1}^{m} x_i Q_i \succeq 0 \tag{6.108}$$

The complexity status of the problems (F) and (G) is a fundamental open issue in the theory of semidefinite programming.

For the standard bit model of computation it is known that either these problems belongs to the complexity class NP $\cap$ co NP, or they are not in NP $\cup$ co NP.

For the real model of computation problems (F) and (G) are known to be in NP $\cap$ co NP, but the question whether they can be solved in polynomial time remains open (see [Ram 95]).

Below we shall describe main results obtaining in [PK 97].

The main results: (i) Any feasible system (6.107) has a solution in the Euclidean ball $B = \{X| \ \ ||X|| \leq R\}$, where $\log R = ln^{O(\min\{m,n^2\})}$ and l is maximum bitlength of the input coefficients. Moreover, the same bound applies to (6.108): any feasible system (6.108) has a solution x such that $\log ||x|| = ln^{O(\min\{m,n^2\})}$.

One can state lower bounds on the discrepancy of infeasible systems (6.107) and (6.108).

(ii) If (6.107) is infeasible then for any symmetric positive semidefinite matrix $X \in B$, $-\log \max_{i=1,\dots,m}\{A_i \cdot X - b_i\} = ln^{O(\min\{m,n^2\})}$. The corresponding result for the infeasible system (6.108) is that for any X satisfying the upper bound of (i), the minimum eigenvalue λ_n of $Q_0 + x_1 Q_1 + \ldots + x_m Q_m$ is negative and $-\log(-\lambda_n) = ln^{O(\min\{m,n^2\})}$. Due to thses results solving (F) with the ellipsoid method requires $lmnO(\min\{m, n^2\})$ arithmetic operations with $\text{ln}^{O(\min\{m,n^2\})}$ bit numbers.

This result was improved by the use of the Renegar decision method [Re 92b] along with the derandomized version [CMa 93] of Clarkson's random algorithm [Clar 95], [ASha 90].

(iii) Problem (F) can be solved in $mnO(\min\{m, n^2\})$ arithmetic operations over $ln^{O(\min\{m,n^2\})}$ bit numbers.

In particular, (F) can be solved in strongly polynomial time for any fixed number of variables or constraints. Note also that for $n = const$, the required number of arithmetic operations grows linearly with m.

(IV) Problem (G) can be solved in $O(mn^4) + nO(\min\{m, n^2\})$ arithmetic operations over $ln^{O(\min\{m,n^2\})}$ bit numbers.

The results (iii) can be extended to the optimization version of the problem (F): find

$$\alpha^* = \inf\{(D, X) \mid (A_i, X) \leq b_i,\ i = 1, \ldots, m,\ X \succeq 0\}, \tag{6.109}$$

where D is a given $n \times n$ integral symmetric matrix.

Specifically, in addition to testing the feasibility of (6.109), each of the following problems can also be solved in $mn^{O(\min\{m,n^2\})}$ operations over $ln^{O(\min\{m,n^2\})}$-bit numbers, namely:

Boundedness. Determine whether the objective function is bounded from below on the set of feasible solutions.

Attainment. Determine whether the infimum is attained, that is, whether (6.109) has an optimal solution.

Finally, for the optimization version of (G) with a given integral m-vector d:

$$find\ \ \beta^* = \inf\{(d, x) | Q_0 + x_1 Q_1 + \cdots + x_m Q_m \succeq 0,\ x \in R^m\}, \tag{6.110}$$

the above four problems can be solved in $O(mn^4) + n^{O(\min\{m,n^2\})}$ operations over $ln^{O(\min\{m,n^2\})}$-bit numbers.

6.7 SEMIDEFINITE PROGRAMMING MODELS IN AUTOMATIC CONTROL

The typical models in automatic control are described by differential inclusions (DI):

$$\frac{dx}{dt} = \dot{x} \in F(x(t), t),\ x(0) = x_0 \tag{6.111}$$

where F is a set-valued function on $R^n \times R_+$. Any $x(\cdot) : R^+ \to R^n$ that satisfies (6.111) is called a solution or trajectory of DI (6.111). In general, there are many solutions of the DI (6.111). It is interesting to investigate various properties satisfied by all solutions of a given DI, for example, the stability properties.

DI given by

$$\dot{x} \in convF(x(t), t) \tag{6.112}$$

is called the relaxed version of DI (6.111). Every trajectory of DI (6.111) is also a trajectory of relaxed DI (6.112). There are so-called Relaxation Theorems in which general conditions are formulated which guarantee that the reachable or attainable sets of the DI (6.111) and its relaxed version coincide, i.e. for every $T \geq 0$

$$\{x(T) \mid x \text{ satisfies DI } (6.111)\} = \{x(T) \mid x \text{ satisfies DI } (6.112)\}$$

We shall assume in all examples given further that the corresponding differential inclusions have the convex right-hand parts.

For beginning, consider one of the extremal problems arising in the Lyapunov's theory of stability for linear systems of differential equations.

Let

$$\frac{dx}{dt} = Ax, \; x \in R^n, \; x(0) = x_0, \tag{6.113}$$

be a linear system of differential equations. The problem of stability for such systems is formulated as follows: whether the trajectory of the system (6.113) converges to zero vector as $t \to \infty$? The Lyapunov method gives us the possibility (see [Ly 47], [Lu 57]) to construct the necessary and sufficient conditions of stability for system (6.113): the system (6.113) is stable if and only if there exists a quadratic strongly convex function

$$V(\xi) = (P\xi, \xi), \; (P \succ 0) \tag{6.114}$$

which decreases along the trajectory of (6.113) when $x(0) \neq 0$. Such function V is called the quadratic Lyapunov function for the system (6.113). Since

$$\frac{d}{dt}V(x) = ((A^tP + PA)x, x),$$

the necessary and sufficient conditions for the quadratic stability of the system (6.113) can be expressed as the following: there exists $P \in S_n$ such, that $P \succ 0$ and

$$A^tP + PA \prec 0 \tag{6.115}$$

An equivalent condition is existing for matrix $Q \in S_n$ such, that

$$Q \succ 0, \; QA^t + AQ \prec 0 \tag{6.116}$$

Q and P are linked by equality: $Q = P^{-1}$. Indeed, let $x = Qy$, $Q = P^{-1}$. Then

$$\begin{aligned} ((A^T P + PA)x, x) &= ((A^T P + PA)Qy, Qy) = \\ (A^T PQy, Qy) + (PAQ, Qy) &= ((QA^T + AQ)y, y), \end{aligned}$$

The appropriate $n \times n$ symmetric matrix P for the stable system (6.113) can be chosen by nonunique way. The Lyapunov's method can be used to obtain some estimates of the behavior of the system (6.113) trajectory.

Example 1. Estimates for the integral stability [KI 85]. Using Lyapunov function one can obtain the following estimate for the system:

$$\frac{dx}{dt} = Ax, \; x \in R^n : \tag{6.117}$$

$$|x(x_0, t)| \leq \sqrt{\frac{\lambda_1(P)}{\lambda_n(P)}} \, |x_0 \exp(-\frac{\lambda_n(C)}{2\lambda_1(P)} t), \tag{6.118}$$

where P is a solution of the Lyapunov equation

$$A^T P + PA = -C; \tag{6.119}$$

$C \succ 0$ is a fixed positive definite matrix, $|\cdot|$ is a symbol of the norm in R^n; $x(x_0, t)$ is the solution of the system (6.117); $x = x_0$ when $t = 0$. Using the estimate (6.118) we may receive integral estimate for $x(x_0, t)$:

$$I(P) = \int_0^\infty |x(x_0, t)| dt \leq 2|x_0| (\sqrt{\frac{\lambda_1(P)}{\lambda_n(P)}})^3 \tag{6.120}$$

In the paper [KI 85] it is shown that the most precise estimate of the form (6.120) is get by the Lyapunov function for which

$$P^* = \arg\min_{P \in G} \frac{(\lambda_1(P))^{\frac{3}{2}}}{\sqrt{\lambda_n(P)} \lambda_n[-(A^T P + PA))]}, \tag{6.121}$$

where G is the set of symmetric positive definite matrices P, for which $C = -(A^T P + PA)$ is positive definite. The corresponding Lyapunov function $V^*(x) = (P^* x, x)$ is called **optimal in integral sense**. Function

$$\varphi(P) = \frac{\lambda_1(P)^{3/2}}{\sqrt{\lambda_n(P)} \lambda_n[-(A^T P + PA)]} \tag{6.122}$$

is continious, almost differentiable on open convex cone G. Let us transit from minimization of the function $\varphi(P)$ to the minimization of function $\psi(P) = 2\ln\varphi(P)$: to find

$$\begin{aligned} \psi(P^*) &= \min\psi(P), \text{ where} && (6.123)\\ \psi(P) &= \min[3\ln\lambda_1(P) - \ln\lambda_n(P) - 2\ln\lambda_n[-(A^TP + PA)]] && (6.124) \end{aligned}$$

Since $\varphi(P) = \varphi(\alpha P)$ for any $\alpha > 0$, we may consider that $\lambda_1(P) = 1$ and, thus reduce the problem (6.124) to the maximization of $\psi_1(P)$, where

$$\psi_1(P) = \ln\lambda_n(P) + 2\ln\lambda_n[-(A^TP + PA)]$$

subject to the constraints:

$$\lambda_1(P) \leq 1;\ P \succ 0;\ C = -(A^TP + PA) \succ 0 \tag{6.125}$$

(It is clear that the constraint $\lambda_1(P) = 1$ we can change on $\lambda_1(P) \leq 1$). Since $\lambda_n(P)$ is a concave function, $-\ln\lambda_n(P)$ is a convex function on cone $P \succ 0$. The constraints (6.125) are convex. Hence, problem of minimization $(-\psi_1(P))$ subject to the constraints (6.125) is the problem of convex programming with semidefinite constraints:

$$\min[-\ln\lambda_n(P) - 2\ln\lambda_n[-(A^TP + PA)]]; \tag{6.126}$$

$$\lambda_1(P) \leq 1; \tag{6.127}$$

$$P \succ 0;\ -(A^TP + PA) \succ 0. \tag{6.128}$$

It is equivalent to the convex programming problem in usual form: (6.126),(6.127) and instead of (6.128) we can write two convex inequalities:

$$\lambda_n(P) > 0;\ \lambda_1(A^TP + PA) < 0. \tag{6.129}$$

Let D be the intersection of convex sets D_1 cutted by inequalities (6.129) (we assume that D_1 is nonempty) and D_2 determined by inequality (6.127). Let P tends to the boundary point of D_1 locating in D_1. Then $\psi_1(P)$ tends to $-\infty$. So optimal solution P^* is achieved at the boundary of D_2, i.e. $\lambda_1(P^*) = 1$. The problem (6.126)–(6.128) can be reduced to unconstrained minimization of nonsmooth penalty function

$$q_s(P) = -ln\lambda_n(p) - 2ln\lambda_n[-(A^TP + PA)] + s^+(\lambda_1(P^*)), \tag{6.130}$$

where

$$s^+(x) = \begin{cases} sx, & \text{if } x > 0 \\ 0, & \text{if } x \le 0 \end{cases}$$

s is a positive number large enough. In order to determine appropriate values for penalty parameter s, consider ε-perturbed problem (6.126), (6.128) with inequality:

$$\lambda_1(P) \le 1 + \varepsilon \tag{6.131}$$

instead of the inequality (6.127).

Let $P^*(\varepsilon)$ be an optimal solution of perturbed problem, $P^*(0) = P^*$. It is clear, that $P^*(\varepsilon) = P^*(1+\varepsilon)$. So

$$\frac{d\psi_1(P^*(\varepsilon))}{d\varepsilon}(0) = 3 \lim_{\varepsilon \to 0} \frac{\ln(1+\varepsilon)}{\varepsilon} = 3.$$

Thus, for $s > 3$ the optimal solution of the convex nonsmooth problem (6.130), (6.128) coincides with optimal solution of the problem (6.126)—(6.128). For solving (6.130), (6.128) one may use r-algorithm.

The notion of the Lyapunov quadratic stability function was generalized on the class of linear differential inclusions (LDI):

$$\frac{dx}{dt} = A(t)x, \; A(t) \in \Omega, \; x(0) = x_0, \tag{6.132}$$

where $x \in R^n$, Ω is a subset of $n \times n$ matrices, i.e. $\Omega \subset R^{n^2}$. The LDI (6.132) is called **stable** if arbitrary trajectory of the system (6.132) $x(t)$ tends to the origin when $t \to \infty$. The sufficient condition for the stability of LDI system (6.132) is so-called **quadratic stability**, i.e. existing of symmetric positive definite matrix P ($P \succ 0$) such that quadratic form

$$V_t(x) = (Px(t), x(t))$$

decreases along any possible trajectory of the LDI (6.132). The problems arising in the theory of quadratic stability of LDI (6.132) often reduce the semidefinite programming (see [BBFG 94]).

Example 2. Polytopic LDI's (PLDI's). PLDI's are described by

$$\frac{dx}{dt} = A(t)x; \; A(t) \in conv\{A_i\}; \; i \in I \tag{6.133}$$

where $\{A_i\}$ is a finite set of matrices $n \times n$. In this case the sufficient condition for determination of stability (6.115) is equivalent to the system of positive definite costraints:

$$P \succ 0; \; A_i^T P + PA_i \prec 0, \; i \in I \tag{6.134}$$

for Lyapunov's matrix P. We can use also the conditions:

$$Q \succ 0;\ QA_i^T + A_iQ \prec 0,\ i \in I,\ Q = P^{-1} \qquad (6.135)$$

Consider the function

$$\varphi(P) = \max_{i \in I}\{\min\{-\lambda_n(P_i);\ \lambda_1(A_i^T P + PA_i)\}\} \qquad (6.136)$$

Since $-\lambda_n(P_i)$ and $\lambda_1(A_i^T P + PA_i)$ are convex functions of $P \in S_n$, $\varphi(P)$ is a convex function. If $\varphi^* = \min \varphi(P) \geq 0$ then the system (6.136) has no solution. Otherwise $\varphi^* = \varphi(P^*) < 0$ and P^* or any other $\overline{P}$, $\varphi(\overline{P}) < 0$, can be used for construction of true Lyapunov quadratic function for the system (6.133).

For finding φ^* one may use one of the mentioned above methods of nondifferentiable optimization, for example, r-algorithm.

Example 3. Studying robust stability for parametric family of linear differential equations (polytopic set of parameters). Let us present an example of so-called robust stability problems. Consider the family of linear differential equations:

$$\frac{dx}{dt} = A(y)x,\ y \in Y \subseteq R^m,\ Y = conv\{\{y^{(i)}\}, i \in I\} \qquad (6.137)$$

where $A(y) \in R^{n \times n}$, $\forall y \in Y$, $x \in R^n$, the entries of $A(y)$ are linear functions of parameter y. (We may interprete the family of systems (6.137) also as one system with uncertain determination of its coefficients).

Consider the set of matrices $A[Y] = \{A(y)\}_{y \in Y} = conv\{A(y_i)\}_{i \in I}$. We try to find $P \succ 0$ to construct universal quadratic Lyapunov function that give the proof of the stability for all systems of the family (6.137) at once. Such P must satisfy to the inequalities (6.134), where $A(y_i) = A_i$, $i \in I$.

Thus, as in a previous example, we obtain the convex programming problem similar to the problem of minimizing $\varphi(P)$ (see (6.136)) with the same post-solution analysis.

Taking into account that conditions of the sort (6.134) are only sufficient for the robust stability of the family (6.133), and if we get a negative result after minimization of $\varphi(P)$ ($\varphi(P^*) \geq 0$), we may continue our analysis dividing parametric set Y in several parts $Y_j, j \in J$, which cover Y in sum:

$$Y \subseteq \cup_{j \in J} Y_j$$

and then use the Lyapunov method for each $Y_j, j \in J$.

Example 4. Invariant ellipsoids. For $Q \succ 0$ let $\mathcal{E}$ denote the ellipsoid centered in the origin:

$$\mathcal{E} = \{y \in R^n | (Q^{-1}y, y) \leq 1\} \tag{6.138}$$

The ellipsoid $\mathcal{E}$ is called **invariant** for the LDI (6.132) if for every trajectory of this LDI, $x(t_0) \in \mathcal{E}$ implies $x(t) \in \mathcal{E}$ for all $t \geq t_0$ [BBFG 94]. This is true if and only if Q satisfies: $QA^t + AQ \preceq 0$ for all $A \in \Omega$ or equivalently $A^T P + PA \preceq 0$) for all $A \in \Omega$, where $P = Q^{-1}$. Thus, the existence of invariant ellipsoid is strictly linked with the existence of quadratic Lyapunov function. When the set of possible initial points $x(t_0)$ is a polytope described by its vertices $\{v_j\}_{j \in J}$

$$\mathcal{P} = conv\{v_j\}_{j \in J}$$

it is naturally to find the smallest in certain sense invariant ellipsoid containing $\mathcal{P}$. The condition that $\mathcal{P} \subset \mathcal{E}$ is equivalent to the system of inequalities:

$$(Q^{-1}v_j, v_j) \leq 1, \; j \in J, \text{or} \tag{6.139}$$

$$(Pv_j, v_j) \leq 1, \; j \in J, \tag{6.140}$$

We can use different criteria for refining what is "the smallest" feasible ellipsoid.

a) The criterion of the minimal n-dimentional volume. We know that volume $\mathcal{E}$ equal to $\omega_n \sqrt{\det Q}$. Hence we can formulate the problem of convex programming:

$$\text{minimize } (-\log \det P), P \succ 0$$

subject to (6.140) and additional inequalities which express the properties of P, as matrix of quadratic Lyapunov function for LDI (6.134). For example when we meet polytopic LDI, $A(t) \in convA_i, i \in I$, we obtain the following problem of convex programming:

$$\text{minimize } (-\log \det P) \tag{6.141}$$

$$(Pv_j, v_j) \leq 1, \; j \in J, \tag{6.142}$$

$$\lambda_n(P_i) \geq 0, \; \lambda_1(A_i^t P + PA_i) \leq 0, \; i \in I \tag{6.143}$$

b) The another typical criterion is minimum of diameter $d(\mathcal{E})$ of $\mathcal{E}$,

$$d(\mathcal{E}) = 2\sqrt{\lambda_1(Q)} = \frac{2}{\sqrt{\lambda_n(P)}} \tag{6.144}$$

We can use as objective function instead of (6.144) the convex function $(-\lambda_n(P))$ preserving all necessary constraints (6.142), (6.143).

How to solve the problem of the type (6.141)–(6.143)? Let us consider the approximate problem in which the constraints (6.143) are changed by the more weak constraints

$$\lambda_1(A_i^T P + PA_i) \leq \varepsilon, \ (i \in I) \tag{6.145}$$

where ε is a sufficiently small positive number. This constraint is equivalent to the following one:

$$\frac{1}{\varepsilon}\lambda_1(A_i^T P + PA_i) \leq 1, \tag{6.146}$$

Note that P enters in the left-hand side of the constraints (6.146) linearly and uniformly, similar as in constraints (6.142). Therefore if we change the right-hand side of constraints (6.142), (6.146) from 1 to $1 + \delta$ ($\delta > 0$) then optimal value of the problem (6.141), (6.142), (6.146) decreases on the value $n\ln(1+\delta)$ and one can obtain the following result similar to Theorem 77:

Theorem 87 *Consider a penalty function in the form of maximum function:*

$$\begin{aligned} \varphi_N(P,\varepsilon) \;\; = \;\; & -\ln\det P + N\max\{0, \max_{i\in I}\{\max[(Ka_i, a_i) - 1]\}, \\ & \max_i[\frac{1}{\varepsilon}\lambda_1(A_i^T P + PA_i) - 1]\} \end{aligned} \tag{6.147}$$

If $N \geq n$, then the problem of minimization of $\varphi_n(P,\varepsilon)$ on the class of positive definete symmetric matrices is fully equivalent to the problem (6.141), (6.142), (6.146).

To find a feasible solution of (6.141), (6.142), (6.146) one may take more strict constraints:

$$\begin{aligned} & \lambda_1(A_i^T P + PA_i) \leq -\varepsilon, \varepsilon > 0 \text{ or} \\ & \frac{1}{\varepsilon}\lambda_1(A_i^T P + PA_i) \leq -1, i \in I \end{aligned} \tag{6.148}$$

Let $P^*(\varepsilon)$ and $P^*(-\varepsilon)$ be the corresponding solutions of the problems (6.141), (6.142), (6.146) and (6.141), (6.142), (6.148). Taking in attention that the left-hand sides of the constraints (6.146) and (6.148) are convex functions of P, we can obtain the feasible solution $\overline{P}$ more good than $P^*(-\varepsilon)$, namely,

$$\overline{P} = \frac{1}{2}(P^*(-\varepsilon) + P^*(\varepsilon)).$$

It is easy to verify, that $\lambda_1(A_i^T\overline{P} + \overline{P}A_i) \leq 0,\ i \in I$. If for some $\varepsilon > 0$ the problem (6.141), (6.142), (6.148) is infeasible, but $P^*(\varepsilon)$ exists, then we can either go to the problems of the form (6.147) with more small ε or try to find a feasible solution of the system of inequalities (6.142), (6.143) by other methods (see, for example, [NN 94]) in a small neigborhood of $P^*(\varepsilon)$.

Consider the partial case of the problem (6.141), (6.142), (6.146) when $|I| = 1$.

In this case we have usual system of linear differential equations:

$$\frac{dx}{dt} = Ax \tag{6.149}$$

and some initial points: $x(t_0) \in M = \{v_1, \ldots, v_m\}$. Let the system (6.149) be stable. Then the trajectories $\{x_i(t)\}, t \geq t_0$, which start from the points v_j are bounded. The invariant ellipsoid $\mathcal{E}^*$ of minimal volume exists that contains all the trajectories $\{x_j(t)\}_{j\in J}$. This ellipsoid has to contain all initial points $v_j, j \in J$.

Let P^* be the matrix of minimum volume invariant ellipsoid. To find P^* we may consider the problem:

$$-\log\det P \longrightarrow \text{minimize} \tag{6.150}$$

$$(Pa_k, a_k) \leq 1,\ k = 1, \ldots, m \tag{6.151}$$

$$P \succ 0, \lambda_1(A^TP + PA) \leq 0 \tag{6.152}$$

and one may use the mentioned above penalty function method for solution of the problem (6.150)–(6.152).

Example 6. Largest invariant ellipsoid contained in a polytope.

Let the polytope $\mathcal{P}$ represent the allowable (or safe) operating region for the system. The maximal in certain sense invariant ellipsoid contained in the polytope can be interpreted as region of safe initial conditions, i.e. initial conditions for which we can guarantee that the state of the system always remains in the safe operating region.

We know, that for solution the problems of the largest inscribed ellipsoids with center in the origin it is convenient to describe polytope $\mathcal{P}$ in the following form:

$$\mathcal{P} = \{y \in R^n | (a_k, y) \leq 1, k \in K\}. \tag{6.153}$$

For ellipsoid $\mathcal{E}$ inscribed in $\mathcal{P}$ we have

$$(Qa_k, a_k) \leq 1 \tag{6.154}$$

where $\varepsilon = \{y : (Q^{-1}y, y) \leq 1\}$. Formula (6.154) describes the system of linear inequalities in variables Q.

When we use the criterion of maximal volume of invariant ellipsoid $\mathcal{E}$, then we may work with convex objective function

$$\text{to minimize } (-\log\det Q)$$

subject to the constraints (6.154), $Q \succ 0$ and specific constraints which determine the properties of quadratic Lyapunov functions.

So we obtain the following optimization problem for polytopic linear differential inclusions (PLDJ)

$$\frac{dx}{dt} = A(t)x; A(t) \in \Omega; \Omega = \operatorname{conv}\{A_i\}_{i \in I}$$

when safe operating region is given in the form of polytope $\mathcal{P}$ (see(6.148):

$$\text{to mimimize } (-\log\det Q), Q \succ 0 \tag{6.155}$$

$$(Qa_k, a_k) \leq 1 \tag{6.156}$$

$$QA_i^{(t)} + A_i Q \preceq 0, i \in I \tag{6.157}$$

This problem can be reduced to the convex programming problem:

$$\text{to minimize } (-\log\det Q), \tag{6.158}$$

$$(Qa_k, a_k) \leq 1 \tag{6.159}$$

$$\lambda_n(Q) \geq 0;\ \lambda_1(QA_i^T + A_i Q) \leq 0, i \in I. \tag{6.160}$$

The form of the last problem is fully equivalent to the form of problem (6.142)–(6.144), but we have to remember that P and Q have polarly inverse sense:

$$P \sim Q^{-1}.$$

So, for solution of the problem (6.158)–(6.160) one may use the same method as for solution the problem (6.142)–(6.144).

In the book [BBFG 94] there are many other instances of reducing important problems arising in automatic control to the semidefinite programming problems. One can see that such problems can be naturally considered also as problems of nondifferentiable optimization.

6.8 NONCONVEX QUADRATIC PROGRAMS AND SEMIDEFINITE PROGRAMMING

Consider a quadratic optimization problem: find

$$q^* = \inf Q_0(x) \tag{6.161}$$

where

$$Q_0(x) = [(K_0 x, x) + (c_0, x) + d_0], \ x \in E^n$$

subjects to constraints

$$Q_i(x) = (K_i x, x) + (c_i, x) + d_i \le 0, \ i = 1, \dots, m \tag{6.162}$$

where $K_0, K_i, i = 1, \dots, m$ are arbitrary symetric $n \times n$ matrices.

Let F be the set of feasible points satisfying (6.162). If $F = \emptyset$ we set $q^* = +\infty$.

One may form Lagrange function $L(x, u)$ with vector of Lagrangian multipliers $u = \{u_1, \dots, u_m\}$:

$$L(x, u) = Q_0(x) + \sum_{i=1}^{m} u_i Q_i(x) = K_L(x, u) + c_L(x, u) + d_L(u),$$

where

$$K_L(x, u) = (K_L(u)x, x), \ K_L(u) = K_0 + \sum_{i=1}^{m} u_i K_i;$$

$$c_L(x, u) = (c_L(u), x), c_L(u) = c_0 + \sum_{i=1}^{m} c_i u_i;$$

$$d_L(u) = d_0 + \sum_{i=1}^{m} u_i d_i.$$

Consider $\psi(u) = \inf_x L(x,u)$. Denote by Ω^+ the set $\{u \in R^n | u \geq 0; \psi(u) > -\infty\}$. Let $\psi^* = \sup_{u \in \Omega^+} \psi(u)$. If Ω^+ is empty set $\psi^* = -\infty$. It is clear, that if $K_L(\overline{u})$ is positive definite, $\overline{u} \in \Omega^+$ and $\psi(\overline{u}) = L(x(\overline{u}), \overline{u})$, where $x(\overline{u})$ is the minimum point of $L(x, \overline{u})$ on x. Denote by Ω^{++} the set $\{\overline{u} \in R^n | \overline{u} \in \Omega^+; K_L(\overline{u}) \succ 0\}$ and by $\overline{dom\psi}$ the set $\{\overline{dom\psi}\} \cap \{u \geq 0\}$. Using these notations one can express ψ^* in the case of nonempty $\overline{dom\psi}$ by formula:

$$\psi^* = \sup_{\overline{dom\psi}} \psi(u).$$

Let $\psi^* = \psi(u^*)$, where u^* is interior point of $dom\psi$. The concave function $\psi(u)$ is differentiable at the u^*, and $x(u^*)$ is a feasible point of the problem (6.161)–(6.162). So $\psi^* = q^*$ in this case and $x(u^*)$ is an optimal point of the quadratic problem.

If $\psi^* > q^*$ one may try to improve dual bound introducing additional quadratic constraints that are consequencies of original constraints (6.162).

Example 1. Consider a simple quadratic problem: find

$$f^* = \inf_{x \geq 0} (Kx, x), \; x \in E^n \tag{6.163}$$

where K is an arbitrary $n \times n$ symmetric matrix. f^* may attain only two values: 0 or ∞. Let K have integer entries, and bit length of input for the problem (6.163) is measured by usual way. It is known that the problem (6.163) is NP-complete.

If K is not positive semidefinite, the dual Lagrangean approach results in the trivial lower estimate $-\infty$.

Let us modify the problem by adding the constraints

$$x_i x_j \geq 0, \; \forall (i,j) \tag{6.164}$$

which are immediate consequences of the constraint $x \geq 0$.

New Lagrange function

$$L(x,u) = (Kx,x) - \sum_{i=1}^{n} u_i x_i - \sum_{i,j=1}^{n} u_{ij} x_i x_j = ((K-U)x, x) - \sum_{i=1}^{n} u_i x_i,$$

where $U = \{u_{ij}\}_{i,j=1}^{n}$.

Let $\psi(u) = \inf_x L(x,u)$, $u = \{\{u_i\}_{i=1}^n \{u_{ij}\}_{i,j=1}^n\}$. The dual bound $\psi^* = \sup_{u \geq 0} \psi(u) \leq f^*$. One can see that $\psi^* = 0$ if and only if there exists $\overline{U} \geq 0$ such that matrix $K - \overline{U}$ is positive semidefinite, otherwise $\psi^* = -\infty$. So, obtaining dual lower bound for the problem (6.163), (6.164) is reduced to the semidefinite programming problem: find $\overline{U} \geq 0$ such that $K - \overline{U} \succeq 0$.

Note that if $n \leq 4$ $\psi^* = f^*$ for any K. If $n \geq 5$ there exist K such that $\psi^* = -\infty$, but $f^* = 0$ [SSt 89].

Example 2. Let us consider the problem: to find the minimum of $K_0(x) = (Kx, x)$, $x \in E^n$ subject to linear constraints $l_i(x) \leq 0$, $i = 1, \ldots, m$. If $K_0(x)$ is nonconvex, then we have the trivial dual bound $\Psi^* = -\infty$. In order to obtain better bounds it is possible to generate the quadratic constraints by multiplying the pairs of linear constraints: from $l_i(x) \leq 0$, $l_j(x) \leq 0$ we get $l_i(x) l_j(x) \geq 0$. Additional quadratic constraints allow us to affect the quadratic part of the Lagrange function. It is proved in [Shor 87] and [SSt 89] that the corresponding dual estimate for the modified quadratic problem is not worse than those which we can obtain by the linearization of the concave part of $K_0(x)$ as it performed by Pardalos-Rosen algorithm (see [PRos 87]).

One is concerned with a canonical form QP(quadratic problem):

$$\textit{minimize } (c^T, y) \text{ subject to } y \in F$$

where

$$\begin{aligned} F &\equiv \{\mathbf{y} \in H : \mathbf{y}^T \mathbf{P}_k \mathbf{y} \leq 0 (k = 1, 2, \ldots, m)\}, \\ H &\equiv \{\mathbf{y} = (y_0, y_1, \ldots, y_n)^T \in R^{1+n} : y_0 = 1\}, \\ c &\equiv \begin{pmatrix} \gamma \\ \mathbf{d} \end{pmatrix} \in R^{1+n}, \\ \mathbf{P}_k &\equiv \begin{pmatrix} \pi_k & \mathbf{q}_k^T/2 \\ \mathbf{q}_k/2 & \mathbf{Q}_k \end{pmatrix} \in S^{1+n} (k = 1, 2, \ldots, m), \\ \pi_k &\in R, \mathbf{q}_k \in R^n, \mathbf{Q}_k \in S^n (k = 1, 2, \ldots, m). \end{aligned} \tag{6.165}$$

Define

$$\left. \begin{aligned} C &\equiv \begin{pmatrix} \gamma & d^T/2 \\ d/2 & 0 \end{pmatrix} \in S^{1+n}, \\ \hat{\mathcal{G}} &\equiv \{Y \in S_+^{1+n} : Y_{00} = 1,\ P_k \cdot Y \leq 0\ (k = 1, 2, \ldots, m)\}. \end{aligned} \right\}$$

We now introduce an SDP: to find

$$\varphi^* = \inf(C, Y) \text{ subject to } Y \in \hat{\mathcal{G}}. \tag{6.166}$$

The SDP problem (6.166) can be considered as SDP relaxation of the problem (6.165). In [FK 97] it is proven that the dual Lagrangian estimate ψ^* for the QP (6.165) is not worse than SDP estimate φ^* (i.e. $\psi^* \geq \varphi^*$) and if general enough conditions of regularity are satisfied, $\varphi^* = \psi^*$.

SDP relaxation method may be regarded as a straightforward application of the Lovasz-Schrijver SDP relaxation method [LS 91] for 0-1 IPs (integer programs) to the QP. In the last few years, the interior-point methods originally proposed for linear programs ([Kar 84], etc.) were extended to SDPs ([Al 95], [BGFB 94], [Fre 94], [HRVW 96], [NN 94], [BV 95], etc.) The extension of the interior-point methods to SDPs has greatly contributed to the recent remarkable development of the SDP relaxation for various combinatorial optimization problems and nonconvex QPs ([Al 95], [GW 95], [GLS 81], [GLS 86], [GLS 88], [HPRW 95], [PRW 95], [Ram 93], [RP 96], [PRes 96], etc.) Among others, Ramana [Ram 93] studied the SDP relaxation for a general nonconvex QP having a quadratic objective function and quadratic inequality constraints, and Poljak, Rendl, and Wolkowicz [PRW 95] studied several relaxation methods, including the SDP relaxation, for 0-1 QP and their relations.

Many examples of using superfluous quadratic constraints for improving dual bounds in combinatorial and multiextremal optimization problems are represented in Chapters 7, 8 and 9 of this book.

In the article [Nes 97] the special problems of binary programming and their positive semidefinite (PSD) relaxations are considered.

Let we have two quadratic discrete optimization problems in the following form:

find

$$f^*(A) \equiv \max(Ax, x), x = (x_1, \ldots, x_n), x_i \in \{-1; 1\}, i = 1, \ldots, n, \qquad (6.167)$$

where A is an arbitrary symmetric $(n \times n)$-matrix. The corresponding minimization problem is to find

$$f_*(A) = \min(Ax, x), x_i \in \{-1; 1\}, i = 1, \ldots, n. \qquad (6.168)$$

Clearly $f^*(-A) = -f_*(A)$.

The PSD-relaxations of the mentioned problems are defined as the following: to find

$$S^*(A) \equiv \max_{\substack{X \succeq 0 \\ d(X) = 1_n}} (A, X)_M \equiv \min_{D(\xi) \succeq A} (1_n, \xi), \ \xi \in R^n \qquad (6.169)$$

$$S_*(A) = \min_{\substack{X \succeq 0 \\ d(X) = 1_n}} (A, X)_M = \max_{A \succeq D(\xi)} (1_n, \xi), \ \xi \in R^n \tag{6.170}$$

($(X, Y)_M$ is $n \times n$ matrix inner product, $d(X)$ is diagonal vector of matrix X, $D(\xi)$ is the diagonal matrix with diagonal vector ξ).

For any function of one variable $f(\tau)$ and $X \in M_n$ we denote $f[X]$ the matrix with the entries $f(x_{ij})$. For example, $[X]^k$ denote the matrix with entries x_{ij}^k, $i, j \in 1, \ldots, n$. Denote $E_u(f(u))$ the average value of the function $f(u)$, $u \in R^n$ on the n-dimensional unit sphere. Let $m^*(A) = \max_{-1_n \leq x \leq 1}(Ax, x)$, where $A \succeq 0$ is a symmetric matrix.

For $a \in R^n$ we denote by $\sigma(a) \in R^n$ the vector with components $sgn(a_i)$; set $sgn(0) = 1$. Let $V = (v_1, \ldots, v_n) \in M_n$.

In [Nes 97] the following results are obtained.

Theorem 88

$$\begin{aligned} m^*(A) &= \max_{\substack{\|v_i\| = 1, i = 1, \ldots, n \\ \|u\| = 1}} (A\sigma(V^T u), \sigma(V^T u)) \\ &= \max E_u\left((A\sigma(V^T u), \sigma(V^T u))\right) \\ &= \max_{\substack{X \succeq 0 \\ d(X) = 1_n}} \frac{2}{\pi} (A, \arcsin[X])_M . \end{aligned}$$

For problems (6.167), (6.168) let

$$S_\beta(A) = \beta S^*(A) + (1 - \beta) S_*(A) \text{ for } \beta \in [0, 1]$$

Theorem 89 *[Nes 97] For any matrix A (possibly indefinite)*

$$S_*(A) \leq f_*(A) \leq S_{1-\alpha}(A) \leq S_\alpha(A) \leq f^*(A) \leq S^*(A),$$

where $\alpha = \frac{2}{\pi}$.

Corollary 5 *Let* $\alpha = \frac{2}{\pi}$ *and* $\beta = \frac{(1-\alpha)^2 - 2}{3\alpha - 1}$. *Then the value* $S_\beta(A)$ *approximates* $f^*(A)$ *with the relative accuracy* $\mu = \frac{\pi - 2}{6 - \pi} < \frac{2}{5}$:

$$|S_\beta(A) - f^*(A)| \leq \mu(f^*(A) - f_*(A)).$$

One may compare this result with the result of Goemans and Williamson [GW 95] obtained for max-cut problem with positive edge weights (see Chapter 8 of our book). The max-cut problem for graphs $G(V, E)$ with n vertices can be reduced to the quadratic optimization problem with binary variables:

$$f^* = \max \frac{1}{2} \sum_{j<i} a_{ij}(1 - x_i x_j), \quad x_i \in \{-1, 1\}, i = 1, \ldots, n.$$

In [GW 95] the polynomial-time algorithm is given, which guarantee the following relative precision of obtained by the algorithm approximate feasible solution $\overline{x}$:

$$f(\overline{x}) \geq 0.878 f^*.$$

The relative accuracy that given by the Corollary is the following :

$$S_\beta(A) \geq (1 - \mu) f^*(A),$$

where $1 - \mu = \frac{8-2\pi}{6-\pi} \approx 0.6$.

Note that improving of estimates obtained by PSD-relaxations or by dual bounds may be achieved on the base of application of superfluous constraints (see Chapters 7-9 of this book).

7

THE ROLE OF ELLIPSOID METHOD FOR COMPLEXITY ANALYSIS OF COMBINATORIAL PROBLEMS

7.1 POLYNOMIAL EQUIVALENCE OF THE OPTIMIZATION AND SEPARATION PROBLEMS

Many combinatorial optimization problems can be reduced to the *LP* problems using the results from the field called "polyhedral combinatorics". The main goal of polyhedral combinatorics is to represent the convex envelope of feasible points of the given combinatorial problem in the form of a system of linear equalities and inequalities. Obtained by this way *LP* problems often have an exponentially growing number of constraints. We know that at each step of ellipsoid method for *LP* we have to use no more than one constraint from the set of constraints which are not fulfilled at current point. In many cases the problem of finding such a constraint can be formulated in the form of a new combinatorial (in some sense polar to the original) optimization problem (so-called separation problem).

Certainly, one may try to investigate a pair of such problems separately. The great merit belongs to M.Grötschel, L.Lovász, A.Schrijver for creating a general approach based on simultaneous consideration of relations between both polar problems for determination of numerical complexity of a wide variety of extremal combinatorial problems. By special implementation of the central cut ellipsoid method it was proven in [GLS 81] that the complexity of the mentioned above pairs of problems is polynomially equivalent. These results are obtained by using the "oracle" ideas in the theory of algorithmic complexity.

Let us remind shortly some special features of the oracle approach (see [KSch 80],[YN7
For many general enough problems of optimization it is difficult to estimate the length of input information because the formulation of the problem may contain such general objects as "convex function", "compact set", "n-dimensional polytope" and so on. In such situation there is no convenient encoding procedure for approximate description of an individual problem and it is not clear how to estimate and compare the quality of different algorithms that may be proposed for solution of the problem.

In the theory of so-called information algorithmic complexity the notion of **oracle** is used informally. We can think of an oracle as a subroutine R, which is called by the main program for some argument x and which returns some true answer $R(x)$ that can be used as additional portion of information by the main program. Sometimes we are not even interested in the work done in the main program, and the "running time" of an algorithm is measured by the number of calls of an oracle from the beginning to the positive signal of stopping criterion.

Denote by $\mathcal{K}$ a class of compact convex bodies. Every convex body $K \in \mathcal{K}$ is given as quintuple (K, n, a_0, r, R), where $n \geq 2$ is the dimension of K, $a_0 \in Q^n$ is a rational point, r, R are positive rational numbers, $r < R$, and $S(a_0, r) \subseteq K \subseteq S(a_0, R)$ (here $S(a, \rho)$ denotes the ball of radius ρ with center a).

We can formulate some algorithmic problems in connection with K.

1 The Strong Optimization Problem (SOPT).

Given a vector $c \in R^n$, find a vector $y \in K$ that maximize (c, x) on K, or assert that K is empty.

2 The Strong Separation Problem (SSEP).

Given a vector $y \in R^n$, decide whether $y \in K$, and if not, find a hyperplane that separates y from K; more exactly, find a vector $c \in R^n$ such that $(c, y) >$ $\max\{(c, x) | x \in K\}$.

3 The Weak Optimization Problem (WOPT).

Given a vector $c \in Q^n$ and a rational number $\varepsilon > 0$, either

(i) *find a vector* $y \in Q^n$ *such that* $y \in S(K, \varepsilon)$ *and* $(c, x) \leq (c, y) + \varepsilon$ *for all* $x \in S(K, -\varepsilon)$
(i.e., y is almost in K and almost maximize (c, x) *over the points deep in K), or*

(ii) *assert that* $S(K, -\varepsilon)$ *is empty.*

4 The Weak Separation Problem (WSEP).

Given a vector $y \in Q^n$ *and a rational number* $\delta > 0$*, either*

(i) *assert that* $y \in S(K, \delta)$*, or*

(ii) *find a vector* $c \in Q^n$ *with* $\|c\|_\infty = 1$ *such that* $c^T x \leq c^T y + \delta$ *for every* $x \in S(K, -\delta)$
(i.e. find an almost separating hyperplane).

An input of the weak optimization problem for $\mathcal{K}$ is the code of some member $K \in \mathcal{K}$, a vector $c \in Q^n$, and a number $\varepsilon > 0$. The length of the input is at least $n + |\log r| + |\log R| + |\log \varepsilon|$.

By using the modification of the central cut ellipsoid method the following theorem was proved [GLS 81].

Theorem 90 *Let* $\mathcal{K}$ *be a class of convex bodies. There is a polynomial-time algorithm to solve the weak separation problem for the members of* $\mathcal{K}$*, if and only if there is a polynomial-time algorithm to solve the weak optimization problem for the members of* $\mathcal{K}$*.*

This theorem is based on the following results, linked with special implementation of the ellipsoid method. Let $K \subset R^n$ be a compact convex set, $S(a_0, r) \subseteq K \subseteq S(a_0, R)$, (c, x) be a linear objective function and $\varepsilon > 0$. Without loss of generality, assume that $\varepsilon < r$, $\|c\| \geq 1$, and $n \geq 2$. Assume that there is a subroutine (oracle) SEP to solve the weak separation problem for K. This means that given a vector $y \in Q^n$ and $\delta > 0$, SEP either concludes that $y \in S(K, \delta)$ or yields a vector d such that

$$\max\{(d, x) | x \in K\} \leq (d, y) + \delta.$$

To solve the weak optimization problem on K we run the following algorithm. Let

$$N = 4n^2 \left\lceil \log \frac{2R^2||c||}{r\varepsilon} \right\rceil, \tag{7.1}$$

$$\delta = \frac{R^2 4^{-N}}{300n}, \tag{7.2}$$

and

$$p = 5N. \tag{7.3}$$

Define a sequence of vectors $x_0, x_1, \ldots$ and a sequence of positive definite matrices $A_0, A_1, \ldots$ as follows. Let $x_0 = a_0$ and $A_0 = R^2 I$. Assuming that x_k, A_k are defined, we run the subroutine SEP with $y = x_k$ and δ. If it concludes that $x_k \in S(K, \delta)$ we say that k is a **feasible index** and set $a = c$. If SEP yields a vector $d \in R^n$ such that $||d|| \geq 1$ and

$$\max\{(d, x) | x \in K\} \leq (d, x_k), \tag{7.4}$$

then we call k an **infeasible index** and set $a = -d$. Next define

$$b_k = \frac{A_k a}{\sqrt{(A_k a, a)}}, \tag{7.5}$$

$$x_k^* = x_k + \frac{1}{n+1} b_k, \tag{7.6}$$

$$A_k^* = \frac{2n^2+3}{2n^2}(A_k - \frac{2}{n+1} b_k b_k^T). \tag{7.7}$$

(With $\frac{n^2}{n^2-1}$ instead of $\frac{2n^2+3}{2n^2}$, we would get the smallest on volume ellipsoid including the appropriate half of E_k. Here we take in (7.7) larger factor to consider the rounding errors.) Further let

$$x_{k+1} \approx x_k^*, \text{ and } A_{k+1} \approx A_k^*, \tag{7.8}$$

where the sign $\approx$ means that the left hand side is obtained by rounding the right hand side to p binary digits behind the decimal point, ensuring that A_{k+1} is symmetric. Let

$$E_k = \{x \in E^n | (B_k(x - x_k), x - x_k) \leq 1\}, \tag{7.9}$$

where $B_k = A_k^{-1}$.

Lemma 7 *Let v denote the n-dimensional volume. Then*

$$\frac{v(E_{k+1})}{v(E_k)} < e^{-\frac{1}{5n}} \tag{7.10}$$

(One may compare this inequality with the inequality (3.32) for the original ellipsoid method without taking into account rounding errors). Set

$$f_k = \max\{(c, x_j) | 0 \leq j < k, x_j \text{ is feasible } \}$$

$$K_k = K \cap \{x | (c, x) \geq f_k\}.$$

It is proven, that

$$E_k \supseteq K_k, \text{ for } k = 0, 1, \ldots, N. \tag{7.11}$$

As a main result, we have the following

Theorem 91 *Let j be a feasible index for which $(c, x_j) = \max_k\{(c, x_k) | 0 \leq k < N, x_k$ is feasible $\}$. Then*

$$(c, x_j) \geq \max\{(c, x) | x \in K\} - \varepsilon.$$

Thus, it was proved that there exists an algorithm for solving the weak optimization problem for class $\mathcal{K}$ of compact convex bodies, which demands polynomial (relatively to the length L of input) number of steps of applications subroutine SEP with parameter δ such that $|\log \delta|$ is bounded by polynomial function of L. The number of additional operations linked with using the ellipsoid method is bounded also by polynomial in L. So, the polynomial-time solvability of the weak separation problem implies polynomial-time solvability of weak optimization problem.

The proof of inverse part of Theorem 90 is based on polar transformation of convex bodies.

Without loss of generality, assume that $a_0 = 0$. Let K^* be the polar of K, i.e.,

$$K^* = \{u | (u, x) \leq 1 \text{ for every } x \in K\}. \tag{7.12}$$

It is well-known that K^* is a convex body, $(K^*)^* = K$, and

$$S(0, 1/R) \subseteq K^* \subseteq S(0, 1/r). \tag{7.13}$$

If $\mathcal{K}$ is a class of convex bodies with $a_0 = 0$, let $\mathcal{K}^* = \{K^* | K \in \mathcal{K}\}$.

Lemma 8 *The separation problem for a class $\mathcal{K}^*$ of convex bodies with $a_0 = 0$ is polynomially solvable iff the optimization problem is polynomially solvable for the class $\mathcal{K}$.*

Proof. (I) The "if" part. Let $K^* \in \mathcal{K}^*$, $v \in R^n$ and $\varepsilon > 0$. Using the optimization subroutine for K, with objective function v and error εr, we get a vector $z \in R^n$ such that $d(z, K) \leq \varepsilon r$, and

$$(v, z) \geq \max_x \{(v, x) | x \in K\} - \varepsilon r. \tag{7.14}$$

Now if $(v, z) \leq 1$, then $(v, x) \leq 1 + \varepsilon r$ and hence $v_0 = \frac{1}{1+\varepsilon r} v \in K^*$. Therefore $||v_0|| \leq 1/r$, whence $d(v, K^*) \leq ||v - v_0|| \leq \varepsilon$.

On the other hand, if $(v, z) > 1$, then z is a solution of the separation problem for K^*. In fact, let $z_0 \in K$ be such that $||z - z_0|| \leq \varepsilon r$. Then for every $u \in K^*, ||u|| \leq 1/r$, and so

$$(z, u) = (z - z_0, u) + (z_0, u) \leq ||u|| \cdot ||z - z_0|| + 1 \leq \varepsilon + (z, v), \tag{7.15}$$

which proves that z is a solution of the separation problem for K^*.

(II) The "only if" part follows from the "if" part of Theorem 90, which we have already proved.□

Since $(\mathcal{K}^*)^* = \mathcal{K}$, Lemma 8 immediately implies the "if" part of the theorem: if the optimization problem is polynomially solvable for $\mathcal{K}$, then the separation problem is polynomially solvable for $\mathcal{K}^*$. But then, the optimization problem is polynomially solvable for $\mathcal{K}^*$ and so, using the lemma again, it follows that the separation problem is polynomially solvable for $\mathcal{K}$. Thus, Theorem 90 is proved.

7.2 JUSTIFICATION OF THE POLYNOMIAL-TIME SOLVABILITY OF SOME COMBINATORIAL PROBLEMS

From polyhedral point of view many combinatorial problems can be formulated as maximization of linear function on *convS*, where S is the finite set of feasible solutions. The set *convS* is a polytope so it can be described by a system of linear inequalities. However, the number of these inequalities may grow very fast when the size of the problem is increased.

The Theorem 90 (obtained by the use of the ellipsoid method) about the link of the complexity of optimization and the corresponding separation problems gives us the chance to prove the polynomial-time solvability of the given problem considering the dual or the polar problem. Sometimes the dual (polar) problem is "good" (belongs to the class P). Then the initial problem also belongs to the class P. If the dual or polar problem is known to be NP-hard, then the initial problem is NP-hard too. In the book [GLS 88] one may find many examples of using Theorem 90 for different classes of combinatorial problems. Let us consider some of them.

Minimum (r, s)-cut problem. Let $D = (V, A)$ be a directed graph: V is a node set; A is an arc set (with "capacity function" $c(\cdot) : A \to R^+$). Two chosen vertices: a "source" $r \in V$ and a "sink" $s \in V \backslash \{r\}$ are given. For a subset of nodes W we denote the set of arcs of D entering W by $\delta^-(W)$ and leaving W by $\delta^+(W)$, while we write $\delta^-(v)$ and $\delta^+(v)$ instead of $\delta^-(\{v\})$ and $\delta^+(\{v\})$. Moreover, if $x \in R^A$ and $\overline{A} \subseteq A$ is a set of arcs, then

$$x(\overline{A}) := \sum_{a \in \overline{A}} x_a.$$

The **maximum flow problem** is formulated as follows:

$$\max[x(\delta^+(r)) - x(\delta^-(r))] \tag{7.16}$$

subject to the constraints

$$x(\delta^-(v)) - x(\delta^+(v)) = 0 \text{ for all } v \in V \backslash \{r, s\} \tag{7.17}$$

$$0 \leq x_a \leq c_a \text{ for all } a \in A \tag{7.18}$$

Every vector $x \in R^A$ satisfying the constraints (7.17), (7.18) is called an (r, s)-flow subject to c or just a **flow** (feasible flow from r to s), and its value is

$$f(x) = x(\delta^+(r)) - x(\delta^-(r)).$$

Let us consider the dual problem to the LP problem (7.16)–(7.18). By introducing variables y_a for all $a \in A$ and z_v for all $v \in V$, we can write the dual problem as the following:

$$\min \sum_{a \in A} c_a y_a \tag{7.19}$$

$$z_w - z_v + y_a \geq 0 \text{ for all } a = (v, w) \in A \tag{7.20}$$

$$z_r = 1; \; z_s = 0 \tag{7.21}$$

$$y_a \geq 0 \text{ for all } a \in A \tag{7.22}$$

The dual problem (7.19)–(7.22) has a well known combinatorial interpretation. If W is a node set with $r \in W$ and $s \notin W$, then we call the arc set $\delta^+(W)$ ($= \delta^-(V \backslash W)$) a **cut separating r from s**, or just an (r,s)-cut. The name derives from the fact that if the arc set $\delta^+(W)$ is removed from D, then all directed paths (dipaths) from r to s are cuts. Since every (r,s)-flow x satisfies the capacity constraints, we have

$$x(\delta^+(W)) \leq c(\delta^+(W))$$

for all cuts separating r from s. From the conservation law (7.17) for all (r,s)-cuts we have

$$x(\delta^+(W)) - x(\delta^-(W)) = x(\delta^+(r)) - x(\delta^-(r)).$$

Thus, the maximum flow value through the capacitated network is not larger than the minimum capacity of a cut separating r from s. In 1956 Ford and Fulkerson [FFul 56] and Elias, Feinstein and Shannon [EFS 56] proved a famous theorem as follows:

Theorem 92 (Max-Flow-Min-Cut) *The maximum value of a flow from r to s is equal to the minimum capacity of a cut separating r from s.*

The minimum (r,s)-cut problem can be formulated directly. For this aim let us formulate the combinatorial version of the Max-Flow-Min-Cut theorem for the integer capacity function $c : A \rightarrow Z^+$.

Theorem 93 *For any digraph $D = (V, a)$, any two different nodes $r, s \in V$ and any capacity function $c : A \rightarrow Z^+$, the minimum of a (r,s)-cut is equal to the maximum number of (not necessary distinct) (r,s)-dipaths such that each arc a is contained in at most c_a of these dipaths.*

So, for $c \geq 0$ the minimum (r,s)-cut problem is equivalent to the following LP problem

$$\min(c, x) \tag{7.23}$$

$$x(P) \geq 1 \text{ for all } (r,s)\text{-dipaths } P \subseteq A, \tag{7.24}$$

$$x(a) \geq 0 \text{ for all } a \in A \tag{7.25}$$

One can see that the minimum cut problem (7.23)–(7.25) can be reduced to the shortest dipath problem using the ellipsoid method. In fact, the shortest dipath problem is just the separation problem for the polyhedron given by the inequalities (7.24), (7.25). So, we can solve the minimum cut problem in polynomial time using the shortest dipath algorithm as a separation subroutine.

In more detail, let $\overline{x} = \{\overline{x}_a\}_{a \in A} \geq 0$ be a current point. Find the shortest dipath from r to s for arc weighted digraph with $\overline{w}(a) = \overline{x}_a$ ($\overline{w}(a)$ is the length of arc a). Let $\overline{P} = P(\overline{w})$ be the shortest dipath. If $x(\overline{P}) \geq 1$, then $\overline{x}$ is feasible, and in the central cut ellipsoid method we use the objective function gradient c for constructing cutting hyperplane through $\overline{x}$; otherwise, $x(\overline{P}) < 1$, and we use the gradient of the linear function

$$x(\overline{P}) = \sum_{a \in \overline{P}} x_a$$

for constructing cutting hyperplane through the current center $\overline{x}$.

Note that the similar ellipsoid-type algorithm can be constructed for the minimum (r,s)-cut problem in the case of undirected graphs with nonnegative capacities of edges.

Remark. Of course the ellipsoid algorithm is not the best way for solving the min-cut problems, there are more efficient polynomial-time algorithms (see, for example, [Or 88], [GTT 90]). But it is very easy to prove the polynomial-time solvability of the mentioned algorithm using general Theorem 90.

The polynomial-time solvability of the min-cut problem can be easily understood if one takes into account that the matrix of the system of linear inequalities (7.20)–(7.22) is **totally unimodular**. A matrix M is called totally unimodular if its every square submatrix has determinant 0, +1, or −1. In particular, each entry of M is 0, +1, or −1. It is easy to show that if M is a totally unimodular matrix and b is an integral vector, then for every objective function (c, x) the LP problem

$$\max\{(c, x) | Mx \leq b\}$$

has an integral optimum solution (if optimum solution exists). Besides if polyhedron $P = \{x | Mx \leq b\}$ has vertices, all vertices are integral. Totally unimodular matrices arise in many graph extremal problems, because the incidence matrix $M(D)$ of a digraph $D = (V, A)$ is totally unimodular. $M(D)$ is a matrix whose rows are indexed by the nodes and whose columns are indexed by the arcs of D, and an entry M_{va} is equal to 1 if node v is the head of a, is equal to -1, if v is the tail of a, and is 0 otherwise.

Seymour [Sey 80] gave a characterization of totally unimodular matrices showing that all such matrices can be constructed by applying simple operations to incidence matrices of digraphs and two special 5×5 matrices. Truemper [Tru 87] proposed the fastest known polynomial-time algorithm, which determines whether a matrix is totally unimodular or not.

One may consider the existence of polynomial-time algorithms for integer LP problems with totally unimodular matrices as a direct consequence of the existence of such algorithms for usual LP problems (for example, Khachiyan's ellipsoid method). There are many other instances of combinatorial problems where the ellipsoid method can be used to prove their polynomial-time solvability.

Below we give the second illustration of the application of Theorem 90.

A matching in graph $G = (V, E)$ is a set of edges M that meet each vertex at most once. The **maximum matching problem** is to find a matching that contains as many edges as posible. If graph G has an even number of vertices, $|V| = 2k$, then the matching with cardinality k is called **perfect**. The perfect matching meets each vertex exactly once.

Let $w = \{w_{ij}\}_{(i,j)\in E}$ be the vector of edge weights. For any $S \subseteq V$, let $\delta(S)$ denote the set of edges with exactly one end in S, and let us call such a set of edges a **cut**. We abbreviate $\delta(\{v\})$ by $\delta(v)$ for any $v \in V$. Let $x = \{x_{ij}\}_{(i,j)\in E}$; $x_{ij} = 1$, if edge (i, j) belongs to the cut, and $x_{ij} = 0$ otherwise.

The problem of finding the maximum weighted perfect matching may be formulated as the following:

$$\text{maximize } f_w(x) = \sum_{(i,j)\in E} w_{ij} x_{ij}, x \in \{0, 1\}^{|E|} \tag{7.26}$$

subject to the constraints

$$x \geq 0; \tag{7.27}$$

$$x(\delta(V)) = 1 \text{ for all } v \in V; \tag{7.28}$$

(here $x(M) := \sum_{(i,j)\in M} x_{ij}$ for any $M \subseteq E$).

Let $E(S)$ denote the set of edges with both ends in S. The so-called blossom inequalities may be introduced:

$$x(E(S)) \leq \frac{1}{2}(|S| - 1) \text{ for all } S \subseteq V \text{such that}$$
$$3 \leq |S| \text{ and } |S| \text{ is odd} \tag{7.29}$$

Edmonds in [Ed 65a], [Ed 65b] proved the following theorems.

Theorem 94 *The convex hull $PM(G)$ of the incidence vectors of the perfect matchings of $G = (V, E)$ is the solution set of the system of linear equalities and inequalities (7.27)-(7.29).*

Theorem 95 *The convex hull $M(G)$ of the incidence vectors of all matching is given by the following system of linear inequalities:*

$$x \geq 0; \tag{7.30}$$

$$x(\delta(V)) \leq 1 \textit{ for all } v \in V; \tag{7.31}$$

$$x(E(S)) \leq \frac{1}{2}(|S| - 1) \textit{ for all } S \subseteq V \textit{ such that}$$
$$|S| \geq 3, \textit{ and } |S| \textit{ is odd} \tag{7.32}$$

Thus, due to Theorems 94 and 95 the problems of finding the maximum weighted perfect (arbitrary) matching can be reduced to the LP problems of maximizing linear function (w, x) on polytopes $PM(G)$ $(M(G))$.

Consider a polytope $M(G)$. The number of constraints of the form (7.31) may grow exponentially with increasing of the size of graph G. But for solving the separation problem at the point $y \notin M(G), y_{i,j} \geq 0$ for all $(i, j) \in E; y(\delta(V)) \leq 1$, for all $v \in V$, we need to find only one constraint from (7.31) that is not satisfied at the point y.

Note that the problem of finding the maximum weighted matching in a graph can be easily reduced to finding a maximum perfect matching for another graph with added edges of zero capacity.

Therefore consider the LP problem (7.26)–(7.29) for the maximum weighted perfect matching. The separation problem corresponding to the inequalities (7.29) is linked to so-called **minimum odd cut problem.**

Definition. *Given an undirected graph $G = (V, E)$ with $|V|$ even and a capacity function $c : E \to Q^+$, find an odd cut (i.e. the cut associated with decomposition $V = S \cup (V \backslash S)$ with $|S|$ odd) of minimum capacity.*

Let us find a vector $y \in R^E$ such that $y \geq 0, y(\delta(V)) = 1$ for all $v \in V$ (see (7.27), (7.28)). In order to check that y satisfies (7.29), we must solve **the minimum odd cut problem** with the capacity function $c(i,j) = y_{ij}$ for all $(i,j) \in E$.

Let a set $W^*(y)$ be optimal solution of the last problem. Then if $x(\delta(W^*(y)) \geq 1$, then y is a feasible solution of the original LP problem. If $x(\delta(W^*(y)) \leq 1$ then one can build the cutting hyperplane $(x(\delta(W^*(y)), x - y) = 0$ with gradient $x(\delta(W^*(y))$ and use this gradient at the current step of the central cut ellipsoid method.

Due to Theorem 90 the polynomial-time algorithm exists for the minimum odd cut problem. The first such algorithm was proposed by Padberg and Rao [PR 82]. In this algorithm the Gomory-Hu method was used for solving the so-called Minimum Cut Problem with nonnegative capacity function: given a graph $G = (V, E)$ and a capacity function $c : E \to Q^+$, find a cut of minimum capacity. The Gomory-Hu algorithm solves the mentioned problem by applying no more than $|V| - 1$ times the sequential solution of $[r, s]$-cut problems (see [GLS 88]).

The method for solving the minimum odd cut problem is based on the following idea. Let $\mathcal{U} \neq \emptyset$ be a subset of V such that $\delta(\mathcal{U})$ is the minimum capacity cut for graph $G = (V, E)$. If $|\mathcal{U}|$ is odd then we are done. Otherwise one can use the following result (see [GLS 88]):

Lemma 9 *There exists a minimum capacity odd cut $\delta(W)$ in G with $\mathcal{U} \subseteq W$ or $W \subseteq \mathcal{U}$.*

The lemma shows that if $|\mathcal{U}|$ is even then the problem of finding a minimum capacity odd cut can be reduced to two smaller problems. Namely, let $G^{(1)}$ and $G^{(2)}$ be the graphs obtained from G by contracting of $\mathcal{U}$ and $V \backslash \mathcal{U}$ respectively and $y^{(1)}$ and $y^{(2)}$ be the corresponding weight functions. Find a minimum capacity odd cuts in $G^{(1)}$ and minimum capacity odd cut in $G^{(2)}$. The smaller of these two cuts will correspond to a minimum capacity odd cut in G.

Since finding of minimum capacity cut $\delta(\mathcal{U})$ for graph G can be done in polynomial time by sequential solution of no more then $|V|$ minimum (r,s)-cut problems, the mentioned above algorithm requires polynomial time for obtaining minimum odd cut in a graph. Therefore one can solve the separation problem for the perfect matching polyhedron in polynomial time. And hence (see Theorem 90) the weighted perfect matching problem can be solved in polynomial time.

So, we have a chain of combinatorial problems:(1) minimum weighted matching→(2) minimum cut in graph→(3) minimum (r,s)-cut→(4) shortest (r,s)-paths. All these problems are polynomially solvable and one can prove this result essentialy by using the ellipsoid method.

7.3 DUAL BOUNDS FOR MAXIMUM WEIGHTED STABLE SETS IN GRAPHS

In this chapter we will consider some extremal graph problems. Let $G = (V, E)$ be an undirected graph with vertex set $V = \{1, \ldots, n\}$ and edge set $E \subseteq V \times V$. Denote by $A_G = (a_{ij})_{n \times n}$ the adjacency matrix of G:

$$a_{ij} = 1 \text{ if } (i,j) \in E; \; a_{ij} = 0 \text{ if } (i,j) \notin E$$

The **complement graph** of $G = (V, E)$ is $\overline{G} = (V, \overline{E})$ with the same vertex set and $\overline{E} = \{(i,j) | i \neq j; (i,j) \notin E\}$. The graph $G = (V, E)$ may be **vertex weighted**, if for every $i \in V$ the weight w_i is given.

Let $w = \{w_i\}_{i \in V}$. For a subset $S \subseteq V$ we define $W(S)$ (weight of S):

$$W(S) = \sum_{i \in S} w_i.$$

We call $G(S) = (S, E \cap S \times S)$ the subgraph induced by S.

A graph G is **complete** if all its vertices are pairwise adjacent. A **clique** C is a subset of V such that $G(C)$ is complete. The **maximum clique** problem is to find a clique of maximum weight. An independent set (stable set) is a subset of V, whose elements are pairwise nonadjacent. The maximum (weight) independent set problem is to find an independent set of maximum cardinality (of maximum weight). The size of a maximum independent set is the **stability number** of G denoted by $\alpha(G)$. The maximum weight of independent set is denoted by $\alpha_w(G)$.

It is easy to see that S is a clique of G if and only if S is an independent set of $\overline{G}$. So, any result obtained for one of the above problems can be reformulated for another problem.

The maximum independent (stable) set and maximum clique problems are known to be NP-complete for the class of arbitrary graphs. Furthermore, it is shown that there is $\varepsilon > 0$ such that, unless $P = NP$, no polynomial time algorithm can approximate the maximum independent set size (or maximum clique size) within a factor n^ε (see [GJ 79]).

Thus, the maximum independent set and maximum clique problems on arbitrary graphs are indeed difficult problems to solve.

But for some classes of graphs with special structure the polynomial-time algorithms for solving maximum weight independent set (MWIS) and maximum weight clique problems were constructed. Below we shall consider such two well known classes of graphs: **t-perfect** and **perfect graphs**.

The definition of these classes is tightly linked with the properties of integer polytopes for the corresponding stable set problems.

The weighted maximum stable set problem in graphs can be formulated as the following 0–1 problem: to find

$$\alpha_w(G) = \quad \max(w, x),\ x = (x_1, \ldots, x_n), \tag{7.33}$$

$$x_i + x_j \leq 1 \quad \text{for all } (i, j) \in E \tag{7.34}$$

$$x_k \in \{0, 1\} \quad \text{for all } k \in V \tag{7.35}$$

We introduce the stable set polytope

$$\text{STAB(G)} := \text{conv}\{\chi^S \in R^V | S \subseteq V \text{ stable set }\} \tag{7.36}$$

defined as the convex hull of the incidence vectors of all stable sets of vertices of G. $\alpha_w(G)$ is equal to the maximum value of linear function (w, x) on convex polytope STAB(G). Of course, it is very useful to represent the STAB(G) by a system of linear inequalities. Unfortunately, in general case, it is a very hard problem. Therefore we consider some partial cases.

The linear relaxation of the problem (7.33)–(7.35) is to find:

$$\begin{aligned} \alpha_1(G,V) &= \max(w,x) && (7.37)\\ x_i + x_j &\leq 1, \forall (i,j) \in E && (7.38)\\ 0 \leq x_k &\leq 1,\ k = 1,\dots,n && (7.39) \end{aligned}$$

Theorem 96 *[GLS 88]. The inequalities (7.38), (7.39) give full description of STAB(G) if and only if G is bipartite. In this case the matrix of the inequalities (7.38), (7.39) is totally unimodular, and vertex optimal solutions of the LP problem (7.37)–(7.39) are always 0–1 solutions. Hence, for bipartite graphs the problem (7.33)–(7.35) can be solved in polynomial time.*

The minimal graphs for which inequalities (7.38), (7.39) are not sufficient to describe STAB(G) are the odd circuits. In fact, if $G(V,E)$ is an odd circuit–$|V| = 2m+1$, the point $x_i = \frac{1}{2}, i \in V$ satisfies the inequalities (7.38), (7.39) but does not belong to STAB(G). So we can propose a new class of linear inequalities valid for STAB(G), so-called **odd circuits constraints**.

$$\sum_{i \in V(C)} x_i \leq \frac{|V(C)|-1}{2} \quad \text{for each odd circuit C} \qquad (7.40)$$

($V(C)$ is the set of vertices that lie in the circuit C).

Let us call the graph t-perfect if (7.38), (7.39) and (7.40) are enough to describe STAB(G). Chvátal [Chv 75] was the first who called attention to this class of graphs. This class includes some known subclasses. For example:

(a) Bipartite graphs.

(b) Almost bipartite graphs, i.e. the graphs $G(V,E)$ that have one vertex v such that $G(V\backslash\{v\})$ is bipartite.

(c) Series-parallel graphs. A graph is series-parallel if it can be obtained from a forest by repeated applications of the following operations: adding an edge, parallel to an existing edge, and replacing the edge by a path.

(d) Nearly bipartite planar graphs. These are planar graphs in which at most two faces are bounded by an odd number of edges.

In general case we do not know whether the problem of checking the t-perfectness is in NP or in P. Despite this fact a maximum weight stable set in a t-perfect graph can be found in polynomial time by using slight modification of the ellipsoid method (see [GLS 88]). This problem is reduced to the LP-problem (7.37)–(7.39), (7.40) with possibly exponentially many odd circuits constraints (7.40). But for obtaining cutting plane in the ellipsoid method it is enough to have one constraint of type (7.40) that is not satisfied for given $\overline{x}$. The search of such a constraint is equivalent to finding the shortest weight odd circuit. For the last problem there exist polynomial time algorithms [GLS 88].

For the maximum weight independent set problem it is natural to study the clique system of linear inequalities:

$$\sum_{i \in Q} x_i \leq 1 \text{ for all cliques } Q \subseteq V \tag{7.41}$$

Let STAB(G,Q) be a set of x, satisfying (7.41) and $x_j \geq 0$ for all $j \in V$. The description of polytope STAB(G,Q) may be very complex because the number of cliques grows exponentially when n tends to infinity.

We need some definitions.

The **clique number** $w(G)$ is the maximum cardinality of a clique in G. It is clear that $\alpha(G) = w(\overline{G})$.

A partition of V into stable sets (cliques, respectively) is called a **coloring** (**clique covering**, respectively) of G. The coloring number (respectively, clique covering number) is the smallest number of stable sets in a coloring(respectively, cliques in a clique covering) of G, and is denoted by $\chi(G)$ (respectively, $\overline{\chi}(G)$). It is clear that

$$\chi(G) = \overline{\chi}(\overline{G}).$$

Each stable set in G has no more than one representative in each clique, hence, we have the inequality:

$$\alpha(G) \leq \overline{\chi}(G) \tag{7.42}$$

Similary we obtain

$$w(G) \leq \chi(G). \tag{7.43}$$

Berge called a graph G **perfect** if the equality

$$w(G') = \chi(G') \tag{7.44}$$

holds for every induced subgraph G' of G. The first Berge conjecture about perfect graphs was the following [Berg 61], [Berg 62]: *The complement of a perfect graph is perfect again.* This was proved by Lovász in 1972 [Lo 72]. In

1962 Berge also proposed the **Strong Perfect Graph Conjecture.** *The graph is perfect if and only if it or its complement, does not contain an odd circuit of length at least five as an induced subgraph.* This conjecture is still unsolved.

For us the most interesting property of perfect graphs is the following (see [Lo 79], [Lo 82]):

Theorem 97 *Graph G is perfect if and only if*

$$STAB(G)=STAB(G,Q).$$

So, in the case of perfect graphs the problem of finding $\alpha_w(G)$ can be reduced to the linear programming problem:

$$\max \sum_{i \in V} w_i x_i$$

subject to the constraints (7.41), (7.39).

But the number of clique constraints (7.41) may grow exponentially with growing $|V|$. Therefore the LP relaxation of the maximum weight independent set problem does not lead to the polynomial algorithm for the class of perfect graphs. Polynomial-time algorithms were obtained by using quadratic-type models and the corresponding semi-definite estimates for $\alpha_w(G)$.

In the works [Lo 79], [Lo 82] some upper bounds for $\alpha_w(G)$ were obtained, which are exact in the case of perfect graphs.

Let $M_w(G)$ be the class of symmetric matrices $A_w = \{a_{ij}\}_{i,j=1}^{n}$ with entries:

$$a_{ij} = \begin{cases} \sqrt{w_i w_j} & \text{for } (i,j) \notin E \\ x_{ij} & \text{for } (i,j) \in E \end{cases}$$

Let $S \subseteq V$ be an independent set of G, $P_w(S) \in R^n$ be the vector with coordinates $\sqrt{w_k}$ for $k \in S$, and zero – for $k \notin S$. Then

$$\frac{(A_w P_w(S), P_w(S))}{||P_w(S)||^2} = \frac{\left(\sum_{k \in S} w_k\right)^2}{\sum_{k \in S} w_k} = \sum_{k \in S} w_k.$$

Let $\lambda_{\max}(A)$ be the maximal eigenvalue of symmetric matrix A. Using the Rayleigh formula, one can obtain:

$$\lambda_{\max}(A_w) = \max_{P, P\neq 0} \frac{(A_w P, P)}{\|P\|^2} \geq \sum_{k\in S} w_k$$

for arbitrary independent set S and any $A_w \in M_w(G)$.

So,

$$\nu_w(G) = \min_{\{A_w : A_w \in M_w(G)\}} \lambda_{\max}(A_w) \geq \alpha_w(G).$$

Thus, $\nu_w(G)$ is an upper bound for $\alpha_w(G)$.

Other formulations for $\nu_w(G)$ are given below (see [Lo 79], [Lo 82]).

(i) Let $B(G)$ be a class of symmetric positive-semidefinite matrices $B = \{b_{ij}\}_{i,j=1}^n$ such that $\sum_{k=1}^n b_{kk} = 1$, and $b_{ij} = 0$ for $(i,j) \in E$. Then

$$\nu_w^{(1)}(G) = \max_{b\in B(G)} \sum_{i,j=1}^n \sqrt{w_i w_j}\, b_{ij} \tag{7.45}$$

is the upper bound for $\alpha_w(G)$, moreover,

$$\nu_w^{(1)}(G) = \nu_w(G).$$

(ii) Another expression for upper bound of $\alpha_w(G)$ can be obtained in the following way. Let $M[G,w]$ be a class of a symmetric positive-semidefinite matrices $C = \{c_{ij}\}$ of size $n \times n$ (n is the number of vertices in G) with zero elements c_{ij} corresponding to nonadjacent pairs of vertices $i, j (i \neq j)$ and with diagonal elements $c_{kk} = w_k$, $k = 1, \ldots, n$.

Let $C \in M[G,w]$, $y \in R^n$. Consider the following problem: find

$$\varphi(C) = \min_y (Cy, y) \tag{7.46}$$

under the constraint: $\sum_{i=1}^n w_k y_k = 1$.

Let $A \subseteq V$ be an independent subset in G, and $y^A \in R^n$ be a vector with entries: $y_i^A = \frac{1}{\sum_{k\in A} w_k}$ for $i \in A$ and $y_i^A = 0$ for $i \notin A$.

Then $(Cy^A, y^A) = \sum_{i\in A} w_i (y_i^A)^2 = \frac{1}{\sum_{i\in A} w_i}$ and $\sum_{k=1}^n y_k^A = 1$, so $\varphi(C) \leq \frac{1}{\sum_{i\in A} w_i}$ for any $C \in M[G,w]$ and any independent set A in G.

So, $\varphi(C) \leq \frac{1}{\alpha(G,w)}$. Since $\varphi(C) \geq 0$, $\alpha(G,w) \leq \frac{1}{\varphi(C)}$. Denote by $\varphi^*(G,w)$ the following value:

$$\varphi^*(G,w) = \sup_{C \in M[G,w]} \varphi(C) \tag{7.47}$$

Thus, we obtain a new upper bound for $\alpha(G,w)$:

$$\overline{\nu}(G,w) = \frac{1}{\varphi^*(G,w)} \tag{7.48}$$

It was proven (see [GLS 88]) that $\overline{\nu}(G,w) = \nu_w(G)$.

Now we shall show that Lovasz estimate $\nu(G,w)$ can be obtained as a usual dual bound for a quadratic-type formulation of the maximum weighted stable set problem. Consider this problem for $G = (V,E), V = \{1,\ldots,n\}$, in the following form: find

$$\alpha_w(G) = \max \sum_{i=1}^{n} w_i x_i \tag{7.49}$$

subject to the constraints:

$$x_k^2 - x_k = 0, \; k = 1,\ldots,n, \tag{7.50}$$

$$x_i x_j = 0, \; \text{for all } i,j \in E. \tag{7.51}$$

Define the Lagrange function

$$L(x,\lambda) = \sum_{i=1}^{n} w_i x_i - \sum_{i,j \in E} \lambda_{ij} x_i x_j - \sum_{k=1}^{n} \lambda_k (x_k^2 - x_k),$$

where $\lambda = (\{\lambda_{ij}\}_{(i,j)\in E}, \{\lambda_k\}_{k \in V})$ is the vector of Lagrange multipliers.

$$L(x,\lambda) = \sum_{i=1}^{n} (w_k + \lambda_k) x_i - (A(\lambda)x, x), \tag{7.52}$$

where $A(\lambda)$ is the symmetric matrix with entries:

$$a_{ij}(\lambda) = \begin{cases} \lambda_k, & \text{for } i = j = k; k = 1,\ldots,n \\ \lambda_{ij}, & \text{for all } (i,j) \in E \\ 0, & \text{otherwise} \end{cases}$$

Let $\Omega^+ = \{\lambda : A(\lambda) \succ 0\}$. The convex function $\psi(\lambda) = \max_\lambda L(x,\lambda)$ is defined on Ω^+. Let $\psi_w^* = \inf_{\lambda \in \Omega^+} \psi(\lambda)$. ψ_w^* be the dual upper bound for $\alpha_w(G)$.

Let $A(\lambda) \in \Omega^+$. Consider

$$\varphi(\lambda_k) = \max_{x_k}[(w_k + \lambda_k)x_k - \lambda_k x_k^2] = \frac{(w_k + \lambda_k)^2}{4\lambda_k}, \quad k = 1, \ldots, n. \tag{7.53}$$

Since $\min_{\lambda_k} \varphi(\lambda_k) = \varphi(w_k) = w_k$, one may fix diagonal elements of $A(\lambda)$:

$$\lambda_k = w_k, \; k = 1, \ldots, n.$$

Note that

$$\lambda_{ij} = 0 \text{ for } (i,j) \notin E, \; i \neq j.$$

Then

$$\psi_w^*(G) = \inf_{\overline{\lambda} \in \overline{\Omega}^+} \max_x \left[2(w,x) - (\overline{A}(\overline{\lambda})x, x)\right],$$

where $\overline{\lambda} = \{\lambda_{ij}\}_{(i,j) \in E}$, $\overline{\Omega}^+$ is the set of positive definite matrices $A(\lambda)$ with fixed diagonal elements:

$$a_{kk}(\overline{\lambda}) = w_k, \; k = 1, \ldots, n.$$

Let us use the following proposition.

Lemma 10 *If A is a positive definite matrix, $b \in R^n$, and*

$$\delta = \max_x [2(b,x) - (Ax,x)] \tag{7.54}$$

$$\delta' = \min_x [(Ax,x) : (b,x) = 1], \tag{7.55}$$

then $\delta \cdot \delta' = 1$.

Proof. The maximum point x in (7.54) equal $x^* = A^{-1}b$, and $\delta = (b, A^{-1}b)$. Using the Lagrange method one can obtain: $\delta' = \frac{1}{(b, A^{-1}b)}$. So, $\delta\delta' = 1$ and Lemma 10 is proved.

Let $b = w$, $\overline{A}(\overline{\lambda}) = A$. Then due to Lemma 10

$$\max_x \left[2(w,x) - (\overline{A}(\overline{\lambda})x, x)\right] = \left[\min_x [(\overline{A}(\overline{\lambda})x, x) | \sum (w,x) = 1]\right]^{-1}. \tag{7.56}$$

One can see that the set of matrices

$$\left\{\overline{A}(\overline{\lambda})\right\}_{\lambda \in \overline{\Omega}^+} = M[G, w] \text{ see (7.47))} \tag{7.57}$$

Using (7.56) and (7.57) one can deduce the following theorem:

Theorem 98 *The dual bound $\psi_w^*(G)$ for the quadratic-type problem (7.49)-(7.51) is equal to the upper bound for MWIS given by formula (7.48). So, $\psi_w(G) = \nu_w(G)$.*

Historically, the first upper bounds of semidefinite type for $\alpha_w(G)$ were obtained by L.Lovász in [Lo 79], [Lo 82]. These investigations were initiated by some problems in information theory.

In 1956 a famous scientist Shannon, the father of information theory, studying the problems of interference stable coding, introduced the concept of information capacity of a graph.

Consider a graph $G = (V, E)$ whose vertices are symbols of some alphabet. Two vertices of G are linked by an edge if the corresponding symbols can be mixed at recognition. The maximal number of different one-symbol words which can be sent with the guarantee of their true recognition is equal to the stable number $\alpha(G)$ of G.

Let us denote by $G_1 \circ G_2$ ($G_1 = (V_1, E_1); G_2 = (V_2, E_2)$) the strong product of two graphs, i.e. the graph $G = (V, E)$, which nodes are any ordered pairs $w = \{v_1, v_2\}$ of nodes of graphs G_1 and G_2 ($v_1 \in V_1, v_2 \in V_2$), and two nodes $w^{(1)}, w^{(2)}$ are adjacent if their first and second entries are adjacent or identical in the corresponding graphs.

Determine the k-th degree of a graph G as $G^k = G_0^{k-1} \circ G; G^1 = G$. If nodes of graph G are alphabetic symbols, then any node of graph G^k corresponds to k-symbol word $\{v_1 v_2 \dots v_k\}$ and two nodes of G^k are adjacent if and only if each pair of entries with the same number is equal or adjacent in graph G (or, by other words, the corresponding words may be mixed). The maximal number of nonmixed k-symbol words is $\alpha(G^k)$.

It is easy to deduce that

$$\alpha(G^k) \geq [\alpha(G)]^k. \tag{7.58}$$

Shannon introduced the measure for so-called information capacity of graph G:

$$\theta(G) = \sup_k \sqrt[k]{\alpha(G^k)}.$$

This value can be used for determination of limit speed of communication translations with zero probability of error.

Since $\alpha(G^{k+l}) \geq \alpha(G^k) \cdot \alpha(G^l)$

$$\theta(G) = \lim_{k \to \infty} \sqrt[k]{\alpha(G^k)}.$$

Shannon showed that capacity number

$$\theta(G) \leq \alpha^*(G),$$

where $\alpha^*(G)$ is the result of solution of the problem

$$\alpha^*(G) = \max_{x \in STAB(G,Q)} \sum_{i=1}^{|V|} x_i.$$

From (7.58) we have: $\alpha(G) \leq \theta(G)$. Thus,

$$\alpha(G) \leq \theta(G) \leq \alpha^*(G)$$

The problem of calculating $\theta(G)$ was seemed very hard even for the simplest graphs, for example, for odd circuits. For a long time starting in 1956 the following Shannon's bound for the "pentagon" circuit C_5 could not be improved:

$$\sqrt{5} \leq \theta(C_5) \leq \frac{5}{2} \qquad (\alpha[(C_5)^2] = 5).$$

After 21 years in 1977 the young Hungarian mathematician L.Lovász using his new theory of orthonormal graph representations received the precise result:

$$\theta(C_5) = \sqrt{5}.$$

Let $G(V, E)$ be a graph. An **orthonormal representation** of G is a sequence $(u_i | i \in V)$ of $|V|$ vectors $u_i \in E^N$, (here N is some positive integer) such that $\|u_i\| = 1$ for all $i \in V$ and $(u_i, u_j) = 0$ for all pairs (i, j) of nonadjacent vertices.

Let $(u_i | i \in V), u_i \in E^N$ be an orthonormal representation of G and let a vector $c \in E^N$ with $\|c\| = 1$. Then for any stable set $S \subseteq V$ the vectors $u_i, i \in S$, are mutually orthonormal and

$$\sum_{i \in S} (c, u_i)^2 \leq 1. \tag{7.59}$$

Let $x = \{\chi^S\} \in E^n$ be the incidence vector of stable set $S \subseteq V$. Then from (7.59) we have for each stable set S:

$$\sum_{i \in V} (c, u_i)^2 x_i = \sum_{i \in S} (c, u_i)^2 \leq 1 \tag{7.60}$$

Denote by $STAB(G)$ the convex envelope of all incidence vectors χ^S, S is a stable set of the graph G. Thus, one may consider (7.60) as a valid family of linear inequalities for $STAB(G)$ for any orthonormal representation $(u_i|i \in V)$, and any unit vector c, ($u_i, i \in V; c$ are vectors from E^N). We shall call (7.60) the **orthonormal representation constraints** for $STAB(G)$(shortly, $ORC(G)$).

If Q is a clique of G, let $\{u_i|i \in V\backslash Q\} \cup \{c\}$ be mutually orthogonal arbitrary unit vectors. Set $u_j = c$ for $j \in Q$. Then the ORC constraint (7.60) is converted into the clique constraint of the form (7.41), determined by Q:

$$\sum_{i \in Q} x_i \leq 1 \text{ for any clique } Q. \tag{7.61}$$

For any graph $G(V, E)$,let us denote

$$STAB(G, Q) = \{x \in E^v | x \geq 0 \text{ and } x \text{ satisfies (7.61) }\}$$

$$TH(G) = \{x \in R^V | x \geq 0 \text{ and } x \text{ satisfies } ORC(G) \text{ (7.60)}\}$$

$TH(G)$ as an intersection of infinitely large number of convex sets is a convex set. It is shown above that

$$STAB(G) \subseteq TH(G) \subseteq STAB(G, Q).$$

For perfect graphs $STAB(G, Q) = STAB(G)$, hence, $TH(G) = STAB(G)$, and $TH(G)$ is a polytope. In general, $TH(G)$ is not a polytope when G is not perfect. The most important fact for us is that one can optimize any linear objective function over $TH(G)$ in polynomial time with given accuracy, using some kinds of upper bounds for $\alpha'_w(G)$.

Given a graph $G = (V, E)$ and weights $w \in R^V_+$. Let us formulate the main problem: find

$$\nu(G, w) := \max\{(w, x)|x \in TH(G)\} \tag{7.62}$$

In the book [GLS 88] there are some reformulations of this problem with the same resulting value.

I. To find

$$\nu_1(G, w) = \min_{\{c, \{u_i\}\}} \max_{i \in V} \frac{w_i}{(c, u_i)^2} \tag{7.63}$$

where $\{u_i|i \in V; u_i \in R^N\}$, ranges over all orthonormal representations of G, $c \in E^N$ over all unit vectors of E^N. (The quotient in (7.63) may have indefinite value (the case 0/0). If $w_i = 0$, then we always set $\frac{w_i}{(c,u_i)^2} = 0$. If $w_i > 0$ but $(c, u_i) = 0$ then we set to this quotient the value $+\infty$).

II. To find

$$\nu_2(G,w) = \min_A\{\lambda_1[A+W],\ A \in M^\perp\} \tag{7.64}$$

where $M^\perp$ is the subset of symmetric matrices $|V|$ by $|V|$ in which all a_{ij} for adjacent i,j are free variables, all other entries equal to zero; $W = \{w_{ij}\}_{i,j=1}^{|V|}$, $w_{ij} = \sqrt{w_i w_j}$.

III. To find

$$\nu_3(G,w) := \max_B\{(B\overline{w},\overline{w})|B \in M \cap \overline{S}_n^+ \text{ and } TrB = 1\}, \tag{7.65}$$

where $M = \{B = \{b_{ij}\} \in S^n; b_{ij} = 0$ for all (i,j) adjacent in $G\}, \overline{w} = \{\sqrt{w_i}\}_{i=1}^{|V|}$.

IV. To find

$$\nu_4(G,w) := \max_{\{d,\{v_i\}\}} \sum_{i\in V}(d,v_i)^2 w_i \tag{7.66}$$

where $\{v_i|i \in V\}$ with $v_i \in R^N$ range over all orthonormal representations of the complementary graph $\overline{G}$ and $d \in E^N$ over all vectors of Euclidean norm 1.

It is proven (see [GLS 88]) that

Theorem 99 *For every graph $G(V,E)$ and every $w \in E_+^n$*

$$\nu(G,w) = \nu_1(G,w) = \nu_2(G,w) = \nu_3(G,w) = \nu_4(G,w).$$

The proof is based on the deduction of series of inequalities:

$$\nu \le \nu_1 \le \nu_2 \le \nu_3 \le \nu_4 \le \nu.$$

Since $STAB(G) \subseteq TH(G) \subseteq STAB(G,Q)$,

$$\alpha_w(G) = \max_{x\in STAB(G)}(w,x) \le \nu(G,w) \le \max_{x\in STAB(G,Q)}(w,x) = \nu_L(G,w),$$

where $\nu_L(G,w)$ is obtained by reducing boolean problem (7.33)–(7.35) to the LP problem with cliques constraints (7.41). In the case of perfect graphs $STAB(G) = STAB(G,Q)$, so we can formulate the following result:

Theorem 100 *If G is a perfect graph then*

$$\alpha_w(G) = \nu(G, w) = \nu_L(G, w). \tag{7.67}$$

In [Lo 79] Lovasz proved the following result. Let $\nu(G) = \nu(G, \overline{w})$, when all weights $w_i = 1, i \in V$.

Theorem 101 *Let $G = G_1 \circ G_2$. Then*

$$\nu(G) = \nu(G_1) \circ \nu(G_2). \tag{7.68}$$

This multiplicity property implies that $\sqrt[k]{\alpha(G^k)} \leq \sqrt[k]{\nu(G^k)} = \nu(G)$, so $\alpha(G) \leq \theta(G) = \lim_{k\to\infty} \sqrt[k]{\alpha(G^k)} \leq \nu(G)$.

Thus, $\nu(G)$ is an upper bound for Shannon number $\theta(G)$, that is not worse than "clique" LP estimate.

Lovasz presented in [Lo 79] many other useful properties of estimates $\nu(G)$. Particularly, he obtained the following formula for $\nu(C_{2k+1})$, where C_{2k+1} is an stable set of odd cycle of size $2k+1$:

$$\nu(C_{2k+1}) = \frac{n \cos \frac{\pi}{2k+1}}{1 + \cos \frac{\pi}{2k+1}}$$

Since $\cos \frac{\pi}{5} = \frac{1}{4}(1+\sqrt{5})$, $\nu(C_5) = \sqrt{5}$.

Thus, Lovasz proved that Shannon capacity number for C_5 $\theta(C_5) = \sqrt{5}$.

Let us return to the dual quadratic upper bounds $\nu_w(G)$ for $\alpha_w(G)$. We want to improve these estimates. For this aim we may use the "superflous" inequalities in the problem formulation. For example,

$$x_i x_j \geq 0, \text{ for all (some) non-adjacent pairs } (i, j); \tag{7.69}$$

or

$$x_k(x_i + x_j) \leq x_k, \text{ for any } k \text{ and } (i, j) \in E, \ (i, j, k \in V); \tag{7.70}$$

The corresponding modification of Lagrange functions may lead sometimes to more precise dual bound for $\alpha_w(G)$ without essential complication of calculations. So, if we add to the constraints (7.50), (7.51) the constraints of the form (7.70), we then take into attention not only clique constraints (7.41) but also the odd circuit constraints (7.40).

Indeed, consider a circuit C of odd length with $(2m+1)$ vertices $\{i_1, i_2, \ldots, i_{2m+1}\}$ and edges (i_r, i_{r+1}), $(r = 1, \ldots, 2m)$ and (i_{2m+1}, i_1). Since $x_{i_1} \cdot x_{i_2} = 0$ and $x_{i_{2m+1}} \cdot x_{i_1} = 0$, one can deduce from (7.70) that

$$x_{i_1}(\sum_{r=1}^{2m+1} x_{i_r}) = x_{i_1}(\sum_{r=2}^{2m} x_{i_r}) = \sum_{p=1}^{m-1} x_{i_1}(x_{i_{2p+1}} + x_{i_{2p+2}}) \leq (m-1)x_{i_1}.$$

Similarly, we obtain

$$x_{i_k}(\sum_{r=1}^{2m+1} x_{i_r}) \leq (m-1)x_{i_k} \text{ for all } k = 1, \ldots, 2m+1.$$

Hence, $(\sum_{r=1}^{2m+1} x_{i_r})^2 - (m-1)\sum_{r=1}^{2m+1} x_{i_r} \leq 0$, and we also obtain odd circuits inequalities $\sum_{i \in C_{2m+1}} x_i \leq m-1$ for all circuits C_{2m+1} of length $(2m+1)$ $(m \geq 2)$ as consequences of inequalities (7.70).

Similarly, let $K \subseteq V$ be an arbitrary clique. Hence, $x_i x_j = 0$ for $i, j \in K$. In combination with (7.50) one can obtain:

$$(\sum_{i \in K} x_i)^2 = \sum_{i \in K} x_i^2 = \sum_{i \in K} x_i,$$

and so

$$0 \leq \sum_{i \in K} x_i \leq 1 \text{ for each clique } K.$$

So, clique inequalities (7.41) may be considered as prime consequences of inequalities (7.50), (7.51).

Consider MWIS problem in the quadratic-type form (7.49)–(7.51) with superflous constraints (7.69), (7.70). The corresponding Lagrange function $L_2(x, \lambda)$ can be written as the following expression:

$$L_2(x, \lambda) = \sum_{i=1}^{n} w_i x_i - \sum_{\{i,j\} \in E} \lambda_{ij} x_i x_j - \sum_{k=1}^{n} \lambda_k (x_k^2 - x_k) -$$
$$- \sum_{(i,j,k) \in M(G)} \lambda_{ijk} x_k (x_i + x_j - 1) + \sum_{(i,j) \in \overline{E}} \lambda_{ij}^{(1)} x_i x_j \qquad (7.71)$$

Here: $\lambda = \{\{\lambda_{ij}\}_{i,j\in E}, \{\lambda_k\}_{k=1}^n, \{\lambda_{ijk}\}_{(i,j,k)\in M(G)}, \{\lambda_{ij}^{(1)}\}_{i,j\in \overline{E}}\}$, where $M(G) = \{(i,j,k) : k \in V,\ (i,j) \in E\}, \overline{E} = \{(i,j) : i \neq j;\ (i,j) \notin E\}$.

Let

$$\varphi_2(\lambda) = \sup_x L_2(x,\lambda);\ \varphi_2^* = \inf_{\lambda\in M^+} \varphi_2(\lambda), \qquad (7.72)$$

(here $M^+ = \{\lambda : \{\lambda_{ijk}\}_{i,j,k\in M(G)} \geq 0;\ \{\lambda_{ij}^{(1)}\}_{i,j\in\overline{E}} \geq 0\}$.) It is clear that $\alpha_w(G) \leq \varphi_2^* \leq \nu_w(G)$, and for some graphs G and weights w $\varphi_2^* < \nu_w(G)$. In particular, $\varphi_2^* = \alpha_w(G)$ not only for the class of perfect graphs, but also for the class of t-perfect graphs.

Even if we use only one trivial family of superflous inequalities (7.69) we can improve the dual upper bounds for $\alpha_w(G)$ considerably in some cases. The corresponding Lagrange function $L_1(x,\lambda)$ can be constructed as follows:

$$L_1(x,\lambda) = \sum_{i=1}^n w_i x_i - \sum_{(i,j)\in E} \lambda_{ij} x_i x_j + \sum_{k=1}^n \lambda_k (x_k^2 - x_k) + \sum_{(i,j)\in\overline{E}} \lambda_{ij}^{(1)} x_i x_j$$

(here $\lambda = \{\{\lambda_k\}_{k=1}^n, \{\lambda_{ij}\}_{i,j\in E}, \{\lambda_{ij}^{(1)}\}_{i,j\in\overline{E}}$). Let

$$\begin{aligned} \varphi_1(\lambda) &= \sup_x L_1(x,\lambda); \\ \varphi_1^*(G,w) &= \inf_{\lambda\in\Omega^+} \varphi_1(\lambda), \end{aligned} \qquad (7.73)$$

where $\Omega^+ = dom\varphi_1 \cup \{\lambda : \lambda_{ij}^{(1)} \geq 0\}$.

There are instances of graphs G for which the upper bound φ_1^* for $\alpha(G)$ is much better than $\nu(G)$. For example, let vertices of graph $G_6 = (V,E)$ correspond to the integer numbers from 0 to 63, written in binary codes of length 6, and two vertices v_1, v_2 are joined by edge if Hamming distance $d(v_1,v_2)$ is no more than 3. It is easy to show that $\alpha(G_6) = 4$ (for instance the maximum stable set is : $\{v_1 = (000000),\ v_2 = (111100),\ v_3 = (110011),\ v_4 = (001111)\}$. It was calculated, that $\nu(G_6) = \frac{16}{3}$, but $\varphi_1^*(G_6) = 4$. Our instance is a bright example of the fact that adding of superflous constraints to nonconvex problem may improve the dual (Lagrange) bound considerably.

It was shown (see [GLS 88]) that improved dual estimate $\varphi_1^*(G,w)$ equals to the modifications (A), (B), (C) of Lovász-Shriver upper bounds for $\alpha_w(G)$.

(A) Let $\overline{w}$ be a vector with entries $\overline{w}_k = \sqrt{w_k}$ and one-range matrix $W = \overline{w} \cdot \overline{w}^+$, $M_w^{(+)}(G)$ be a class of $n \times n$ matrices $A(x,w)$ with entries $a_{ij}(x,w) = W_{ij} + x_{ij}$.

Find

$$\overline{\nu}_1(G,w) = \inf_x \lambda_1(A(x,w))$$

under constraints:

$$x_{ij} \geq 0 \text{ for } (i,j) \notin E$$

(B) Let $Y = \{y_{ij}\}_{i,j=1}^n$ be symmetric $n \times n$ matrices.

Find

$$\overline{\nu}_2(G,w) = \sup(Y,W)$$

under constraints:

$$y_{ij} \geq 0 \text{ for } (i,j) \in E;\ Y \succeq 0;\ \sum_{i=1}^n y_{ii} = 1.$$

(C) Let $Z = \{z_{ij}\}_{i,j=1}^n$ be $n \times n$ symmetric matrices.

Find

$$\overline{\nu}_3(G,w) = [\inf_z \inf_x (Zx,x)]^{-1}$$

under constraints

$$z_{ij} \geq 0 \text{ for } (i,j) \notin E;\ Z \succeq 0 \sum_{i=1}^n w_i x_i = 1.$$

The following theorem was proven:

Theorem 102 *For each graph $G = (V,E)$ bounds $\overline{\nu}_1(G,w) = \overline{\nu}_2(G,w) = \overline{\nu}_3(G,w) = \varphi_1^*(G,w)$ (see [GLS 88]).*

7.4 POLYNOMIAL-TIME ALGORITHM FOR PERFECT GRAPHS

For the case when $w_i, i \in V$ are positive integers, the algorithm was proposed in [GLS 81] for finding $\nu_3(G,w)$ based on ellipsoid method. This algorithm is bounded in time by polynomial in $|V|$, $\log\varepsilon$ and $\log T$, where ε is the given precision of determination $\nu_3(G,w)$, T is the maximum among the denominators and numerators of approximation $\sqrt{w_i w_j}$ by rational numbers ω_{ij} so that

$$|\sqrt{w_i w_j} - \omega_{ij}| \leq \frac{\varepsilon}{2n^2}\ (i,j \in (1,\ldots,n)).$$

In the case of perfect graphs $\alpha_w(G)$ and $\nu_2(G, w)$ are equal integer numbers, so we may take $\varepsilon = \frac{1}{2}$. For perfect graphs the algorithm was proposed that calculates $\nu_2(G, w)$ in polynomial time bounded by $|V|$ and $\log ||W||$ with precision $\frac{1}{2}$. Thus, by rounding of obtained approximation of $\nu_2(G, w)$ one can receive the exact integer value $\alpha_w(G)$.

Let us recall that the first polynomial time algorithm for determination of $\alpha_w(G)$ for perfect graphs (see [GLS 81]) is based on calculation of $\nu_3(G, w)$ by ellipsoid method.

Consider an application of ellipsoid algorithm for finding

$$\nu_w(G) = \max[(B\overline{w}, \overline{w}) \mid B \in \mathcal{K}],$$

where $\overline{w} = \{\overline{w}_i\}_{i=1}^n \in R^n$, B denotes a $n \times n$ matrix with entries $\{b_{ij}\}_{j=1}^n$, $\mathcal{K}$ is the subclass of $n \times n$ symmetric positive semidefinite matrices B satisfying the following conditions:

(i) $b_{ij} = 0$ for pairs i, j corresponding to adjacent vertices i, j in G;

(ii)

$$\sum_{k=1}^{n} b_{kk} = 1. \tag{7.74}$$

It is convenient to represent $\mathcal{K}$ as an intersection of a cone of $n \times n$ symmetric semidefinite matrices $K^+ \subset M_n$ (M_n denotes linear space of all $n \times n$ real matrices) and affine subspace $M_1 \subset M_n$ of real symmetric matrices satisfying (i), (ii), i.e.

$$K = M_1 \cap K^+.$$

Futher we shall consider K as a subset in affine subspace M_1. Note that M_1 is isomorphic to $[\frac{n(n+1)}{2} - |E| - 1]$ - dimensional linear space. Since $B \in K$ is positive semidefinite all $b_{ii} \geq 0$, $i = 1, \ldots, n$, and due to (ii),

$$b_{kk} \leq 1, \; k = 1, \ldots, n. \tag{7.75}$$

Consider (i, j), $i \neq j$, $(i, j) \notin E$, and principal submatrix $S_{ij} = \begin{pmatrix} b_{ii} & b_{ij} \\ b_{ij} & b_{jj} \end{pmatrix}$ of the second order.

$$det S_{ij} = b_{ii} \cdot b_{jj} - b_{ij}^2 \geq 0, \text{ so } |b_{ij}| \leq 1.$$

Thus subset $\mathcal{K}$ is contained in sphere $S(0, n)$.

Let $\overline{E} = S(0, n) \cap M_1$. Consider sphere $S(\frac{1}{n}I, \frac{1}{n})$. If $B \in S(\frac{1}{n}I, \frac{1}{n}) \cap M_1$, $B \in \mathcal{K}$. So

$$S(\frac{1}{n}I, \frac{1}{n}) \cap M_1 \subseteq \mathcal{K}.$$

Thus, $\mathcal{K}$ is a centered full dimensional convex well-bounded body in M_1. So we can use ellipsoid algorithm in M_1 for finding $\nu_w(G)$. Since $B = \frac{1}{n}I_n \in \mathcal{K}$, we can use sphere $E_0 = S(\frac{1}{n}I, n + \frac{1}{\sqrt{n}}) \cap M_1$ as initial ellipsoid in M_1 containing $\mathcal{K}$.

Futher we construct sequence of ellipsoids $E_1, E_2, \ldots, E_n, \ldots$ in M_1 containing optimal subset $K^* \subseteq K$, moreover, volumes $V(E_i)$, $i = 1, 2, \ldots$, of these ellipsoids decrease as geometrical progression with ratio $q(n)$ dependent only on n. Let matrix $B_k \in M_1$ be the center of ellipsoid E_k. If $B_k \in K$ we use for constructing E_{k+1} the cutting hyperplane $\overline{P}_k$ in M_1, orthogonal to projection of vector $\{\overline{w}_i \overline{w}_j\}_{i,j=1}^n$ on M_1 (see branch (A) of ellipsoid algorithm). If $B_k \neq K$ we use for constructing E_{k+1} the cutting hyperplane in M_1 passing through B_k and separating K (see branch (B) of general ellipsoid algorithm).

Determination of $\{B_k \in K^+$ or not $\}$ demands $O(n^3)$ arithmetical operations (for example, one may use Cholesky decomposition). If B_k is not positive semidefinite, calculation of cutting hyperplane $\overline{P}_k$ demands $O(n^3)$ operation also.

Let all w_i, $i = 1, \ldots, n$ be positive integers. One can approximate $\overline{w}_i = \{\sqrt{w_i}\}_{i=1}^n$ by rational numbers with given precision $\delta > 0$ and use the modification of ellipsoid algorithm which takes into consideration rounding errors. Detailed analysis shows that there exists a polynomial-time algorithm of determination $\nu_w(G)$ with arbitrary precision $\varepsilon > 0$ (the length of input includes $|\log \frac{1}{\varepsilon}|$) (see [GLS 88]). We know that $\alpha_w(G) = \nu_w(G)$ in the case of perfect graphs G. So if w is an integer vector it is enough to calculate $\nu_w(G)$ with precision $\varepsilon = \frac{1}{2}$ in order to obtain the exact value of $\alpha_w(G)$. Thus we can formulate the following proposition:

Theorem 103 *There exists a polynomial-time algorithm for obtaining maximum weight independent set value $\alpha_w(G)$ in the case of perfect graphs.*

Such an algorithm gives us a possibility to find the corresponding optimal independent set $S^* \subseteq V$ also. Indeed, let $G(V, E)$ be a perfect graph. Then graph $G'_w(V \backslash s, E')$ (induced subgraph of G_w after deleting a vertex $s \in V$) is perfect also. If vertex s belongs to all optimal independent subsets $S^* \subseteq V$ then

$$\alpha_w(G) - \alpha_{w'}(G') = w(s).$$

Otherwise

$$\alpha_w(G) - \alpha_{w'}(G') = 0,$$

and further we can consider subgraph $G'_{w'}$ for obtaining an optimal subset of vertices for graph G_w.

In both cases the problem of finding an optimal weight independent set S^* for graph $G_w(V, E)$ is reduced to finding a similar set of vertices for perfect graph with smaller number of vertices.

7.5 USING R-ALGORITHM FOR OBTAINING UPPER BOUNDS FOR THE MAXIMUM WEIGHTED STABLE SET PROBLEMS

One can use different models to calculate upper bounds $\nu_w(G)$ for $\alpha_w(G)$.

If we want to use the formula (7.64) for calculation $\nu_2(G, w)$ we have the non-smooth convex unconstrained matrix optimization problem with $|E|$ unknown parameters which affinely enter in matrix elements.

This problem can be formulated as follows: find

$$x = \{x_{ij}\}_{(i,j)\in E}$$

to minimize

$$\varphi(x) = \lambda_1 \left(W + A(x)\right),$$

where $W = \left\{\sqrt{w_i w_j}\right\}_{i,j=1}^{n}$, $A(x)$ - $n \times n$ symmetric matrix with entries

$$a_{ij}(x) = \begin{cases} x_{ij} & , \text{if} \quad (i,j) \in E \\ 0 & , \text{if} \quad (i,j) \notin E. \end{cases}$$

We recommend to use r-algorithm for solving this problem. At each step of the algorithm we need to calculate the subgradient of minimized function (the largest matrix eigenvalue of the current matrix). For this aim we must solve the corresponding spectral problem and find the maximal eigenvalue and associated with its eigenvector.

Let $x = \overline{x}$, and $v(\overline{x}) = \{v_k(\overline{x})\}_{k=1}^n$ be a normalized eigenvector associated with the maximal eigenvalue of matrix $W + A(\overline{x})$. Then

$$g_\varphi(\overline{x}) = \{v_i(\overline{x})v_j(\overline{x})\}_{(i,j)\in E}\,. \tag{7.76}$$

When multiplicity of $\lambda_1\,(W + A(\overline{x}))$ is more than 1 then $v_k(\overline{x})$ is determined nonuniquely, so $\overline{x}$ is a point of discontinuity of g_φ, but for realization of r-algorithm one may use arbitrary $v(\overline{x})$ in formula (7.76).

The results of numerical experiments for calculating $\nu_w(G)$ using formula (7.64) are tabulated in Table 7.1. We consider the graph G_2^{64} with vertex set $V, |V| = 64$, and weights equal 1, corresponding to binary coding words of the length 6, the edges are corresponded to pairs of vertices with Hamming distance not exceeding 3. Taking into account that graph G_2^{64} has nontrivial group of automorphisms the problem of finding Lovasz estimate $\nu(G_2^{64})(\nu^*(G_2^{64}))$ reduces to minimization of convex matrix function in 3 variables (in 6 variables).

$$\nu(G_2^{64}) = \min_{\{u_1,u_2,u_3\}} \lambda_1(A(u))$$

where

$$a_{ij}(u) = \begin{cases} u_k & , \text{if } k = h_{ij}(u) \le 3 \\ 1 & , \text{otherwise} \end{cases}$$

where $h_{ij}(u)$ is Hamming distance among words i and j;

$$\overline{\nu}_1(G_2^{64}, W) = \min_{u\in E^6} \lambda_1(B(u)),$$

where

$$b_{ij}(u) = h_{ij}(u);$$

$$u \in E^6;\ u_4, u_5, u_6 \ge 0$$

In Table 7.1 we describe the behavior of minimizing processes for $r(\alpha)$-algorithm. Parameters of $r(\alpha)$-algorithm are the following: $\alpha = 2, h_0 = 1.0, q_1 = 0.95, \overline{m} = 3, q_2 = 1.1, \varepsilon_x = 10^{-5}$. In column N_λ we give the number of eigenvalues of matrices $A(u)$ and $B(u)$ which values deviate from maximal eigenvalue less than 0.01 (0.0001–in brackets).

If $\frac{n(n+1)}{2} - |E| < |E|$, one may prefer to use the formula (7.65) for calculation $\nu(G, w) = \nu_3(G, w)$.

Table 7.1

Estimate $\overline{\nu}_1(G_2^{64}, W)$ ($\overline{\nu}_1^*(G_2^{64}, W)=4$)			Estimate $\nu(G_2^{64})$ ($\nu^*(G_2^{64}) = 16/3$)		
iter	$f(x_{iter})$	N_λ	*iter*	$f(x_{iter})$	N_λ
0	64.00000	0(0)	0	64.00000	0(0)
5	7.609600	14(14)	5	6.914880	14(14)
10	8.147414	0(0)	10	5.937764	0(0)
15	4.777998	0(0)	15	5.403329	0(0)
20	4.197216	19(19)	20	5.336638	20(19)
25	4.085952	19(19)	25	5.333766	21(19)
30	4.038819	19(19)	29	5.333389	21(19)
35	4.014916	19(19)	–	–	–
40	4.011060	34(14)	–	–	–
45	4.003007	50(14)	–	–	–
50	4.001437	51(19)	–	–	–
55	4.000639	51(34)	–	–	–
60	4.000307	51(15)	–	–	–
61	4.000280	51(29)	–	–	–

Formula (7.65) for calculating $\nu_3(G, w)$ gives us the other problem of convex programming in $\frac{|V||V+1|}{2} - |E|$ variables, namely, the minimization of linear function of matrix subject to one linear constraint on diagonal elements and one nonlinear nonsmooth, but convex constraint in the form of positive semidefinite condition. Let us formulate the problem (7.65) in the following form:

$$\text{maximize } (B(x), W), \tag{7.77}$$

subject to constraints

$$\operatorname{tr}(B(x)) = 1 \tag{7.78}$$

$$\lambda_n(B(x)) \geq 0, \text{ or } \lambda_1(-B(x)) \leq 0 \tag{7.79}$$

where

$$x = \{x_{ij}\}_{(i,j)\in E}$$

$$(B(x))_{ij} = \begin{cases} 0 & \text{, if } (i,j) \text{ are adjacent} \\ x_{ij} & \text{, for other entries} \end{cases}$$

We can take into account the costraint (7.78) by excluding x_{nn}.

$$x_{nn} = 1 - \sum_{i=1}^{n-1} x_{ii}. \tag{7.80}$$

We may include the constraint (7.79) in objective function using nonsmooth penalty function:

$$\text{maximize } [(B(x), W] + c \min\{0, \lambda_n(B(x))\}], \tag{7.81}$$

where c is some constant large enough.

The problem (7.81) is reduced to the unconstrained convex nonsmooth minimization problem and may be solved by r-algorithm.

Note that problem (7.65) is a typical problem of semidefinite programming. After substitution (7.80), we obtain the problem in $(n-1)$ variables $\overline{x} = (x_1, \ldots, x_{n-1})$ and corresponding variable matrix $\overline{B}(\overline{x})$. To account for semidefiniteness one can use the barrier penalty function of the form $\log\det \overline{B}(\overline{x})$ with small parameter, that tends to zero in the process of optimization. So for solution of the problem (7.77)–(7.79) it is convenient to use a modification of interior point method based on introducing barrier penalty function with a certain rule of changing penalty multiplier (see [NN 94]).

Note that ellipsoid method converges slowly. Therefore we do not recommend this method for solving practical problems.

In test experiments [SSt 89] we used the following form of Lovász's upper bound $\nu_w(G)$:

$$\nu_w(G) = \inf_{\lambda \in \Lambda^+} \sup_x L_w(x, \lambda), \tag{7.82}$$

where

$$L_w(x, \lambda) = 2\sum_{j=1}^{n} w_j x_j - (A(\lambda)x, x), \tag{7.83}$$

Λ^+ is a set of symmetric positive semidefinite matrices $A(\lambda)$ with entries

$$a_{ij}(\lambda) = \begin{cases} w_j & , \quad i = j; \\ 0 & , \quad (i,j) \notin E,\ i \neq j; \\ \lambda_{ij} & , \quad (i,j) \in E. \end{cases}$$

Function

$$\psi(\lambda) = \sup_x L_w(x, \lambda)$$

is a convex function in $m = |E|$ variables. In nontrivial cases $\psi(\lambda) = +\infty$ for almost all boundary points of the set Ω^+ and the behaviour of $\psi(\lambda)$ near the boundary of Ω^+ is similar to the behaviour of interior barrier penalty functions. It gives us possibility to use for minimization of $\psi(\lambda)$ a certain modification of r-algorithm with special step-size regulation (see [SSt 89]) which ensures that all points $A_k = A(\lambda_k)$, $k = 1, 2, \ldots$, are interior points of the set Ω^+. We used the Cholesky factorization for checking of positive definitness of matrices $A(\lambda)$

Numerical experiments were carried out to illustrate the practical effectiveness of the described above approach for finding $\nu_w(G)$. Graphs $G(V, E)$ with a number of vertices $|V| = 100$ and a number of edges $|E| = 200$ were generated stochastically. Weights $\{w_i\}$ were chosen as uniformly distributed numbers from the interval $[20; 40]$. Let $[f_k]$ denote the rounded value of $\psi(\lambda_k)$ obtained after k iterations of r-algorithm for one of the test examples:

k	20	40	60	80	100	120	140	160	180	200
$[f_k]$	1349	1317	1297	1281	1276	1269	1266	1265	1264	1264

Table 7.2

Table 7.2 illustrates the rate of convegence of r-algorithm.

We can find an explicit maximum weight stable set in a perfect graph as follows. Compare $\alpha_w(G)$ and $\alpha_w(G')$, where G' and W' arise from G and W by removing vertex 1 from G and the first component w_1 from w. If $\alpha_w(G) - \alpha_{w'}(G') = w_1$, then vertex 1 enters in all optimal independent sets, and we account for this fact, otherwise we replace G by induced subgraph G'. Next we try to remove vertex 2 similarly, and so on. At the end we are left with a collection of vertices forming a maximum weighted independent set in G.

Consider the third way of calculation $\nu(G, w)$ as a dual estimate for quadratic-type problems (7.49)–(7.51):

$$\nu(G, w) = \psi^* = \inf_{\lambda \in \Omega^+} \psi(\lambda),$$

where

$$\psi(\lambda) = \min_x \left[\sum_{k=1}^{n} 2 w_k x_k - (A(\lambda)x, x) \right], \tag{7.84}$$

$$A(\lambda) = \{a_{ij}(\lambda)\}_{i,j=1}^n, \; a_{i,j}(\lambda) = \begin{cases} \lambda_{ij} & , (i,j) \in E; \\ w_k & , \text{when } i = j = k; \\ 0 & , \text{otherwise,} \end{cases}$$

$\Omega^+ = \{\lambda : A(\lambda) \in S_n^+$ (the set of symmetric positive definite matrices) $n \times n\}$.

The solution $x(\overline{\lambda})$ of problem (7.84) for $\overline{\lambda} \in \Omega^+$ can be obtained by solving the system of linear equations:

$$A(\overline{\lambda})x(\overline{\lambda}) = w,$$

i.e.

$$x(\overline{\lambda}) = [A(\overline{\lambda})]^{-1}w \tag{7.85}$$

Thus, $\psi(\overline{\lambda}) = ([A(\overline{\lambda})]^{-1}x(\overline{\lambda}), x(\overline{\lambda}))$. The subgradient $g_\psi(\overline{\lambda})$ can be obtained by formula:

$$g_\psi(\overline{\lambda}) = \{x_i(\overline{\lambda})x_j(\overline{\lambda})\}_{(i,j)\in E},$$

if $\overline{\lambda} \in \Omega^+$.

But when $\overline{\lambda}$ tends to the boundary of Ω^+ det $A(\lambda)$ tends to zero, and calculation of $x(\lambda)$ becomes unstable. So, we must use a regularization technique to prevent the unstability in calculations, that may arise near the boundary of Ω^+. For example, we recommend to add the objective function a term in form of barrier penalty function $-\log\det A(\lambda)$ with a small parameter $\varepsilon > 0$, i.e. to consider the problem of minimizing

$$\psi_\varepsilon(\lambda) = [(A^{-1}(\lambda)w, w)] + \varepsilon \ln\det A(\lambda)$$

in the region Ω^+.

Let λ be an interior point of Ω^+ and $(\lambda + \eta)$ belongs to the boundary of Ω^+. Then $\psi_\varepsilon(\lambda + t\eta)$ tends to $+\infty$ when t tends to 1 from below. For minimizing $\psi_\varepsilon(\lambda)$ one may use the modification of r-algorithm with a special subroutine to control stepsize of approximate minimization in given direction, which guarantees that a new point of the minimizing sequence belongs to region Ω^+. Testing that $\overline{\lambda} \in \Omega^+$ may be done by Cholesky's decomposition of matrix $A(\overline{\lambda})$.

The suggested technique of finding dual quadratic bounds for MWIS problems can be easily extented to the so-called set packing problems:

$$\min_x \sum_{j=1}^n w_j x_j, \; (w_j \geq 0; j = 1, \ldots, n), \tag{7.86}$$

subject to constraints

$$\sum_{j=1}^{n} a_{ij} x_j = 1, \; i = 1, \ldots, m \tag{7.87}$$

$$x_j \in \{0,1\}; a_{ij} \in \{0,1\}, \; j = 1, \ldots, n; i = 1, \ldots, m \tag{7.88}$$

It is known (see [MTS 86]) that if system (7.87)–(7.88) is feasible, there exists large enough $\theta > 0$ such that the problem (7.86)–(7.88) has the same set of optimal solutions as the following **MWIS** problem:

$$\max_x \left[\theta m + \sum_{j=1}^{n} \overline{w}_j x_j \right], \; w_j \geq 0, \; j = 1, \ldots, n \tag{7.89}$$

$$\sum_{j=1}^{n} a_{ij} x_j \leq 1, \; i = 1, \ldots, m; \tag{7.90}$$

$$x_j \in \{0,1\}; \; a_{ij} \in \{0,1\}, \; j = 1, \ldots, n; i = 1, \ldots, m, \tag{7.91}$$

where $\overline{w}_j = \theta \sum_{i=1}^{m} a_{ij} - w_j; j = 1, \ldots, n.$

The lower bound for the appropriate θ is given by the following theorem:

Theorem 104 *Let x^* be an optimal solution for initial problem (7.86)–(7.88). Then if $\theta \geq \sum_{j=1}^{n} w_j x_j^*$, then the sets of optimal solutions of the problems (7.86)–(7.88) and (7.89)–(7.91) coincide.*

So, the boolean problem (7.86)–(7.88) can be replaced by a nonconvex quadratic problem (with a large enough parameter θ):

$$\max_x \left[\theta m + \sum_{j=1}^{n} \overline{w}_j x_j \right] \tag{7.92}$$

subject to constraints

$$x_i x_j = 0 \tag{7.93}$$

for pairs (i,j) such that there exists k with $a_{ki} \cdot a_{kj} = 1, k = 1, \ldots, m;$

$$x_k^2 - x_k = 0; \; k = 1, \ldots, n. \tag{7.94}$$

One may use Lagrange function for finding dual bounds of the packing problem (7.92)–(7.94). For this aim we can use the same algorithms that we have constructed for finding $\nu_w(G)$.

8

SEMIDEFINITE PROGRAMMING BOUNDS FOR EXTREMAL GRAPH PROBLEMS

8.1 THE MAXIMUM CUT PROBLEM

One of the problems that is naturally reduced to a quadratic boolean programming is the maximum cut problem (the max-cut problem).

Let $G = (V, E)$ be a weighted graph with n vertices and weights w_{ij} associated with the edges $l = \{i, j\}$ (if $(i, j) \notin E$, we set $w_{ij} = 0$). For each partition $V = S \cup (V \backslash S)$ of the vertex set $1 \leq |S| < |V|$, we define the value

$$c(S) = \sum_{i \in S, j \in V \backslash S} w_{ij} \tag{8.1}$$

The value of the maximum cut $\{S^*, V \backslash S^*\}$ of G, denoted by $mc(G)$, is the number defined by

$$mc(G) = \max_{S \subset V} c(S) = c(S^*).$$

Note that an ordinary (unweighted) graph is identical to a weighted graph where the weights are 1 for edges and 0 for non-edges. The max-cut problem is NP-complete [GJ 79]. Yannakakis [Yan 78] proved that it preserves this property even for class of graphs with degrees of vertices not exceeding 3. The mathematical model of the max-cut problem can be represented in boolean variables but due to its symmetric form related to the sets S and $V \backslash S$, it is better to use the variables with feasible set $\{1, -1\}$. Let $|V| = n$ and let $y = \{y_i\}_{i \in V}$ be a n-dimentional vector of variables $y_i \in \{-1, 1\}$. Variables y_i have such interpretation:

$$y_i = \begin{cases} 1, & \text{for } i \in S \\ -1, & \text{for } i \in V \backslash S, \end{cases}$$

where $\emptyset \subset S \subset V$. Then the maximum cut problem is reduced to maximization of $f(y)$, where

$$f(y) = \frac{1}{8} \sum_{(i,j) \in E} w_{ij}(y_i - y_j)^2 \tag{8.2}$$

subject to the constraints

$$y_i^2 - 1 = 0, \ \forall i \in V \tag{8.3}$$

After trivial transformations we obtain:

$$f(y) = \frac{1}{4} \sum_{(i,j) \in E} w_{ij} - \frac{1}{4} \sum_{(i,j) \in E} w_{ij} y_i y_j. \tag{8.4}$$

Note that in formulas (8.2), (8.4) each $e \in E$ is considered twice: as (i, j) and as (j, i).

Let $A = \{a_{ij}\}$ be an incidence matrix of graph $G(V, E)$. Then the max-cut problem can be reduced to minimization of $f_1(y)$, where:

$$f_1(y) = \sum_{(i,j) \in E} w_{ij} y_i y_j = \sum_{i \in V} \sum_{j \in V} w_{ij} a_{ij} y_i y_j \tag{8.5}$$

subject to (8.3). Note that $f(y) = \frac{1}{4} \sum_{i \neq j} w_{ij} - \frac{1}{4} f_1(y)$. Let $\overline{W} = \{\overline{w}_{ij}\}_{i,j \in V}$, $\overline{w}_{ij} = a_{ij} w_{ij}$, $\forall (i, j)$. The Lagrange function $L(y, u)$, $(u \in E^n$ where $n = |V|)$ of the problem (8.5), (8.3) has the form

$$L(y, u) = f_1(y) + \sum_{i \in V} u_i(y_i^2 - 1) = ((\overline{W} + diag(u))y, y) - \sum_{i \in V} u_i.$$

Let $\Omega^+(\overline{\Omega}^+) = \{u : \overline{W} + diag(u) \in S_n^+ \ (\in \overline{S}_n^+)\}$, where $S_n^+(\overline{S}_n^+)$ is the class of positive definite (semidefinite) symmetric matrices $n \times n$. One can obtain dual bound Ψ for the problem (8.5), (8.3):

$$\Psi^* = \sup_{u \in \Omega^+} \Psi(u), \quad \text{where } \Psi(u) = \inf_y L(y, u) \tag{8.6}$$

Since $x(u) = \arg\min_y L(y, u)$ equals zero for $u \in \Omega^+$,

$$\Psi(u) = -\sum_{i \in V} u_i \text{ for } u \in \Omega^+. \tag{8.7}$$

So the supremum u^* in (8.6) is obtained at the boundary of Ω^+, i.e. the matrix $A + diag(u^*)$ has a zero minimal eigenvalue. Consider the problem of finding $(-\Psi^*)$:

$$\min \sum_{i=1}^{n} u_i \tag{8.8}$$

subject to the constraint

$$\lambda_n(\overline{W} + diag(u)) \geq 0. \tag{8.9}$$

Taking into consideration that the dual problem (8.8), (8.9) is convex, one can use the nonsmooth penalty function method (see Theorem 27) for solving this problem: minimize

$$S_N(u) = \sum_{i \in V} u_i - N \min\left[0, \lambda_n(\overline{W} + diag(u))\right], \tag{8.10}$$

where $N > 0$ is a penalty multiplier. Consider the ε-perturbed problem: find

$$\Psi_1^\varepsilon = \min \sum_{i=1}^{n} u_i^\varepsilon, \; u^\varepsilon = \{u_i^\varepsilon\}_{i=1}^n, \tag{8.11}$$

subject to the constraint

$$\lambda_n(\overline{W} + diag(u^\varepsilon)) \geq \varepsilon. \tag{8.12}$$

This problem is equivalent to the problem:

$$\min \sum_{i=1}^{n} u_i^\varepsilon$$

subject to

$$\lambda_n(\overline{W} + diag(u^\varepsilon - J_n \varepsilon)) \geq 0,$$

or

$$\min \sum_{i=1}^{n} (\overline{u}_i + \varepsilon)$$

subject to

$$\lambda_n(\overline{W} + diag(\overline{u})) \geq 0,$$

where $\overline{u}_i = u_i^\varepsilon - \varepsilon$, $J_n = \{1, \ldots, 1\}$. Therefore $(\Psi_1^\varepsilon)^* = \Psi_1^* + n\varepsilon$. Using Theorem 27 one can see that if $N > n$ the problem (8.8), (8.9) is fully equivalent to the minimization of the penalty function (8.10).

The subgradient of the penalty function $S_N(u)$ (see 8.10) can be calculated by the formula:

$$g_{S_N}(u) = \begin{cases} J_n - N g_{\lambda_n}(u) & , \text{if } \lambda_n(\overline{W} + u) < 0 \\ J_n & , \text{otherwise,} \end{cases}$$

where J_n is the vector with all entries equal 1; $g_{\lambda_n} = \{s_i^2(u)\}_{i=1}^n$, where $s_i(u)$ are the entries of a normalized eigenvector $s(u)$ associated with $\lambda_n(\overline{W} + u)$. Thus, we have the formulas for calculating subgradients of nonsmooth convex function $S_N(u)$. We recommend to use r-algorithm for minimization of $S_N(u)$.

Consider the following heuristic algorithm of finding "good" feasible solutions for the max-cut problems. Let $\{u(k)\}_{k=0}^{\infty}$ be the minimizing sequence, generated by r-algorithm in the process of minimizing $S_N(u)$. Then for all k

$$-S_N(u(k)) \leq f_1(y^*).$$

Remember that

$$f(y) = \frac{1}{4}\left(\sum_{i,j} a_{ij}w_{ij} - f_1(y)\right),$$

so

$$f(y^*) \leq \frac{1}{4}\left(\sum_{i,j} a_{ij}w_{ij} + S_N(u(k))\right) \tag{8.13}$$

for all k.

Let u^* be the optimal solution of the dual problem (8.8),(8.9) and let $R(u^*)$ be the subspace of eigenvectors associated with eigenvalue $\lambda_n(\overline{W} + diag(u^*)) = 0$. If we could find a vector $\tilde{y} \in R(u^*)$ with entries ± 1, such vector would be feasible for the primal problem (8.5), (8.3) and its objective value could be equal to the lower bound. So, such vector $\tilde{y}$ could be the optimal for the original problem. However, such feasible vectors can be found only in rare cases.

In [GW 94] there was proposed a way of generating "good" feasible solutions for the max-cut problem using the eigenvectors associated with $\lambda_n(\overline{W} + diag(\overline{u}))$ where $\overline{u}$ is an approximation of u^*. Let $s(\overline{u}) = \{s_i(\overline{u})\}_{i=1}^n$ be an eigenvector associated with $\lambda_n(\overline{W} + diag(\overline{u}))$. Write its entries in nonincreasing order:

$$s_{i_1}(\overline{u}) \geq s_{i_2}(\overline{u}) \geq \ldots \geq s_{i_n}(\overline{u}).$$

For different k, $1 \leq k \leq n$, construct the partition of V:

$$V = S(k) \cup (V \backslash S(k)),$$

where $S(k) = \{i_1, \ldots, i_k\}$, and find $\max_k c(S(k)) = c(S(\overline{k}))$ (see (8.1)). Choose a feasible vector $y(s(\overline{u}))$ corresponding to this partition:

$$y(s(\overline{u})) = \begin{cases} 1, & \text{for } i \in S(\overline{k}) \\ -1, & \text{for } i \in V \backslash S(\overline{k}) \end{cases}$$

In many cases $y(s(\overline{u}))$ is a good approximation of the optimal solution. So, in the process of minimization of $S_N(u)$ by r-algorithm one may use $u(k)$ at each step k not only for obtaining an upper bound for $f(y^*) = mc(G)$, using formula (8.13), but also for obtaining a feasible integral solution $y(u(k))$ by the described above way. One can use the record value of the objective function for generated feasible solutions as lower bound for $mc(G)$. The results of numerical experiments are represented in [SB 95] (see also Tables).

Now we consider an alternative approach for solving the max-cut problem in which the so-called Laplacian matrix is used [DPo 93].

Definition 19 *The Laplacian matrix $L(G)$ of the weighted graph $G(V, E)$ is the symmetric matrix $n \times n$ with entries*

$$l_{ij} = \begin{cases} -w_{ij} & , \quad \text{for } i \neq j \\ \sum_{j \neq i} w_{ij} & , \quad \text{for } i = j \end{cases}$$

so, for the ordinary G

$$l_{ij} = \begin{cases} -1 & , \quad \text{for } (i,j) \in E \\ d_i & , \quad \text{for } i = j \\ 0 & , \quad \text{otherwise} \end{cases}$$

i.e. $L(G) = diag(d) - A$, where A is the adjacency matrix and d is the vector of vertex degrees.

Let $v = \{v_i, \ldots, v_n\}$ be a n-dimensional vector. Call a vector v with $\sum_{i=1}^{n} v_i \geq 0$ a **correcting vector**. In [DPo 93] the following estimate for every correcting vector v is obtained:

$$mc(G) \leq \frac{1}{4} n \lambda_1 (L(G) + diag(v)), \tag{8.14}$$

where $\lambda_1(M)$ denotes the maximum eigenvalue of matrix M. Using this inequality we can receive the upper bound $\varphi(G)$ for $mc(G)$:

$$\varphi(G) = \min_{v} \frac{1}{4} n \lambda_1 (L(G) + diag(v)), \tag{8.15}$$

subject to

$$\sum_{i=1}^{n} v_i = 0 \tag{8.16}$$

A vector v^* realizing the minimum in (8.15), (8.16), will be called **an optimum correcting vector**. Let us show that the bound $\varphi(G)$ can be obtained as dual bound for the problem: find

$$mc(G) = \frac{1}{4} \max \sum_{i,j} w_{ij}(x_i - x_j)^2, \tag{8.17}$$

subject to

$$x_i^2 - 1 = 0, \quad i \in V = \overline{1,n} \tag{8.18}$$

$$(x, x) = n \tag{8.19}$$

Consider the Lagrange function $L(x, v)$ with multiplier vector $v = \{v_i\}_{i=1}^n$:

$$\begin{aligned} L(x,v) &= \frac{1}{4} \max \sum_{i,j} w_{ij}(x_i - x_j)^2 + \frac{1}{4} \sum_{i=1}^{n} v_i(x_i^2 - 1) \\ &= \frac{1}{4} \left[\sum_{i,j} (-2w_{ij}x_i x_j) + \sum_i \left(\sum_j w_{ij} + v_i \right) x_i^2 \right] - \frac{1}{4} \sum_{i=1}^{n} v_i \\ &= \frac{1}{4} \left[(L(G) + diag(v)x, x) - \sum_{i=1}^{n} v_i \right] \end{aligned}$$

Let

$$\begin{aligned} \varphi(v) &= \max_{\{x:(x,x)=n\}} L(x,v) \\ &= \frac{1}{4} n\lambda_1(L + diag(v)) - \frac{1}{4} \sum_{i=1}^{n} v_i \end{aligned} \tag{8.20}$$

(we take into account the Rayleigh formula: $\lambda_1(M) = \max_{(y,y)=1}(My, y)$ for symmetric matrix M). One can see that (8.20) is invariant to the shift of v :

$$v \to v + J_n\Delta \text{ for arbitrary } \Delta,\ J_n = \{1, \ldots, 1\} \in E^n.$$

Thus, we can set: $\sum_{i=1}^n v_i = 0$, and obtain an upper bound $\varphi(G)$ for $mc(G)$ in the previous form (by solving the problem (8.15), (8.16)). By the other shift of v we can get $v = \overline{v}$ in (8.20) such that

$$\lambda_1(L(G) + diag(\overline{v})) = 0,$$

and

$$\varphi(\overline{v}) = -\frac{1}{4} \sum_{i=1}^{n} \overline{v}_i,$$

So, for finding $\varphi(G)$ we obtain the convex problem of the type similar to (8.8), (8.9): find

$$\varphi(G) = -\frac{1}{4}\max_{v}\sum_{i=1}^{n} v_i \tag{8.21}$$

subject to

$$\lambda_1(L(G) + diag(v)) \leq 0 \tag{8.22}$$

and one can observe that

$$\varphi(G) = \frac{1}{4}\sum_{(i,j)\in E} w_{ij} - \frac{1}{4}\Psi_1(u^*)$$

(see formula (8.4)).

A trivial upper bound for $mc(G)$ is the sum of all nonnegative weights of edges. It is proven in [DPo 93] that

$$\varphi(G) \leq \sum_{w_l \geq 0} w_l \tag{8.23}$$

Definition 20 *A weighted graph is called* **exact** *if it satisfies* $mc(G) = \varphi(G)$.

There are several interesting families of exact graphs, but in general the problem of recognition of "exactness" for weighted graphs is NP-complete. Let

$$f(v) = \frac{1}{4}n\lambda_1(L + v) \tag{8.24}$$

The function f is convex, Lipschitzian and attains its minimum exactly once unless all the weights are null [DPo 93]. The convexity of f simplifies the possible forms of the optimal correcting vector if the weighted graph G has a nontrivial group of automorphisms $\Gamma(G)$.

Theorem 105 *If vertices i and j belong to the same orbit of* $\Gamma(G)$, *then* $v_i^* = v_j^*$. *If G is vertex-transitive, then all* $v_j^* = 0$ *(here* $v^* = \{v_i\}_{i=1}^{n}$ *is the optimal correcting vector) [DPo 93].*

One can use the previous theorem for calculating $\varphi(G)$ in simple cases.

a) Ordinary complete graphs K_N :

$$\varphi(K_{2k}) = k^2 = mc(K_{2k}).$$

$$\varphi(K_{2k+1}) = \frac{1}{4}(2k+1)^2 > k(k+1) = mc(K_{2k+1}).$$

b) Ordinary cycles C_N, $N = 3, 4, 5, \ldots$

$$\varphi(C_{2k}) = 2k = mc(C_{2k})$$

$$\varphi(C_{2k+1}) = \left(1 + \cos\frac{\pi}{2k+1}\right)\frac{2k+1}{2} \geq 2k = mc(C_{2k+1})$$

For C_5

$$\varphi(C_5) = \frac{5}{2}(1 + \cos\frac{\pi}{5} = \frac{5\sqrt{5}}{8}(1+\sqrt{5}) > mc(C_5) = 4. \quad (8.25)$$

Note that for an odd N values $\frac{\varphi(C_N)}{mc(C_N)}$ achieve its minimum at $N = 5$, and

$$\min_N \frac{\varphi(C_N)}{mc(C_N)} = \frac{(\sqrt{5}+1)5\sqrt{5}}{32} \approx 1.13 \quad (8.26)$$

i.e.

$$\varphi(C_N) \leq mcG \quad (8.27)$$

Note that if G is exact then there is an eigenvector x^* with entries ± 1 associated with eigenvalue $\lambda_1(A + V^*)$, where $V^* = diag(v^*)$.

Consider some useful properties of $\varphi(G)$ proved in [DPo 93].

Definition 21 *Let $V^{(1)}$ and $V^{(2)}$ be the vertex sets of weighted graphs $G^{(1)}$ and $G^{(2)}$ respectively (the sets $V^{(1)}$ and $V^{(2)}$ may intersect). Let $n^{(1)}, n^{(2)}$ be the number of vertices and let $w^{(1)}, w^{(2)}$ be the weight functions of $G^{(1)}$ and $G^{(2)}$ respectively. The* **amalgam** *of graphs $G^{(1)}$, $G^{(2)}$ is a weighted graph G with vertex set $V = V^{(1)} \cup V^{(2)}$ and the weight function defined by:*

$$w_{ij} = \begin{cases} w_{ij}^{(1)} + w_{ij}^{(2)} & , \text{ if } i \text{ and } j \text{ are both in } V^{(1)} \cap V^{(2)}, \\ w_{ij}^{(1)} & , (i,j) \text{ are both in } V^{(1)} \text{ but not both in } V^{(2)}, \\ w_{ij}^{(2)} & , (i,j) \text{ are both in } V^{(2)} \text{ but not both in } V^{(1)}, \\ 0 & , \text{ otherwise.} \end{cases}$$

Theorem 106 *If G is amalgam of $G^{(1)}$ and $G^{(2)}$, then*

$$mc(G) \le mc(G^{(1)}) + mc(G^{(2)});$$

$$\varphi(G) \le \varphi(G^{(1)}) + \varphi(G^{(2)}).$$

If the graphs $G^{(1)}$ and $G^{(2)}$ have at most one common vertex, then

$$mc(G) = mc(G^{(1)}) + mc(G^{(2)});$$

$$\varphi(G) = \varphi(G^{(1)}) + \varphi(G^{(2)}).$$

This theorem can be easily proved by representation of $\varphi(G)$ as dual bounds.

Let $G = (V, E)$ be a graph and $F \subseteq E$ be an edge subset. The vector $x^F \in R^E$ with $x_e^F = 1$ if $e \in F$ and $x_e^F = 0$ if $e \notin F$ is called the incidence vector of F. The polytope $P_B(G) := conv\{x^F \in R^E | (V, F)$ is a bipartite subgraph of $G\}$ is called the **bipartite subgraph polytope** of G. It is clear that for positive edge weights $c(e)$, $e \in E$, every optimum basic solution of the linear program

$$\max(c, x), \quad x \in P_B(G) \tag{8.28}$$

corresponds to a cut.

Consider the trivial inequalities:

$$0 \le x_e \le 1 \tag{8.29}$$

The inequalities (8.29) determine $P_B(G)$ completely if and only if G is bipartite. Consider any odd cycle C in a graph G. It is obvious that full edge set of C cannot belong to a cut. So, one can formulate for an incidence vector x of a cut the following odd cut inequalities:

$$x(C) := \sum_{e \in C} x_e \le |C| - 1, \quad C \text{is an odd cycle in } G \tag{8.30}$$

Definition 22 *A graph $G(V, E)$ that has the property: $P_B(G) = \{x \in R^E | x$ satisfies all inequalities (8.29) and (8.30) $\}$ is called* **weakly bipartite**. *A graph that is not weakly bipartite is called* **strongly nonbipartite**.

Thus, in the case of weakly bipartite graphs the max-cut problem with positive weights of edges reduces to LP problem but the number of inequalities of the type (8.29) may increase as exponential function of the size of a graph. To overcome this difficulty for obtaining a polynomial time algorithm of solving the max-cut problem in the mentioned above case one can use the ellipsoid method. Certain steps of this method require to solve the problem of finding the odd cycle of minimum weight for checking the feasibility of the inequalities (8.29). The search of such a cycle can be reduced to at most n applications of the algorithm of finding the shortest even path in a weighted graph.

Consider the polynomial time algorithm of finding a shortest path from a vertex to other one with an even number of edges discribed in [GLS 88]. Suppose a graph $G = (V, E)$ with nonnegative edge weights c_e, $e \in E$, and two nodes $i, j \in V$ are given.

Let $G_i(G_j)$ be the graph obtained by removing vertex i (vertex j). Consider graph H_{ij} consisting of the vertex-disjoint union of G_i and G_j where in addition each vertex in G_i (except j) is linked to its copy in G_j by edge. The edges in G_i (respectively in G_j) keep their original weights, the weight of the new edges are set equal zero.

Suppose P is an even length path from j to i in G. We start with the edge e_1 of P incident with j and label the edge in G_i corresponding to e_1. Then we take the edge e_2 of P following e_1 and label the edge in G_j corresponding to e_2. We continue this way by alternately labeling edges in G_i and G_j. For every node v of G not on the path P we label the (new) edge in H_{ij} linking the two copies of v. It is obvious from the construction that the labeled edges in H_{ij} form a perfect matching with the same weight as P. On the other hand, from every perfect matching M in H_{ij} we can obtain an even length path in G from i to j having the same weight by simply skipping all "new" edges contained in M and possibly removing even cycles of weight zero.

It follows from the above discussion that G contains an even length path from i to j if and only if H_{ij} contains a perfect matching, and that a shortest path between i and j with an even number of edges can be computed in polynomial time using any of the existing polynomial matching algorithms, e.g. [Ed 65a].

Moreover, we would like to mention that shortest paths of odd length can be computed similarly. We construct a new graph H'_{ij} from G by taking the vertex disjoint union of G and the graph G_{ij} obtained from G by removing i and j, and add "new" edges having weight zero linking the copies of the vertices different from i and j. Then we apply the perfect matching algorithm as before.

Thus, one can formulate the following result (see [GLS 88]):

Theorem 107 *Given a graph $G = (V, E)$ with nonnegative edge weights, then there is a polynomial algorithm that checks whether G contains a path of even (or odd) length and computes the shortest path of even (or odd) length between every pair of vertices, if such a path exists.*

To solve the max-cut problem for weakly bipartite graph one can use the ellipsoid method.

For any graph $G = (V, E)$ set $P_c(G) = \{x \in R^E | x$ satisfies the inequalities (8.28), (8.29) $\}$.

We shall show that for any $c \geq 0$ the LP problem

$$\max(c, x), \quad x \in P_c(G) \tag{8.31}$$

can be solved in polynomial time. As we know (see Chapter 7) for obtaining this result it is enought to prove that there exists a polynomial time algorithm for solving **separation problem** for $P_c(G)$:

Let $y \in Q^E$. Determine whether $y \in P_c(G)$, and if y is not in $P_c(G)$ find a vector $d \in Q^E$ such that $(d, y) > (d, x)$ for all $x \in P_c(G)$.

Suppose a vector $y \in Q^E$ is given. It is trivial to check whether y satisfies the inequalities (8.28). If not, we have found a violated inequality. Thus, we may assume in the sequel that y satisfies $0 \leq y_e \leq 2$ for all $e \in E$.

For every edge $e \in E$ define a "weight" $w_e := 1 - y_e$. If C is an odd cycle in G, then clearly $y(C) > |C| - 1$ if and only if $w(C) < 1$. This implies that we can check whether an odd cycle constraint (8.29) is violated by computing an odd cycle C^* of minimum weight $w(C^*)$. Namely, if $w(C^*) \geq 1$, then y satisfies all the constraints (8.29); if $w(C^*) < 1$ then $x(C^*) \leq |C^*| - 1$ is the desired cutting plane.

To compute a minimum weight odd cycle we proceed as follows. We pick any edge $ij \in E$ with $w_{ij} < 1$ and compute the shortest (with respect to the weights w_e) path from i to j of even length with the described above method. If the weight of this path plus the weight w_{ij} is less than one, an odd cycle C^* is found with $w(C^*) < 1$. If the sum of the path weight and w_{ij} is at least one, we pick another edge and continue untill all edges with $w_{ij} < 1$ have been considered. Thus, after at most $|E|$ applications of the described above algorithm we have determined whether y is in $P_C(G)$ and if y is not in $P_C(G)$ we obtained a violated inequality of the form (8.29). Clearly, the overall running time of the separation algorithm for $P_C(G)$ is polynomial in the data, thus we have shown

Theorem 108 *There is an algorithm which for any graph $G = (V, E)$ and any $c \in Q^E$ solves the linear program* $\max(c, x)$, $x \in P_C(G)$ *in polynomial time. This algorithm in particular solves the max-cut problem for weakly bipartite graphs in polynomial time.*

We do not claim that the algorithm for the above problem is fast in practice. It remains a challenging problem to find a practically efficient method for the max-cut problem in weakly bipartite graphs.

We do not know how hard the problem of recognizing weakly bipartite graphs is. The only result we have is that the decision problem "Is a given graph weakly bipartite?" is in co-NP.

In closing, we observe that although a minimum weight odd cycle can be computed in polynomial time, we do not know whether it is possible to find an odd hole (cycle without chord) in polynomial time. This would be of interest for checking whether a graph is perfect or not (see Chapter 7).

Let $G = (V, E)$ be a graph, with edges $e = (i, j) \in E$, $\delta(i)$ be a set of vertices incident with i. We call the graph $G_l^{'} = (V', E')$ the contraction of G by edge l, if $V' = (V \backslash i)$, $E' = [(E \backslash \cup (i, k), k \in \delta(i)) \cup (j, k), k \in \delta(i)]$. We say that graph G is contractible to graph $\overline{G}$ if there exists a sequence of contractions $G \to G_1 \to G_2 \to \ldots \to G_r$ such that graph G_r is isomorphic to $\overline{G}$. It is well known that planar graphs are not contractible to K_5. Wagner [Wag 64], [Wag 70] gave a full characterization of graphs not contractible to K_5.

It is known that planar graphs are weakly bipartite, so the polynomial-time algorithm exists for the max-cut problem in the case of planar graphs [OD 72], [Had 75]. This algorithm was generalized for the case of the graphs not contractible to complete graph K_5 (with five vertices) [Barah 83].

Barahona has generalized the polynomial time algorithm for planar graphs on the cardinality version (all edge weights are equal to one) of the max-cut problem to graphs with fixed genus p. Using various transformations and matching techniques it was shown in [Barah 81] that for a graph $G(V, E)$ of genus p a maximum cardinality cut can be found in $O(4^p|V|^5)$ time (see [GLS 88]).

Remark. Note that a surface of genus p is a surface obtained from a sphere by attaching p "handles". Planar graphs have genus zero.

If a graph G can be drawn on a surface of genus p and cannot be drawn on a surface of genus $(p-1)$ such that no two edges intersect, then g is said to have genus p. The genus of a given graph can be obtained with the algorithm of Filotti and Miller [FMi 78].

It is interesting that the max-cut problem for the class of all graphs not contractible to K_6 is NP-complete.

Remark. The notion of contractibility is used in the well known graph colouring **Hadwiger's conjecture**:*If a graph G is not contractible to the complete graph K_m, then there exists a true vertex colouring of graph G by m colours.* The Hadwiger conjecture for $m = 4$ is equivalent to the **four colour conjecture** for planar graph which was proven in 1976 with essential using of might of computers (see [AH 77a]). This proof is not checked carefully for now. Recently Robertson proved Hadwiger's conjecture for $k = 5$ (under the condition that four colour conjecture is true) [AH 77b].

The quadratic dual bound for the max-cut problem in the case of weakly bipartite graphs is not always exact, but if all weights are nonnegative, the following estimate is proven.

Theorem 109 *Let G be a weakly bipartite graph with non-negative weights. Then*

$$\varphi(G) \le \frac{5\sqrt{5}(1+\sqrt{5})}{32} mc(G).$$

We shall give a brief description of the proof. It is known that the max-cut problem for any graph G with non- negative weights can be expressed in terms of edge variables x_e as follows: $\max\{(w, x) | x \in P_B(G)\}$. For weakly bipartite graphs we obtain the equality:

$$mc(G) = \max \sum_{e \in E} w_e x_e \text{ subject to (8.28), (8.29)} \qquad (8.32)$$

Let $C(G)$ be a set of odd cycles in G. Due to duality theory for the LP problem (8.32) there exist a set of non-negative coefficients α_c, $c \in C$, and a set of non-negative coefficients β_e, $e \in E$ such that

$$\beta_e + \sum_{c \ni e} \alpha_c \geq w_e \text{ for every edge } e; \tag{8.33}$$

$$\sum_{e \in E} \beta_e + \sum_{c \in C} \alpha_c(|c| - 1) = mcG \tag{8.34}$$

For an odd cycle $c \in C$ and a positive coefficient α let αc be a weighted cycle with all edges having weight α, and βe denotes the single edge with the weight β. Let H be the weighted graph obtained as the amalgam of the weighted graphs $\alpha_c c$, $c \in C$ and $\beta_e e$, $e \in E$. By (8.33), the weight of each edge of H is at least 1 and G is clearly a subgraph of H. Thus, we have $\varphi(G) \leq \varphi(H)$. Using Theorem 106, (8.34) and the inequality for odd cycles c : (8.27)

$$\varphi(G) \leq \frac{25 + 5\sqrt{5}}{32}(|c| - 1),$$

we receive:

$$\begin{aligned} \varphi(H) &\leq \sum \alpha_c \varphi(c) + \sum \beta_e \varphi(e) \\ &\leq \frac{25 + 5\sqrt{5}}{32}\left(\sum \alpha_c(|c| - 1) + \sum_e \beta_e\right) \\ &= \frac{25 + 5\sqrt{5}}{32} mcG. \end{aligned}$$

Note that $\varphi(G)$ can be efficiently computed for the weighted graphs G with rational weights by the ellipsoid method:

Theorem 110 *([GLS 88]) There exists a polynomial time algorithm, which for a given graph G with rational weights and a rational number $\delta > 0$, computes a rational vector $\overline{v}$ and a rational number $\delta > 0$ such that*

(i) $f(\overline{v}) \leq \overline{\lambda}$;

(ii) $\overline{\lambda} \leq f(v) + \delta$ *for all* v.

(here $f(v) = \frac{1}{4} n \lambda_1(L + diag(v))$*).*

The worst-case ratio of $\frac{\varphi(G)}{mcG}$ for unweighted G is with $G = C_5$, as it was conjectured in [DPo 93] and was proven in [GW 94]:

$$\frac{\varphi(G)}{mc(G)} \leq d = \frac{25+5\sqrt{5}}{32} = 1.131\ldots$$

More exactly, Goemans and Williamson have shown that the solution of (8.15), (8.16) yields a cut whose size is guaranteed to be within $\frac{1}{d} = 0.87\ldots$ of the optimum. This fact affirms good properties of the mentioned above nonsmooth bounds for $mc(G)$. The previous best result guaranteed only 0.5 of the optimum.

8.2 GRAPH PARTITIONING PROBLEMS

The problem of minimizing weighted sums of the k largest eigenvalues of a parametric family of symmetric matrices has many combinatorial applications. The best known of them is the graph partitioning problem. The problem is to divide n nodes of a given graph $G(V, E)$ into k disjoint subsets with specified cardinalities $m_1 \geq \ldots \geq m_k$ in a way to minimize the total number of edges connecting different subsets. The problem is NP-hard and therefore we have no hope to get a computationally "good" algorithm for its precise solution in general case. But Donath and Hoffman proposed in [DH 72], [DH 73] an effective lower bound for this problem that gives us possibility to obtain good enoughapproximate solutions or test the accuracy of the solution that can be received by other relatively simple heuristic methods. Consider the parametric family of symmetric matrices $A(x)$:

$$A(x) = A_0 + \sum_{i=1}^{m} x_i A_i \tag{8.35}$$

where $A_\nu, \nu = 0, 1, \ldots, m,$ are symmetric matrices, $x \in R^m$

with affinely dependent on $x = \{x_1, \ldots, x_m\}$ entries of $A(x)$. Let $S_n^k(x) = \sum_{i=1}^{k} \lambda_i(A(x))$ be a sum of k largest eigenvalues of matrix $A(x)$.We know that $S_n^k(x)$ is a convex function in x. In the case of graph partitioning problem the variables x are contained only in diagonal elements of $A(x)$. The lower bound for the partitioning problem is:

$$\rho^*(A, m) = -\frac{1}{2}s^* \tag{8.36}$$

where s^* is the solution of the following problem [DH 72]: minimize

$$S_n^k(x, m) = \sum_{i=1}^{k} m_i \lambda_i(A_0 + D(x)) \tag{8.37}$$

subject to the constraint $\mathrm{tr} D(x) = 0$

where the nondiagonal elements a_{ij}^0 of the symmetric matrix A_0 equal to one, if the i-th and j-th nodes are connected, and zero-otherwise; the diagonal elements are defined by

$$a_{ii}^0 = - \sum_{j=1;j\neq i}^{n} (a_{ij}^0), \quad i = 1, \ldots, n, \tag{8.38}$$

$D(x)$ is a diagonal matrix with elements

$$d_{ii} = x_i, \quad , i = 1, \ldots, n, \quad \sum_{i=1}^{n} x_i = 0. \tag{8.39}$$

Note that $m_1 \geq m_2 \geq \ldots \geq m_k$, therefore due to Lemma 5 the problem (8.37) is the problem of convex programming.

The lower bound $\rho^*(A, m)$ can be deduced by the following way. Let $V = \cup_{j=1}^{k} V_j$ be a partition Π of the vertex set V of graph $G(V, E)$, $|V_j| = m_j$; $j = 1, \ldots, k$. Let $Z = \{z_{ij}\}$ be the $n \times k$ matrix such that

$$z_{ij} = \begin{cases} 1, & \text{if vertex } i \in V_j \\ 0, & \text{otherwise} \end{cases}$$

Z_j is the j-th column of the matrix Z. One can see, that $\|Z_j\|^2 = m_j$, and $(Z_{j_1}, Z_{j_2}) = 0$, if $j_1 \neq j_2$, so $Z \in M_n^k(m)$ (see the definition of $M_n^k(m)$ in Chapter 6). Let A be an incidence matrix of $G(V, E)$. Then one can see that $\mathrm{tr}(AZZ^t) = 2\bar{s}$, where $\bar{s}$ is the number of edges, whose both ends lie in the same set of the partition Π. The number of edges whose ends lie in different sets of the partition Π can be expressed as

$$\rho(\Pi) = \frac{1}{2}(|E| - \bar{s}) = \frac{1}{2}\mathrm{tr}[A(JJ^t - ZZ^t)], \tag{8.40}$$

where J is a vector with entries 1. Since all diagonal entries of $JJ^t - ZZ^t$ equal to zero, one may change in formula (8.40) A for an arbitrary matrix of the type $A + \mathrm{diag}(y), y \in R^n$. Let us consider the matrix $A_0 = A - \mathrm{diag}(s)$, where $s = \{s_i\}_{i=1}^n$, $s_i = \sum_{j=1}^n a_{ij}$. Note, that $\mathrm{tr}(A_0 JJ^t) = 0$, so

$$\rho(\Pi) = -\frac{1}{2}\mathrm{tr}(A_0 ZZ^t) = -\frac{1}{2}\mathrm{tr}[(A_0 + \mathrm{diag}(x))ZZ^t] \tag{8.41}$$

for an arbitrary $x = \{x_1, \ldots, x_n\}$, $\sum_{i=1}^n x_i = 0$. This equality is true for any feasible partition Π of vertex set V. So, using (8.41), we obtain:

$$\rho(\Pi) \leq -\frac{1}{2} \max_{Y \in M_n^k(m)} [\mathrm{tr}[(A_0 + \mathrm{diag}(x))YY^t]] = -\frac{1}{2}\sum_{i=1}^{k} m_i \lambda_i(A_0 + \mathrm{diag}(x)),$$

under the condition

$$\sum_{i=1}^{n} x_i = 0.$$

Thus, we obtain the estimate of Donath and Hoffman (8.37):

$$\rho(\Pi) \le -\frac{1}{2} \min_{x} \sum_{i=1}^{k} \lambda_i m_i (A_i + \text{diag}(x))$$

subject to the constraint

$$\sum_{i=1}^{n} x_i = 0.$$

The problem (8.37) is a special case of minimization of the weighted sum of largest eigenvalues of parametric family of matrices. How to calculate the subgradients of the function $S_n^k(x, m) = \sum_{i=1}^{k} m_i \lambda_i (A_0 + diag(x))$?

Let us consider a more general case of the matrix functions:

$$S_n^k(x) = \sum_{i=1}^{k} \lambda_i(A(x)), \tag{8.42}$$

where $A(x) = A_0 + \sum_{i=1}^{m} x_i A_i$ (see (8.35). We shall use the described above procedure for calculating subgradient (gradient) of the function $S_n^k(A)$(see Chapter 6). Let $g_{S_n^k}(x) = \{g_{S_n^k}(x)^{(i)}\}_{i=1}^{m}$. One can use the following procedure for calculating $g_{S_n^k}(\overline{x})$:

(i) calculate $A(\overline{x})$;

(ii) solve the eigenvalue and eigenvector problem of the matrix $A(\overline{x}) \in S_n$;

(iii) let $\lambda_1(\overline{x}) \ge \lambda_2(\overline{x}) \ge \ldots \ge \lambda_k(\overline{x})$ be the k largest eigenvalues for matrix $A(\overline{x})$, $\{y_i(\overline{x})\}_{i=1}^{k}$ be the associated orthonormal system of eigenvectors. Then

$$g^i_{S_n^k}(\overline{x}) = (A_i, YY^{(t)}), \quad i = 1, \ldots k \tag{8.43}$$

where g^i is the i-th entry of the vector g.

One can give a more general formulation of the graph partitioning problem (GPP) by introducing the nonnegative weights of edges. As earlier we have a set of integers $m_1 \geq m_2 \geq \ldots \geq m_k$ with $\sum_{j=1}^{k} m_j = n$. Denote by m the k-vector made up of m_j-s.

Let $G = (V, E)$ be a complete graph with $|V| = n$ and each edge $(i, j), i < j$ has weight $w_{ij} \geq 0$. We want to partition a set of vertices V into k subsets such that the j-th subset has a prescribed cardinality m_j; and that sum of the weights of those edges whose endpoints are in different subsets is minimized. Let us denote this minimum value by $\pi_m(G)$. Let $W = \{a_{ij}\}$ be a matrix with $a_{ij} = w_{ij}, i < j$, and $a_{ii} = 0, i = 1, \ldots, n$.

Donath and Hoffman [DH 72] proved the following inequality

$$\pi_m(G) \geq U_m^W(G) = -\frac{1}{2} \min_x \sum_{j=1}^{k} m_j \lambda_j (W + \text{diag}x) \tag{8.44}$$

subject to constraint

$$\sum_{i=1}^{n} x_i = - \sum_{(i,j)\in E} w_{ij}. \tag{8.45}$$

For calculating the lower bounds (see (8.44), (8.45)) for $\pi_m(G)$ We can use subgradient-type methods for calculating the lower bounds (see (8.44), (8.45)) for $\pi_m(G)$ At each iteration of such methods we must solve the eigenvalues and eigenvectors problem for symmetric $n \times n$ matrix $W(\overline{x}) = W + \text{diag}\overline{x}$ and use the formula (8.43) for obtaining subgradients for function

$$\varphi_{k,m}^W(x) = \textstyle\sum_{j=1}^{k} m_j \lambda_j (W + \text{diag}x)$$
$$= \textstyle\sum_{r=1}^{k} (m_r - m_{r+1}) S_{n,r}(W + \text{diag}x); \; m_{k+1} = 0$$

The first practically efficient algorithm for obtaining the lower bounds for graph partitioning problems using the minimization of $S_n^k(x, m)$ (see (8.37)) was proposed in [CDW 75] by Cullum, Donath and Wolfe for equal m_i, $i = 1, \ldots, k$. They noticed that the problem (8.37) is a problem of nondifferentiable optimization and, moreover as a rule, the optimum point is the point of nondifferentiality of the minimized function. Their method is based on idea of smoothing the function $S_n^k(x, m)$ by using information not only about the k largest eigenvalues and corresponding eigenvectors, but also the information about other eigenvalues and eigenvectors, if they are little different from k-th eigenvalue. For smoothed function they proposed to use one of the feasible direction methods. The interior point methods are proposed for solution of the problems of the type (8.37) in series of works ([OW 92], [OW 93], [HY 91], [HY 92]).

We used for this purpose one of the modifications of the subgradient- type method with space dilation in the direction of difference of two successive subgradients (r-algorithm) and obtained good results in test experiments (see Table 8.7).

When n is even, $k = 2$, and $m_1 = m_2 = \frac{n}{2}$, the graph partitioning problem reduces to the so-called graph bisection problem that can be considered as max-cut problem with one additional constraint.

Consider weighted graph $G = (V, E)$ with even number of vertices: $|V| = n = 2s$. The bisection problem for G can be formulated as the following:

$$maximize f(y) = \frac{1}{4}\left(\sum_{(i,j) \in E} w_{ij} - \sum_{i,j \in E} w_{ij} y_i y_j \right) \tag{8.46}$$

subject to constraints:

$$y_i^2 - 1 = 0, \ \forall i \in V = \{1, \ldots, n\} \tag{8.47}$$

$$\sum_{i=1}^{n} y_i = 0 \tag{8.48}$$

(for feasible solution $\bar{y}$ the number of $\bar{y}_i$ having value 1 must be equal to number $\bar{y}_i$ having value -1). one may propose some ways for obtaining upper bounds for the problem (8.46)–(8.48):

(i) use the estimate of the type (8.36), (8.37) with $m_1 = m_2 = s$.

(ii) due to (8.48), $y_n = \sum_{i=1}^{n-1} y_i$. Set $\tilde{y} = \{y_1, \ldots, y_{n-1}\}$.

Consider the problem in variables $\tilde{y}$ which is equivalent to the problem 8.46–8.48:

$$maximize \overline{f}(\tilde{y}) = \frac{1}{4}\left(\sum_{(i,j) \in E} w_{ij} - \sum_{i,j \in E} [w_{ij} y_i y_j - \sum_{i=1}^{n-1} w_{in}(y_i y_n + y_n y_i)] \right. \tag{8.49}$$

subject to constraints:

$$y_i^2 - 1 = 0, \ \ \forall i \in \{1, \ldots, n-1\} \tag{8.50}$$

$$(\sum_{i=1}^{n-1} y_i)^2 - 1 = 0. \tag{8.51}$$

One can construct Lagrange function for the problem (8.49)–(8.51) in the form:

$$L(\tilde{y}, u) = -\overline{f}(\tilde{y}) + \sum_{i=1}^{n-1} u_i(y_i^2 - 1) + u_n[(\sum_{i=1}^{n-1} y_i)^2 - 1]$$

and consider marginal function

$$\psi(u) = \inf_{\tilde{y}} L(\tilde{y}, u).$$

$\psi(u)$ is a convex function, and $dom\psi \subseteq \overline{\Omega}^+$, where $\overline{\Omega} + (\Omega^+) = \{\overline{u} \in R^n : L(\tilde{y}, \overline{u})$ is convex (strictly convex) function in $\tilde{y}$ for fixed $\overline{u}\}$.

Since $L(\tilde{y}, u)$ has no linear in $\tilde{y}$ members

$$\psi(u) = -\frac{1}{4}\overline{W} + \sum_{i=1}^{n} u_i, \text{ if } u \in dom\psi$$

where $\overline{W} = \sum_{i,j=1}^{n} w_{ij}$.

Let $Q(u)$ be matrix of quadratic part of $L(x, u)$. Let $\psi^* = \sup_{u \in dom\psi} \psi(u)$

$$\psi^* = \sup_{u \in \Omega+} \psi(u) = \{\sup \psi(u) : Q(u) \succeq 0\}.$$

Thus upper dual bound φ^* for the maximum bisection problem (8.46)–(8.48) can be calculated by solving the following problem:

$$\text{find } \psi^* = \min[\sum_{i=1}^{n} u_i + \frac{1}{4}\overline{w}] \tag{8.52}$$

subject to constraint:

$$\lambda_n[Q(u)] \succeq 0. \tag{8.53}$$

The convex programming problem (8.52), (8.53) can be solved by application of $r(\alpha)$-algorithm for nonsmooth penalty function

$$\varphi_p(u) = \frac{1}{4}\overline{w} + \sum_{i=1}^{n} u_i + s\lambda_n^+[Q(u)],$$

where

$$\lambda_n^+(u) = \begin{cases} \lambda_n(u), & \text{if } \lambda_n(u) < 0; \\ 0, & \text{if } \lambda_n(u) \geq 0; \end{cases} \quad s > 0$$

is a penalty multiplier.

(iii) the bisection problem (8.46)–(8.48) can be approximately reduced by use of quadratic-type penalty $s(\sum u_i)^2$ for constraint (8.48) to usual max-cut problem with other weights:

$$w_{ij}^c = w_{ij} + s,$$

where s is the penalty multiplier.

The results of numerical tests of finding bounds for maxcut, graph partioning and maximum bisection problems for graphs from the tables 8.1, 8.2 are described in the tables 8.3–8.7. The corresponding programs were prepared by P.I. Stetsjuk (see Chapter 3 of this book).

Comments for the Tables. Parameters of $r(\alpha)$-algorithm are: $\alpha = 2$, $h_0^0 = 1.0$, $q_1 = 0.9$, $\overline{m} = 3$, $q_2 = 1.1$, $\varepsilon_x = 10^{-3}$, starting point is $u_0 = \{\overbrace{1,\ldots,1}^{(\frac{n}{2})times},\overbrace{0,\ldots,0}^{(\frac{n}{2})times}\}$.

We use the following notations:

iter – the number of r-algorithm iterations,

n_f – the number of calculations of value of function and subgradient,

φ_{up} is the upper estimate (see 8.8),

φ_{lo} is the lower heuristic estimate obtained as record value of feasible solutions,

G^{+1} denotes the vertex subset S in the best heuristic feasible solution,

$G^{-1} = V \backslash S$,

φ_r is the record value of the upper estimate,

φ_{r10} is the record value of the estimate obtained at the 10-th iteration,

m_i – is the number of vertices in i-th subset of vertices, $m_1 \geq m_2 \geq \ldots \geq m_k$,

ε_p – is an accuracy of fulfiling of constraint (8.45).

Table 8.1 Peterson graph (10 vertices and 15 edges)

e_{ij}	w_{ij}								
(i,j)	P_1	P_2	P_3	P_4	P_5	P_6	P_7	P_8	P_9
(1,2)	1	5	2	5	1	5	2	5	1
(1,5)	1	1	3	6	8	2	7	2	3
(1,10)	1	3	4	3	1	2	2	2	3
(2,3)	1	2	7	2	3	1	3	6	8
(2,9)	1	2	6	2	5	5	1	5	9
(3,4)	1	3	1	6	7	3	4	3	1
(3,8)	1	4	3	4	2	1	2	1	2
(4,5)	1	1	2	1	2	2	6	2	5
(4,7)	1	1	5	1	3	4	3	4	2
(5,6)	1	5	1	5	9	3	1	6	7
(6,8)	1	5	5	5	4	1	5	1	3
(6,9)	1	3	1	3	1	5	5	5	4
(7,9)	1	2	2	2	3	3	1	3	1
(7,10)	1	1	2	1	8	1	2	1	8
(8,10)	1	1	1	1	1	1	1	1	1

In Tables 8.3 – 8.4 the test results for maxcut problem are given. In Table 8.3 penalty parameter is n: for $P_1, \ldots, P_8$ $n = 10$; for $T_1, \ldots, T_5$ $n = 20$. The additional stopping criterion was used: $\varphi_{up} - \varphi_{lo} < 1$.

In the Tables 8.5 – 8.6 the results for bisection problem are given. Table 8.5 corresponds bisection model (8.46)–(8.48) with penalty coefficients $n, 5n$. Table 8.6 corresponds bisection model (8.49)–(8.51) (variant (ii)).

In the Tables 8.7 test results for graph k partitioning problem (8.44)–(8.45). For constraint (8.45) the penalty coefficient $s = 10$ was used.

One of the possible generalizations of the max-cut problem is the so-called MAX k-CUT problem: a graph $G = (V, E)$is given; to partition V into k subsets: for a partition $\mathcal{P} = \mathcal{P}_1, \mathcal{P}_2, \ldots, \mathcal{P}_k$ of V we let $|\mathcal{P}| = k$ and

$$w(\mathcal{P}) = \frac{1}{2} \sum_{\{(i,j)\in E: i\in\mathcal{P}_l, j\in calP_m, l\neq m\}} w_{ij}.$$

Table 8.2 Graph with 20 vertices and 36 edges (see [SB 95])

e_{ij}	w_{ij}						e_{ij}	w_{ij}				
(i,j)	T_1	T_2	T_3	T_4	T_5		(i,j)	T_1	T_2	T_3	T_4	T_5
(1, 2)	1	7	7	2	6		(10,12)	1	7	1	2	1
(1,18)	1	7	7	3	5		(10,20)	1	9	7	4	2
(2, 3)	1	7	7	5	4		(11,12)	1	1	7	5	5
(2,16)	1	7	9	4	3		(11,13)	1	7	1	6	5
(2,18)	1	7	9	1	3		(12,13)	1	7	7	1	4
(3, 4)	1	7	7	5	4		(13,14)	1	7	7	1	4
(4, 5)	1	1	7	5	5		(13,15)	1	9	1	1	1
(5, 6)	1	7	7	3	6		(13,19)	1	9	1	3	1
(5, 9)	1	1	1	2	1		(13,20)	1	7	9	4	2
(5,19)	1	7	9	2	3		(14,15)	1	9	7	2	2
(5,20)	1	7	9	4	4		(14,16)	1	7	1	3	5
(6, 7)	1	7	7	5	6		(15,16)	1	1	7	2	3
(6, 8)	1	1	1	1	3		(15,20)	1	9	9	4	2
(7, 8)	1	7	7	3	3		(16,17)	1	7	7	2	1
(8, 9)	1	1	7	3	2		(16,18)	1	9	9	3	3
(8,19)	1	7	7	2	1		(16,20)	1	9	7	2	2
(9,10)	1	7	7	1	2		(17,18)	1	7	7	1	1
(10,11)	1	1	7	1	1		(19,20)	1	9	9	3	3

Table 8.3

Test	*iter*(n_f)	φ_{up}	φ_{lo}	G^{-1}	G^{+1}
P_1	4(6)	12.71280	12.0	1,2,4,6,7,8	3,5,9,10
P_2	3(9)	35.70171	35.0	2,4,5,8,9,10	1,3,6,7
P_3	5(13)	40.91025	40.0	2,5,7,8,10	1,3,4,6,9
P_4	2(9)	43.81359	43.0	1,3,6,7	2,4,5,8,9,10
P_5	49(70)	52.76558	51.0	1,2,4,6,10	3,5,7,8,9
P_6	3(11)	34.97500	34.0	2,3,5,6,7,10	1,4,8,9
P_7	49(74)	40.13301	39.0	2,3,5,6,7,10	1,4,8,9
P_8	3(12)	41.75555	41.0	1,3,5,7,8,9	2,4,6,10
P_9	17(31)	52.99566	52.0	1,2,4,6,7,8	3,5,9,10
T_1	14(21)	29.93753	29.0	1,3,5,7,8,10, 13,15,16	2,4,6,9,11,12, 14,17,18,19,20
T_2	52(81)	192.08551	185.0	2,4,6,9,11,12, 14,17,18,19,20	1,3,5,7,8,10, 13,15,16
T_3	56(87)	189.71114	181.0	1,2,4,6,8,10, 12,14,17,20	3,5,7,9,11,13, 15,16,18,19
T_4	37(50)	88.07765	87.0	2,4,6,9,11,14, 17,18,19,20	1,3,5,7,8,10, 12,13,15,16
T_5	48(71)	91.79930	90.0	2,4,6,9,11,12,14, 15,17,18,19,20	1,3,5,7,8,10, 13,16

Table 8.4

Test	Penalty = 2n			Penalty = n		
	iter	φ_r	φ_{r10}	*iter*	φ_r	φ_{r10}
P_1	44	12.50275	13.27069	39	12.50047	12.62278
P_2	42	35.14717	35.80706	46	35.14696	35.19580
P_3	54	40.67444	43.33323	36	40.67427	40.70837
P_4	42	43.00449	43.36042	39	43.00424	43.04024
P_5	67	52.76651	58.06457	38	52.76566	52.92837
P_6	41	34.57070	36.21280	42	34.57037	34.61336
P_7	58	40.13305	45.88946	49	40.13301	40.31695
P_8	42	41.32763	42.25898	41	41.32630	41.39632
P_9	54	52.99161	56.56011	42	52.99118	53.08104
T_1	63	29.87477	31.36940	54	29.87397	30.13975
T_2	74	192.08707	212.81858	55	192.08549	192.44747
T_3	69	189.71210	208.24474	53	189.71115	190.39966
T_4	51	88.07844	92.16235	37	88.07765	88.28446
T_5	59	91.80358	101.15711	49	91.79925	92.70205

Then the problem MAX k-CUT is:

$$\textit{maximize } w(\mathcal{P})$$

subject to

$$|\mathcal{P}| = k \text{ (see [MW 97])}.$$

This problem has an important interpretation as the search for a ground in the anti-ferromagnetic k-state Potts model [Wel 93]. A similar problem faced in trying to colour graph [FJ 95](see also the maximum k-partite problem [NM 90]).

The formulation of MAX k-CUT problem is similar to graph partitioning problem: to divide the set V of a graph $G = (V, E)$ into k disjoint subsets V_i, $i = 1, \ldots, l$ with given cardinality number: m_i ($|V_i| = n_i$) with the aim to maximize sum of weights of edges connecting different subsets.

We know how to obtain the quadratic dual upper bounds for a graph partitioning problem. We can receive these bounds for all combinations of $\{m_i\}$such that $m_1 \geq m_2 \geq \ldots \geq m_l$, $;\sum_{i=1}^{k} m_i = |V|$, $m_i \geq 1$, $i = 1, \ldots, k$, and the minimum of them gives the upper bound for corresponding MAX k-CUT problem.

Table 8.5

Test	$iter(n_f)$	φ_{up}	φ_{lo}	G^{-1}	G^{+1}
P_1	41(46)	12.50045	11.0	1,2,5,7,8	3,4,6,9,10
P_2	7(16)	34.91024	34.0	2,3,6,7,10	1,4,5,8,9
P_3	2(8)	40.84215	40.0	2,5,7,8,10	1,3,4,6,9
P_4	44(68)	41.63351	40.0	1,3,6,7,10	2,4,5,8,9
P_5	35(52)	52.52589	51.0	1,2,4,6,7	3,5,8,9,10
P_6	32(43)	34.14082	33.0	1,4,5,8,9	2,3,6,7,10
P_7	17(32)	39.99753	39.0	1,2,4,8,9	3,5,6,7,10
P_8	11(21)	40.99564	40.0	2,4,6,8,10	1,3,5,7,9
P_9	7(21)	52.92904	52.0	1,2,4,6,7	3,5,8,9,10
T_1	5(9)	29.86129	29.0	2,4,6,9,11, 12,14,17,19,20	1,3,5,7,8,10, 13,15,16,18
T_2	49(78)	190.43549	185.0	2,4,6,9,11, 12,14,18,19,20	1,3,5,7,8,10, 13,15,16,17
T_3	50(73)	189.48545	181.0	1,2,4,6,8,10, 12,14,17,20	3,5,7,9,11,13, 15,16,18,19
T_4	13(27)	87.94716	87.0	2,4,6,9,11, 14,17,18,19,20	1,3,5,7,8,10, 12,13,15,16
T_5	54(83)	91.64590	90.0	2,4,6,9,11,12, 14,18,19,20	1,3,5,7,8,10, 13,15,16,17

Table 8.6

	Penalty = 2n			Penalty = n		
Test	*iter*	φ_r	φ_{r10}	*iter*	φ_r	φ_{r10}
P_1	52	12.50481	13.10771	43	12.50217	12.82710
P_2	40	34.65168	35.54201	34	34.65175	35.84211
P_3	57	40.31385	45.55675	46	40.31385	41.73139
P_4	56	41.62693	45.58307	47	41.62677	43.21611
P_5	43	52.53272	59.95141	39	52.52565	54.71852
P_6	52	34.14219	38.19486	48	34.13868	35.45551
P_7	46	39.97493	41.72035	50	39.97470	41.49242
P_8	43	40.88004	44.17209	46	40.87335	42.44732
P_9	53	52.73087	54.70961	42	52.73382	54.71606
T_1	77	29.40667	31.32538	54	29.40674	30.55627
T_2	93	190.43591	201.92446	48	190.43447	195.96313
T_3	71	189.49221	213.64884	59	189.48520	193.28992
T_4	93	87.77915	98.70850	50	87.77958	89.87398
T_5	63	91.64848	100.38115	61	91.64593	94.85597

Table 8.7

	$k = 2; m_1 = n/2$			$k = 3; m_1 = n/3; m_2 = n/3$		
Test	$iter(n_f)$	$\pi_m(G)$	ε_ρ	$iter(n_f)$	$\pi_m(G)$	ε_ρ
P_1	52(67)	4.9995	.129E-04	50(61)	5.9995	.142E-04
P_2	67(105)	7.5518	.376E-08	52(62)	10.360	.382E-08
P_3	54(84)	6.8379	.316E-07	49(65)	9.9230	.251E-07
P_4	57(85)	10.097	.653E-06	56(76)	12.985	.621E-07
P_5	60(97)	8.6894	.825E-07	53(72)	12.295	.369E-06
P_6	67(99)	8.5483	.599E-07	57(80)	10.966	.883E-08
P_7	64(105)	7.8972	.292E-07	50(67)	10.702	.101E-06
P_8	59(88)	10.516	.177E-07	48(59)	13.561	.665E-06
P_9	58(90)	8.0225	.162E-06	50(65)	12.304	.172E-07
T_1	71(100)	2.8568	.167E-06	57(68)	3.9373	.446E-07
T_2	66(109)	15.461	.216E-06	47(64)	24.820	.513E-07
T_3	69(110)	17.119	.627E-07	52(68)	23.463	.125E-06
T_4	62(89)	7.7247	.151E-06	56(71)	10.446	.463E-07
T_5	63(89)	7.6582	.117E-06	60(81)	10.166	.563E-07

For usual max-cut problem ($k = 2$) Goemans and Williamson [GW 94] proposed approximation algorithm for obtaining feasible primal solution based on the following idea: choose the random hyperplane through the origin and partition the vertex set V according to which side of the hyperplane corresponding vertices of the cube $\{-1, +1\}$ fall. This approximation algorithm seems to be good in average.

In [FJ 95] similar algorithm was proposed for obtaining primal feasible solution for k-cut problem: consider SDP_k problem of semidefinite programming:

$$SDP_k : maximize \frac{k-1}{k} \sum_{i<j} w_{ij}(1 - (v_i, v_j)) \tag{8.54}$$

$$\text{subject to } v_j \in S_n, \ \forall j \tag{8.55}$$

$$(v_i, v_j) \geq \frac{-1}{k-1}, \ \forall i \neq j \tag{8.56}$$

(here S_n is the symplex: $\sum_{j=1}^{n} v_j = 1, \ v_j \geq 0, \forall j$).

In [FJ 95] the following heuristic algorithm is described :

k-CUT:

Step 1 solve the problem SDP_k to obtain vectors $v_1, v_2, \ldots, v_n \in S_n$.

Step 2 choose k random vectors $z_1, z_2, \ldots, z_k$.

Step 3 partition V according to which of $z_1, z_2, \ldots, z_k$ is closest to each v_j, i.e., let $\mathcal{P} = \mathcal{P}_1, \mathcal{P}_2, \ldots, \mathcal{P}_k$ be defined by

$$\mathcal{P}_i = \{j : (v_j, z_i) \geq (v_j, z_{i'}), \forall i \neq i'\}, \text{ for } 1 \leq i \leq k.$$

The most natural way of choosing $z_1, z_2, \ldots, z_k$ is to choose them independently at random from S_n.

Described above algorithm satisfies the following

Theorem 111 *[FJ 95] There exists a sequence of constants α_k, $k \geq 2$ such that if $\mathcal{P}_k^*$ denotes the optimal partition in MAX k-CUT then*

$$E(w(\mathcal{P}_k)) \geq \alpha_k w(\mathcal{P}_k^*),$$

where the α_k *satisfy*

$$(i)\alpha_k > 1 - k^{-1};$$

$$(ii)\alpha_k - (1-k^{-1}) \sim 2k^{-2}\ln k;$$

$(iii)\alpha_2 \geq 0\cdot 878567,\ \ \alpha_3 \geq 0\cdot 800217,\ \ \alpha_4 \geq 0\cdot 850304,\ \ \alpha_5 \geq 0\cdot 874243,\ \ \alpha_{10} \geq 0\cdot 926642,\ \ \alpha_{100} \geq 0\cdot 990625.$

8.3 UPPER BOUNDS FOR THE SIZE OF K-PARTITE INDUCED SUBGRAPHS OF GIVEN GRAPH AND THEIR APPLICATIONS TO COLOURING PROBLEMS

The Lovasz's estimate $\nu(G,w)$ in the form $\nu_2(G,w)$ may be obtained by the following way. Let $G=(V,E)$ be a simple undirected graph without loops, $V=\{1,\ldots,n\}$. Denote by $x(S)=\{x_i(S)\}_{i=1}^n$ the indicator vector of the subset $S\subseteq V$. Let $W=\{\sqrt{w_iw_j}\}_{i,j\in V}$ be a weight matrix, $u=\{u_{ij}\}_{i,j\in E}$;

$$A_G(u)=\{u_{ij},(i,j)\in E;0,\text{ otherwise}\}$$

be a subclass of symmetric matrices $n\times n$. Consider an n-dimensional vector

$$w(S)=\left\{\frac{\sqrt{w_i}}{\sqrt{\sum_{j\in S}w_j}},i\in S;\ 0,\text{ otherwise}\right\}.$$

Note that $\|w(S)\|=1$, and for arbitrary u and stable set S

$$(W+A_G(U)w(S),w(S))=(Ww(S),w(S))=\frac{\sum_{i,j\in S}w_iw_j}{\sum_{k\in S}}=\sum_{i\in S}w_i. \quad (8.57)$$

Let $\nu_G(u)$ be the maximal eigenvalue of $M(u)=W+A_G(u)$. From (8.57) one can obtain:

$$\nu_W(u)=\lambda_1[W+A_G(u)]=\max_{\{z:\|z\|=1\}}(M(u)z,z)\geq(M(u)w(S),w(S))=\sum_{i\in S}w_i$$

for arbitrary u and a stable set S. So, $\nu_W(u)$ is an upper bound for the maximum weight of stable sets for every u.

Thus, $\nu(G,w) = \min_u \lambda_1(W + A_G(u)) \geq \alpha_W(G)$ is a well known estimate of Lovasz $\nu_2(G,w)$ (see Chapter 7).

In similar way one can obtain the upper bound for $\alpha^{(k)}(G)$, where $\alpha^{(k)}(G)$ is the size of the largest induced k-partite subgraph of $G = (V, E)$, i.e. the maximum number of nodes, that can be covered by k subsets of V so that no edge has both ends in any subset.

Let $S_1, S_2, \ldots, S_k$ be k pairwise nonintersecting stable subsets of V. Introduce the family of n-dimensional vectors $\{Y_r\}_{r=1}^k, Y_r = (y_r^{(1)}, \ldots, y_r^{(n)})$,

$$y_r^{(i)} = \begin{cases} \frac{1}{\sqrt{|S_i|}} & , i \in S_r, r = 1, \ldots, k; \\ 0 & , \text{otherwise} \end{cases}$$

Vectors Y_r, $r = 1, \ldots, k$, form an orthonormal system. Denote by $Y(k)$ the $n \times k$ matrix with columns Y_r, $r = 1, \ldots, k$.

Let $A_G(u)$, $u = \{u_{ij}\}_{(i,j)\in E}$, be the parametric family of symmetric matrices $n \times n$ with entries $a_{ij}(u)$

$$a_{ij} = \begin{cases} u_{ij} & , \text{if } (i,j) \in E \\ 1 & , \text{otherwise} \end{cases}$$

It is easy to verify that for arbitrary u, $\mathrm{tr}\big(A(u)Y(k)[Y(k)]^{(t)}\big) = \sum_{r=1}^k |S_r|$. Due to Theorem 84, $\max_{Y\in M_{n,k}} \mathrm{tr}[A(u)YY^{(t)}]$, where $M_{n,k}$ is the class of matrices $n \times k$ with orthonormal system of k columns, is equal to the sum of k largest eigenvalues of matrix $A(u)$.

So,

$$S_{n,k}(A(u)) = \sum_{r=1}^k \lambda_r[A(u)] \geq \sum_{r=1} |S_r|$$

for arbitrary u and any k-partite induced subgraph of G with stable sets $S_1, \ldots, S_k$. Thus,

$$\nu^{(k)}(G) = \min_u S_{n,k}(A(u)) \geq \alpha^{(k)}(G). \tag{8.58}$$

This upper bound for $\alpha^{(k)}(G)$ was first derived by Narasimhan and Manber in [NM 90]. When $k = 1$, one obtains the Lovasz's estimate $\nu_2(G)$ for the stable number $\alpha(G)$.

Calculation of $\nu^{(k)}(G)$ is reduced to the minimization of $|E|$-dimensional convex nonsmooth function of matrix parameters u, which enter into matrix affinely.

If $\alpha^{(k)}(G) = |V| = n$, it means that graph $G = (V, E)$ has a true vertex colouring in k colours. So if $\nu^{(k)}(G) < n$ the graph G cannot be coloured in k colours. Thus, we can use the upper bound $\nu^{(k)}(G)$ for studying some of the colouring problems. In [Gr 67] B.Grunbaum proposed the following conjecture:

The edges of any triangulation of arbitrary oriented surface in R^3 can be coloured by three colours (1,2,3) so, that edges, related to the same triangle in the given triangulation, have to be coloured by different colours.

This proposition is very hard to prove because in the partial case of spheric surface it is equivalent to the famous **four colour conjecture** (we know how many efforts were spent to prove the last conjecture and how disputable the proof elaborated the active use of computers was). One can reduce Grunbaum's conjecture to the perfect matching problem of a special kind.

Let T be a set of triangles of given triangulation of oriented surface S, let $\overline{E}$ be a set of edges in this triangulation. Since each edge is the boundary for two triangles, and any triangle contains three edges

$$\frac{3}{2}|T| = |\overline{E}|,$$

so the number of triangles is even.

Consider the graph $G(V, E)$, in which any node $i \in V$ corresponds to some triangle $t_i \in T$, $|V| = |T|$, and $(i, j) \in E$ if and only if two triangles t_i and t_j have the same boundary edge. It is clear, that graph $G(V, E)$ is homogeneous of degree three with even number of vertices. Grunbaum's conjecture is equivalent to the question: can the set of edges E be decomposed into three nonintersected perfect matching in graph $G(V, E)$.

Let us consider graph $G(\overline{E}, R)$ with vertex set $\{i\}$, $i = 1, \ldots, |\overline{E}|$, that is a line graph for graph $G(V, E)$; $(i, j) \in R$ if and only if the edges of i and j have one common end. The graph $G(\overline{E}, R)$ is the homogeneous graph of degree four with triangle cliques, associated with edges of triangles in triangulation. Each vertex $e \in \overline{E}$ belongs to pair of the triangle cliques in $G(\overline{E}, R)$. We have an equality:

$$|R| = 2|\overline{E}| = 3|T|. \tag{8.59}$$

Thus, Grunbaum's conjecture is reduced to the problem of existance of a true vertex colouring by three colours for graph $G(\overline{E}, R)$.

Consider a graph of the type $G(\overline{E}, R)$ with m triangle cliques, $3m$ vertices and $6m$ edges (due to (8.59)).

Let $n = 3m$. If Grunbaum's conjecture is true, the Narasimhan–Manber upper bound $\nu_3(G)$ must be equal to n for $G = G(\overline{E}, R)$.

Remember that in order to get the Narasimhan–Manber estimate $\nu^{(3)}(G)$ we must solve a complex problem of minimizing a convex nondifferentiable function:

$$\varphi_3(x) = \lambda_1(A(x)) + \lambda_2(A(x)) + \lambda_3(A(x)), \tag{8.60}$$

In a special case of graph $G(\overline{E}, R)$ there is additional complexity linked with nonuniqueness of the set of optimal solutions X^* for the problem of minimizing $\varphi_3(x)$. As a rule, the dimension of bounded convex set X^* is far from full, so in the neighborhood of X^* the surfaces of equal values of function $S_{n,3}(A(x))$ are very compressed in the directions, orthogonal to support of X^*. Such problems may be used as hard tests for procedures of nondifferentiable optimization.

Let $n = 3m$. Consider the incidence matrix $n \times n$ for graph $G_4 = (\overline{E}, R)$:

$$B = \{b_{ij}\}_{i,j=1}^{n}, \quad b_{ij} = \begin{cases} 1, & (i,j) \in R; \\ 0, & \text{otherwise.} \end{cases}$$

The corresponding quadratic form is

$$(Bx, x) = \sum_{i=1}^{2m} (x_{i1} + x_{i2} + x_{i3})^2 - 2\sum_{i=1}^{n} x_{ii}^2,$$

where $i1, i2, i3$ are the numbers of vertices (from $\overline{E}$) corresponding to i-th triangle clique.

Consider matrix

$$\begin{aligned} \overline{B} &= B + 2I_n \\ (\overline{B}x, x) &= \sum_{i=1}^{2m} (x_{i1} + x_{i2} + x_{i3})^2 \end{aligned} \tag{8.61}$$

The rank of positive semidefinite matrix $\overline{B}$ does not exceed $2m$ so the multiplicity of zero eigenvalues is not less than $n - 2m = m$. The maximum eigenvalue of the incidence matrix B is 4, the associated eigenvector is $J = (1, 1, \ldots 1)$. From (8.61) one can observe that the minimum eigenvalue of matrix B is -2 and its multiplicity is not less than m.

For graph $G_4 = (\overline{E}, R)$,

$$\nu^{(3)}(G) = \min_x S_{n,3}(A(x)), \tag{8.62}$$

where

$$A(x) = JJ^t - B + \sum_{(i,j)\in R} x_{ij}A_{ij}, \tag{8.63}$$

and A_{ij} are the matrices with two nonzero entries: $\alpha_{ij} = \alpha_{ji} = 1$. If $\overline{x}$ is such, that all $\overline{x}_{ij} = 1$, $(i,j) \in R$, then $A(\overline{x}) = JJ^t$, so $S_{n,3}(A(\overline{x})) = n$, and also $\lambda_1(A(\overline{x})) = n$; $\lambda_2(A(\overline{x})) = \lambda_3(A(\overline{x})) = 0$.

The structure of the subgradient set of the function $S_n^3(x) = S_{n,3}(A(\overline{x}))$ in the point $\overline{x}$ is very complex, because eigenvalue 0 has multiplicity $(n-1)$ and its eigenvector is any vector, orthogonal to J. Let $G_{S_n^3}(\overline{x})$ be the subgradient set of function S_n^3 in the point $\overline{x}$. If $0 \in G_{S_n^3}(\overline{x})$, then $\overline{x}$ is the point of minimum S_n^3, and $\nu^{(3)}(G) = n$. Otherwise, $\nu^{(3)}(G) < n$, and graph $G_4 = (\overline{E}, R)$ cannot have the vertex 3-colouring. But it is hard to define whether $0 \in G_{S_n^3}(\overline{x})$ or not because even if $0 \in G_{S_n^3}(\overline{x})$ it lies as a rule on the boundary of subgradient set.

Consider another initial point

$$\begin{aligned} \overline{z} &= \{\overline{z}_{ij}\},\ \text{with } \overline{z}_{ij} = -\frac{m}{2} + 1 \\ A(\overline{z}) &= JJ^t - \frac{m}{2}B \end{aligned}$$

It is easy to notice that $A(\overline{z})$ has eigenvector J, besides,

$$A(\overline{z})J = (n - 4 \cdot \frac{m}{2})J = mJ,$$

and there are no less than m mutually orthogonal vectors v_α which are eigenvectors of B corresponding to eigenvalue -2, so

$$A(\overline{z})v_\alpha = (-2) \cdot (-\frac{m}{2})v_\alpha = mv_\alpha.$$

All remaining eigenvectors of $A(\overline{z})$ are associated with eigenvalues less than m.

So, in the point $\overline{z}$ the three largest eigenvalues of the matrix $A(\overline{z})$ equal m, and their sum equals n. If we prove that $0 \in G_{S_n^3}(\overline{z})$, then $\nu^{(3)}(G_4)$ equals n. Otherwise, $\nu^{(3)}(G_4) < n$, and graph $G_4 = (V, E)$ cannot be coloured by three colours. To verify this fact it is enough to obtain by using, for example, r-algorithm with initial point $\overline{z}$ and small initial step $h > 0$ the value of $S_{n,3}(A(x))$ that is less than n.

My student O.A. Berezovskii wrote a program for obtaining Narasimhan–Manber's bounds $\nu_3(G)$. The experiments were carried out with graphs of the type $G(\overline{E}, R)$ and with modified graphs $G(\overline{E}, R')$, where $R' = R \cup D$, D is some nonempty set of additional edges. In all experiments with graphs $G(\overline{E}, R)$, satisfying Grunbaum's conjecture we, naturally, obtained the estimate equal to $|\overline{E}|$. For all graphs of the type $G(\overline{E}, R \cup D)$ with such D, that $G(\overline{E}, R \cup D)$ cannot be coloured by 3 colours, $\nu_3(G)$ was less than $|\overline{E}|$.

Although we had limited experience in solving the above mentioned problems we want to propose the following conjecture which would be very interesting to verify.

Conjecture A. *Let $G(\overline{E}, R)$ be the homogeneous graph of degree four, which corresponds to the triangulation of the oriented surface, i.e. each vertex $\overline{e} \in \overline{E}$ of this graph lies on the some edge of this triangulation and is incident to the four vertices which lie on the other edges of two triangles with common edge e. Let us join some pairs of nonadjacent vertices from $\overline{E}$ by additional set of edges D and consider the constructed this way graph $G(\overline{E}, R \cup D)$. If graph $G(\overline{E}, R \cup D)$ has no vertex colouring by three colours then Narasimhan–Manber estimate $\nu_3(G) < |\overline{E}|$ for graph $G(\overline{E}, R \cup D)$.*

If Conjecture A is true, one can use it to obtain the vertex colouring of graph $G(\overline{E}, R)$ by three colours if such colouring exists. Indeed, let us colour vertices of any triangle of $G(\overline{E}, R)$ by colours 1, 2, 3. We can extend this colouring by many ways. For example, we colour next vertex α by 1. Then we may join this vertex by edges with vertices which have colour 2 and 3. After this operation we obtain new graph G_1 with additional edges. Let us calculate $\nu_3(G_1)$. If $\nu_3(G_1) = |\overline{E}|$, we can continue our colouring. If $\nu_3(G_1) < |\overline{E}|$ we must change the colouring of α, obtain new graph G_1' with additional edges, calculate $\nu_3(G_1')$, etc. By this way one can receive vertex colouring of $G(\overline{E}, R)$ using mentioned above techique sequentially "vertex by vertex".

In partial case of spheric triangulation if Conjecture A is true we would obtain the noncombinatorial algorithm for colouring of any planar graph by 4 colours.

9

GLOBAL MINIMIZATION OF POLYNOMIAL FUNCTIONS AND 17-TH HILBERT PROBLEM

9.1 THE PROBLEM OF REPRESENTATION OF NONNEGATIVE POLYNOMIALS AS SUM OF SQUARES OF POLYNOMIALS

Consider one of the simplest on formulation problem of nonconvex polynomial programming: find infinum $\rho^*(A)$ of

$$K(x) = (Ax, x) \text{ on the cone: } x \geq 0, x = (x_1, \ldots, x_n) \in E^n, \tag{9.1}$$

where A is an arbitrary symmetric matrix $n \times n$, $A = \{a_{ij}\}_{i,j=1}^n$. $\rho^*(A)$ may accept only two values: 0 or $-\infty$. If $\rho^*(A) = 0$, then $x^* = 0$ is the optimal point of the problem (9.1). It is known that class of the problems: "$\rho^*(A) = 0$?" for arbitrary integer (rational) symmetric matrices A of arbitrary size $n \times n$ is NP-complete (see [GJ 79]. Note that for convex $K(x)$ the answer is trivial: $\rho^*(A) = 0$.

Consider the substitutions:

$$x_i = y_i^2, \ i = 1, \ldots, n.$$

Then the problem (9.1) can be represented as an unconstrained minimization problem for homogeneous polynomials $P_A(y)$ of total degree 4 in n variables:

$$\rho^*(A) \quad = \quad \inf P_A(y), \text{ where}$$

$$P_A(y) = \sum_{i,j=1}^{n} a_{ij} y_i^2 y_j^2. \tag{9.2}$$

One of the ways of proving that $\rho^*(A) = 0$ (if it is correct) is to decompose nonnegative polynomial $P_A(y)$ into a sum of squares of other polynomials. Unfortunately, not all nonnegative polynomials of some variables have such decomposition.

As young David Hilbert showed in 1888 [Hi 88] there are nonnegative polynomial functions of several variables which cannot be represented as sum of squares of polynomials. Moreover, when he studied the homogeneous nonnegative polynomials (polynomial forms) $F(x)$ of even total degree $r = 2k$ of n variables and asked: "is it true that if $F(x) \geq 0$ for all x, then $F(x) = \sum p_i^2(x)$, where p_i are real polynomials?" he showed: the answer is fully positive only in such cases:

1) n is an arbitrary number, $r = 2$. This case corresponds to positive semidefinite quadratic forms.

2) $n = 2, r = 2k$ is an arbitrary even number. In this case $F = F(x_1, x_2)$ can be represented as a sum of no more than two squares of forms. This fact is the consequence of the following proposition.

Theorem 112 *Every nonnegative polynomial with elder coefficient d^2 of even degree of one variable $P_{2k}(x_1)$ can be represented as sum of squares of two polynomials.*

Proof. All real roots of such polynomial c_β, if they exist, cannot be simple and have even multiplicity, all complex roots, if they exist, can be joint in pairs of the form $a_\alpha + ib_\alpha$ and $a_\alpha - ib_\alpha$, where $i^2 = -1$, a_α and b_α are real numbers. Let

$$f_1(x_1) = \prod_\alpha [x_1 - (a_\alpha + ib_\alpha)] = P(x_1) + iQ(x_1)$$

$$f_2(x_1) = \prod_\alpha [x_1 - (a_\alpha - ib_\alpha)] = P(x_1) - iQ(x_1)$$

Then

$$\begin{aligned} P_{2k}(x_1) &= d^2 f_1(x_1) \cdot f_2(x_1) \cdot \prod_\beta (x_1 - c_\beta)^{2\rho(\beta)} = \\ &= d^2 \prod_\beta (x_1 - c_\beta)^{2\rho(\beta)} [(P(x_1))^2 + (Q(x_1))^2] \end{aligned} \tag{9.3}$$

(here $2\rho(\beta)$ denotes the multiplicity of real root c_β), so

$$P_{2k}(x_1) = [d(x - c_\beta)^{\rho(\beta)} P(x_1)]^2 + [d(x - c_\beta)^{\rho(\beta)} Q(x_1)]^2,$$

i.e. $P_{2k}(x_1)$ is represented as sum of two squares of polynomials (or one square if all the roots are real).□

The previous proof is not useful for finding a global minimum of $P_{2k}(x_1)$, because it includes the finding of all roots of polynomial $P_{2k}(x_1)$. In general, the last procedure is more complex than finding the global minimum of polynomials.

The nonnegative homogeneous form of two variables $F(x_1, x_2)$ of total degree $r = 2k$ for $x = (x_1, x_2) \neq 0$ can be represented as $x_2^{2k} P(z)$, where $P(z)$ is a nonnegative polynomial of $z = \frac{x_1}{x_2}$. Due to Theorem 112 we obtain for $P(z)$ the decomposition in no more than two squares of polynomials of z, and after the substitution $z = \frac{x_1}{x_2}$ we obtain the corresponding decomposition of $F(x_1, x_2)$ into sum of squares of no more than two polynomials.

Let x_1^* be a point of global minimum of the polynomial $P_{2k}(x_1)$ given by (9.3), and $P^* = P_{2k}(x_1^*)$. Then the polynomial $\overline{P}_{2k}(x_1) = P_{2k}(x_1) - P^*$ is nonnegative, and the point x_1^* is a root of the equation $\overline{P}_{2k}(x_1) = 0$ of even multiplicity, so

$$\overline{P}_{2k}(x_1) = (x_1 - x_1^*)^2 \cdot P_{2(k-1)}(x_1) \tag{9.4}$$

where $P_{2(k-1)}(x_1)$ is a nonnegative polynomial of even degree $2(k-1)$. Now we can give a new proof of the fact that each nonnegative polynomial of even degree $2k$ can be represented as sum of squares of polynomials of degree not exceeding k. The proof can be performed by induction in k. For $k = 1$ a nonnegative polynomial of degree $2k$ has the form:

$$P_2(x_1) = a_2(x_1 - x_1^*)^2 + b,$$

where x_1^* is a global minimum, $a_2 > 0$; $b \geq 0$, so our statement is true.

Let the proposition be true for $r = k - 1$.

Consider a polynomial $P_{2k}(x_1)$ with global minimum value $P^* \geq 0$.

Due to (9.4)

$$P_{2k}(x_1) = \overline{P}_{2k}(x_1) + P^* = (x_1 - x_1^*)^2 P_{2(k-1)}(x_1) + P^*.$$

Since, by induction proposition, $P_{2(k-1)}(x_1)$ can be decomposed into a sum of squares of polynomials, one can see that $P_{2k}(x_1)$ also can be decomposed into a sum of squares of polynomials. (One may compare this proof with that of Theorem 112.)

3) D. Hilbert proved that an arbitrary nonnegative biquadratic form ($r = 4$) of three variables can be represented as sum of no more than three squares of quadratic forms ([Hi 88], [Sw 95]). His proof is rather complex, so, we give more elementary demonstration of the fact, that biquadratic nonnegative form in 3 variables can be decomposed into several squares of quadratic forms.

Let $C^+(3,4)$ be the cone of all nonnegative homogeneous polynomial forms in 3 variables of total degree $m = 4$ imbedded in 15-dimensional linear space $H(3,4)$ of arbitrary homogeneous pólynomials in 3 variables of degree 4. Let $x = (x_1, x_2, x_3)$.

Each polynomial $P(\cdot) \in C^+(3,4)$ can be represented in canonical form as

$$P(x) = P(x_1, x_2, x_3) = a_1x_1^4 + a_2x_2^4 + a_3x_3^4 + \\ b_{12}x_1^3x_2 + b_{21}x_2^3x_1 + b_{13}x_1^3x_3 + b_{31}x_3^3x_1 + b_{23}x_2^3x_3 + b_{32}x_3^3x_1 + \\ c_1x_1^2x_2x_3 + c_2x_1x_2^2x_3 + c_3x_1x_2x_3^2 + d_{12}x_1^2x_2^2 + d_{13}x_1^2x_3^2 + d_{23}x_2^2x_3^2. \quad (9.5)$$

Consider $\overline{P}(\cdot)$ that belongs to the boundary of the cone $C^+(3,4)$. It means that there exists a vector $\overline{x} \neq 0$ such that $\overline{P}(\overline{x}) = 0$ and for all $x \in R^3$ $\overline{P}(x) \geq 0$.

Definition 23 *We say that a triple of linear independent vectors $\{\eta^{(1)}, \eta^{(2)}, \eta^{(3)}\}$ forms P-zero basis if $\eta^{(1)}, \eta^{(2)}, \eta^{(3)} \in R^3$ and $P(\eta^{(i)}) = 0,\ i = 1,2,3$.*

Consider the vectors $e^{(1)} = (1,0,0)$, $e^{(2)} = (0,1,0)$, $e^{(3)} = (0,0,1)$, and let

$$e^{(i)} = \sum_{j=1}^{3} \beta_{ij}\eta_i^j, \quad i = 1,2,3;$$

$$B = \{\beta_{ij}\}_{i,j=1}^3, \quad x = B\overline{x}, \quad \overline{x} = \{\overline{x}_1, \overline{x}_2, \overline{x}_3\}.$$

Let $\overline{P}(\overline{x}) = P(B\overline{x})$.

Note, that if $P(x)$ is a nonnegative homogeneous polynomial in 3 variables of 4-th degree then $\overline{P}(\overline{x})$ is also a nonnegative homogeneous polynomial function in 3 variables of degree 4.

Let the coefficients of polynomial $\overline{P}(\overline{x})$ have the same designation as the corresponding coefficients of the polynomial $P(x)$ (see (9.5)) but with bar.

Theorem 113 *If polynomial $P \in C^{+}(3,4)$ has a P-zero basis $\{\eta^{(1)}, \eta^{(2)}, \eta^{(3)}\}$, then it can be represented as sum of squares of real polynomials.*

Proof. Note, that η^i, $i = 1,2,3$ are minimum points of polynomial $P(x)$ therefore $\overline{e}_1 = (1,0,0)$, $\overline{e}_2 = (0,1,0)$, $\overline{e}_3 = (0,0,1)$ are minimim points of $\overline{P}(\overline{x})$ Since $\overline{P}(\overline{e}_i) = 0$, $\frac{\partial P}{\partial \overline{x}_j}(\overline{e}_i) = 0$, $i = 1,2,3$, $j = 1,2,3$, $\overline{a}_1 = \overline{a}_2 = \overline{a}_3 = 0; \overline{b}_{ij} = 0$ for all i, j, $i \neq j$.

So, $\overline{P}(\overline{x})$ has the following form:

$$\overline{P}(\overline{x}) = \overline{c}_1\overline{x}_1^2\overline{x}_2\overline{x}_3 + \overline{c}_2\overline{x}_1\overline{x}_2^2\overline{x}_3 + \overline{c}_3\overline{x}_1\overline{x}_2\overline{x}_3^2 + \overline{d}_{12}\overline{x}_1^2\overline{x}_2^2 + \overline{d}_{13}\overline{x}_1^2\overline{x}_3^2 + \overline{d}_{23}\overline{x}_2^2\overline{x}_3^2.$$

Let us make the following substitutions:

$$y_1 = \overline{x}_1\overline{x}_2, \quad y_2 = \overline{x}_1\overline{x}_3, \quad y_3 = \overline{x}_2\overline{x}_3.$$

Then $\overline{P}(\overline{x}) = K(y)$, where $K(y) = \overline{c}_1 y_1 y_2 + \overline{c}_2 y_1 y_3 + \overline{c}_3 y_2 y_3 + \overline{d}_{12} y_1^2 + \overline{d}_{13} y_2^2 + \overline{d}_{23} y_3^2$.

$K(y)$ is a quadratic form in variables $y = (y_1, y_2, y_3)$. Since, if $K(y) \not\equiv 0$, there exists $\overline{y} \in R^3$ such that $K(\overline{y}) > 0$. $K(y)$ can be represented in one of the form:

$$(i) \quad K(y) = \sum_{i=1}^{3}[l_i(y_1, y_2, y_3)]^2 \tag{9.6}$$

$$(ii) \quad K(y) = [l_1(y_1, y_2, y_3)]^2 - [l_{ij}(y_i, y_j)]^2 \tag{9.7}$$

where l_{ij} is the nonzero linear form in two variables y_i, y_j, $i, j \in \{1,2,3\}$, $l_i(y_1, y_2, y_3)$ are the linear forms in variables y_1, y_2, y_3, besides in the case (ii) $l_1(\cdot,\cdot,\cdot)$ contains y_k, $k \neq i, k \neq j$ with nonzero coefficient.

One can easy show, that in the case (ii) there exists $\overline{x}^{(0)} \in R^3$ such, that for $y^{(0)} = (\overline{x}_1^{(0)}\overline{x}_2^{(0)}, \overline{x}_1^{(0)}\overline{x}_3^{(0)}, \overline{x}_2^{(0)}\overline{x}_3^{(0)})$ $K(y^{(0)}) < 0$. This fact contradicts to the supposition that $\overline{P}(\overline{x}) \geq 0$ for all $\overline{x} \in R^3$. So, the case (i) only can be realized, and $\overline{P}(\overline{x})$ can be represented as sum of 3 squares of quadratic forms:

$$\overline{P}(\overline{x}) = \sum_{i=1}^{3}[a_{i1}x_1x_2 + a_{i2}x_1x_3 + a_{i3}x_2x_3]^2.$$

Thus, Theorem 113 is proven.

Lemma 11 *Let $P(x)$, $x \in R^n$, be a nonnegative homogeneous polynomial of total degree 4 and let $Z(P)$ be set of zero points of polynomial P:*

$$Z(P) = \{x \in R^n : \; P(x) = 0\}$$

and $LZ(P)$ be linear envelope of $Z(P)$.

If dimension of $LZ(P)$ $dim\, LZ(P) < n$, then there exists a polynomial $\overline{P}(x)$ such that $1 \leq \overline{P}(x) \leq P(x)$ for all $x \in R^n$, and

$$dim\, LZ(\overline{P}) \geq dim\, LZ(P) + 1.$$

Proof. Since $dim\, LZ(P) < n$, one can find nonzero $\eta \in R^n$ orthogonal to $LM(P)$.

Consider $x^* \in \arg\max_{\{x:(\eta,x)=1\}} P(x)$ and let $P(x^*) = c > 0$. The polynomial

$$\overline{P}(x) = P(x) - c(\eta, x)^4$$

satisfies the inequalities

$$0 \leq \overline{P}(x) \leq P(x) \text{ for all } x \in R^n$$

and $x^* \notin LM(P)$, $\overline{P}(x^*) = 0$. Hence, $dim\, LM(\overline{P}) \geq dim\, LM(P) + 1$. Lemma is proven.

Corollary 6 *Every nonnegative homogeneous polynomial of degree 4 in 3 variables $P(x)$ can be represented as sum of a polynomial $\overline{P}(x) \in C^+(3,4)$ with $\overline{P}$-zero basis and sum of squares of polynomials.*

Proof. Set $n = 3$ and use the previous lemma.

Theorem 114 *Each polynomial $P \in C^+(3,4)$ can be represented as sum of squares of polynomials.*

Proof. Since an arbitrary polynomial $\overline{P}(x)$ with $\overline{P}$-zero basis can be represented as sum of squares Theorem 114 can be immediatly deduced from Corollary 6.

Hilbert proved also the main negative result: in all other possible combinations n and $r = 2k$ (besides $n = 2$ or $r = 2; r = 4, n = 3$) there exist nonnegative forms that cannot be represented as sum of squares of other forms [Hi 88].

For example, polynomial in 4 variables of the fourth total degree

$$P_1(x, y, z, w) = w^4 + x^2y^2 + x^2z^2 + y^2z^2 - 4xyzw$$

and polynomial in three variables of the sixth total degree

$$P_2(x, y, z) = z^6 + x^4y^2 + x^2y^4 - 3x^2y^2z^2$$

are both nonnegative, but they cannot be represented as sum of squares of polynomials.

The left-hand side of some known polynomial inequalities $P(x) \geq 0$ can be expressed as sum of squares, and such representation makes them to be obvious.

For example consider the known inequality: for a positive vector $a = \{a_1, \ldots, a_n\}$,

$$F_n(a) = \sum_{i=1}^{n} a_i^n - na_1 \ldots a_n \geq 0$$

(equivalent formulation: arithmetic average is not less than geometric average).

For $n = 6$ we have:

$$\begin{aligned}
&a_1^6 + a_2^6 + a_3^6 + a_4^6 + a_5^6 + a_6^6 - 6a_1a_2a_3a_4a_5a_6 = \\
&= \frac{1}{2}(a_1^2 + a_2^2 + a_3^2)\{(a_2^2 - a_3^2)^2 + (a_3^2 - a_1^2)^2 + (a_1^2 - a_2^2)^2\} + \\
&+ \frac{1}{2}(a_1^2 + a_2^2 + a_3^2)\{(a_5^2 - a_6^2)^2 + (a_6^2 - a_4^2)^2 + (a_4^2 - a_5^2)^2\} + \\
&+ 3(a_1a_2a_3 - a_4a_5a_6)^2
\end{aligned}$$

so $F_6(a)$ is represented as a sum of 9 squares (see [HLP 62]).

Let us remember that in 1900 at the II-nd International Congress of Mathematicians in his famous report "Mathematical problems" (see [Hi 35]) David Hilbert proposed 22 interesting directions of development (or points of growth) for the mathematics of the XX-th century. Great Hilbert wanted to emphasize the riches of ideas, diversity and unity of mathematics. His foresees was completely come true. We shall consider the 17-th Hilbert problem, which was unexpectedly found to be linked tightly with the subject of our book.

D. Hilbert asked: can an arbitrary nonnegative rational function (the quotient of two polynomials) be represented as sum of squares of rational functions? E. Artin in his famous article [Art 27] gave a complete positive answer on this question in generalized algebraic form.

Let K be an algebraic field. It is called a **formally real** if from equality the $\sum a_i^2 = 0$, $a_i \in K$, it follows that $a_i = 0$ for all i. **Ordering** of the field K is the determination of the subset $P \subset K$ (positive in given ordering), which has the following properties:

a) P is closed relatively to summation and multiplication;

b) for each element $a \in K$, $a \neq 0$, either $a \in P$ or $-a \in P$;

c) $0 \notin P$.

It is naturally to write $a > b$ if $a - b \in P$. E.Artin proved that sums of squares in formally real fields are exactly those elements that are positive for arbitrary ordering of K.

Let $k(x_1, \ldots, x_n)$ be a field of rational real functions on real field R. This field is formally real. As a partial case of his general result Artin proved:

Theorem 115 (Artin) *If $f \in k(x_1, \ldots, x_n)$ is a* **nonnegative** *function, then $f = \sum_i g_i^2$, $g_i \in k(x_1, \ldots, x_n)$ (see [Art 27]).*

Thus, Artin solved the 17-th Hilbert problem in a generalized form.

From the Artin theorem one can formulate the following.

Theorem 116 *If a real polynomial $P(x) \geq 0$ for all $x \in R^n$ then there exists a strictly positive polynomial $H(x)$ such that $H(x)P(x)$ can be represented as sum of squares of polynomials.*

Now we consider a form that is strictly positive for a positive x

Theorem 117 (Polya) [Poly 28]. *If a form $F(x_1, \ldots, x_n)$ is strictly positive for $x \geq 0$, $\sum_{i=1}^n x_i > 0$, then it can be represented as quotient $F = \frac{G}{H}$ where G and H are forms with positive coefficients. Particullarly, one can take*

$$H = (x_1 + x_2 + \ldots + x_m)^N.$$

Polya's theorem gives us a good illustration for Artin's theorem. Let us put $x_i = y_i^2$, $i = 1, \ldots, n$. Then $\varphi(y) = F(y_1^2, \ldots, y_n^2)$ and $\varphi(y) > 0$ for all $y \neq 0$. So

$$\varphi(y) = \frac{G(y_1^2, \ldots, y_n^2)}{(\sum y_i^2)^N},$$

where G is the form with nonnegative coefficients. So, $\varphi(y)$ can be represented as quotient of two forms that both are sums of squares. We obtain the partial case of Artin's theorem.

If we set $x_n = 1 - x_1 - \ldots - x_{n-1}$ one can deduce from Polya's theorem the following

Theorem 118 *If an arbitrary (nonhomogeneous) polynomial $P(x_1, \ldots, x_{n-1})$ is positive in a simplex region*

$$x_1, \ldots, x_{n-1} \geq 0,\ x_1 + \ldots + x_{n-1} \leq 1$$

then $P(x)$ can be represented in the form

$$P(x) = \sum_{\alpha} c_\alpha x_1^{a_1} \ldots x_{n-1}^{a_{n-1}} (1 - x_1 - \ldots - x_{n-1})^{a_m}$$

where a_i, $i = 1, \ldots, n$ are integer nonnegative numbers, c_α are positive, (α is an index of the corresponding monomial, $\alpha = \{a_1, \ldots, a_{n-1}\}$)(see [HLP 62]).

The Polya theorem gives us the regular procedure for solving the question: is the form $F(x)$ strictly positive for $x \geq 0$? We may multiply $F(x)$ by $s(x) = \sum_{i=1}^n x_i$ repeatedly and obtain $F_1(x) = F(x)s(x)$, $F_2(x) = F_1(x)s(x)$ and so on. If $F(x) > 0$ for $x > 0$ we obtain after a finite number of steps the form $F_k(x)$ with nonnegative coefficients. But how many steps must we perform? In [HLP 62] the analysis of this problem is given for the following polynomials: $F_\varepsilon^{(n)}(x) = \sum_{i=1}^n x_i^n - (n - \varepsilon) \prod_{j=1}^n x_j$. This polynomial satisfied the conditions of Polya's theorem, and if $N \geq \frac{n^2(n-1)}{2\varepsilon}, \varepsilon > 0$, it guarantees, that the form $G_{\varepsilon,N}(x) = F_\varepsilon^{(n)}(x) \cdot [H(x)]^N$ has only positive coefficients. This estimate shows that, if ε tends to zero, N may tend to infinity, so the above mentioned method of verifying that $F(x) > 0$ if $x > 0$ is not so simple as it seems at first sight. In general the problem of checking that a polynomial function $F(x) \geq 0$ for all $x \in R^n$ is multiextremal and very hard.

9.2 CONSTRUCTING OF QUADRATIC PROBLEM WITH SUPERFLUOUS CONSTRAINTS FOR GLOBAL MINIMIZATION OF ONE-DIMENSIONAL POLYNOMIALS

Let a bounded-from-below polynomial $P(x_1, x_2, \ldots, x_n)$ be given and let P^* be the value of the polynomial at the global minimum point.

By introducing new variables and making use of quadratic substitutions of the form: $x_i^2 = y_i; x_j x_k = z_{jk}$ and so forth we can reduce the minimization problem for the polynomial $P(x_1, \ldots, x_n)$ to a quadratic extremal problem with constraints in the form of equalities. The direct application of the dual estimate technique to this quadratic problem results in nontrivial estimates only in rare cases. But we may modify the quadratic problem by generating new quadratic equalities of the quadratic problem variables, which are consequences of the previous ones. In many cases using the Lagrange function technique to the modified quadratic problems may essentially improve dual estimates. Unexpectedly these facts were found to be in tight connection with the decomposition of nonnegative polynomials into a sum of squares of other polynomials.

To begin with let us consider one-dimensional case.

Let $P_{2k}(x_1)$ be a polynomial in one variable of even degree $2k$ with the leading coefficient equal to 1:

$$P_{2k}(x_1) = x_1^{2k} + a_{2k-1}x_1^{2k-1} + \ldots + a_2 x_1^2 + a_1 x_1 + a_0$$

Such polynomials are bounded from below. Let x_1^* be a point (one of the points) of global minimum, and $P_{2k}(x_1^*) = P^*$ be the minimal value of P_{2k}. We want to find P^* and x_1^* by reducing the unconstrained problem of minimiziation of $P_{2k}(x_1)$ to a quadratic-type problem of nonlinear programming.

Example 1 $k = 2$. Minimize

$$P_4(x_1) = x_1^4 + a_2 x_1^2 + a_1 x_1 \tag{9.8}$$

(for simplicity we set $a_3 = 0$).

Let us introduce the substitution: $x_2 = x_1^2$ or $x_1^2 - x_2 = 0$. We obtain the following quadratic type optimization problem:

$$\text{find } P^* = \min(x_2^2 + a_2 x_1^2 + a_1 x_1) \tag{9.9}$$

subject to the constraint:

$$x_1^2 - x_2 = 0 \tag{9.10}$$

which is equivalent to the problem of minimizing $P_4(x_1)$. We introduce a Lagrange multiplier u and obtain the Lagrange function:

$$\begin{aligned} L(x,u) &= x_2^2 + a_2 x_1^2 + a_1 x_1 + u(x_1^2 - x_2) = \\ &= x_2^2 + (a_2 + u)x_1^2 + a_1 x_1 - u x_2 = \\ &= K_4^u(x) + l_4(x), \end{aligned}$$

where $x = \{x_1, x_2\}$, $K_4^u(x)$ and l_4 are respectevly quadratic and linear forms of the Lagrange function considered as a quadratic function in x which depend on parameter u. Introduce the marginal function:

$$\psi(u) = \inf_x L(x,u) \tag{9.11}$$

For $\overline{u} > -a_2$ the function $L_{\overline{u}}(x) = L(x, \overline{u})$ is strictly convex in x, and optimal point $\{x_1(\overline{u}), x_2(\overline{u})\}$ in (9.11) possesses the values:

$$x_1(\overline{u}) = \frac{-a_1}{2(a_2 + \overline{u})}; \; x_2(\overline{u}) = \frac{\overline{u}}{2},$$

and

$$\psi(u) = L(x(u), u) = -\frac{u^2}{4} - \frac{1}{4} \cdot \frac{a_1^2}{a_2 + u} = -\frac{1}{4}\left(u^2 + \frac{a_1^2}{a_2 + u}\right) \tag{9.12}$$

$\psi(u)$ is a concave function. If $a_1 \neq 0$ $\psi(u)$ tends to $-\infty$ as u tends to a_2 from interior points of dom $\psi = (-a_2; +\infty)$. In this case there exists $u^* \in (-a_2, +\infty)$ such that $\psi^* = \psi(u^*) = \sup_{u \in \mathrm{dom}\psi} \psi(u)$. $u^* \in$ int dom ψ, so $\frac{d\psi(u^*)}{du} = 0; /; x_2(u^*) = x_1^2(u^*)$, $x_1^* = x_1(u^*)$ is a unique global minimum of $P_4(x_1)$, and $P_4(x_1^*) = \psi^*$, i.e. the dual lower bound of the problem (9.9)–(9.10) is exact.

If $a_1 = 0$, the point $u = -a_2$ also belongs to dom ψ.

$$\psi(u^*) = \sup_{\{u: u \geq -a_2\}} \psi(u) = \begin{cases} 0 & \text{, if } a_2 \geq 0 \\ -\frac{a_2^2}{4} = \psi(-a_2) & \text{, if } a_2 < 0 \end{cases}$$

In the first case $x_2(u^*) = 0$, which corresponds to the global minimum $x_1^* = 0$; besides, $\psi^* = P^* = 0$.

In the second case we have two points of global minimum, corresponding to

$$x_2(u^*) = -\frac{a_2}{2} : \; x_1^* = \pm\sqrt{-\frac{a_2}{2}}.$$

We see that $\psi(u^*) = P_4(x_1^*) = -\frac{a_2^2}{4} = P^*$. So in all cases, $\psi^* = \inf_x P_4(x_1)$.

When $a_3 \neq 0$, one can use the substitution $x_1 = y_1 - \frac{a_3}{4}$ and pass to the previous case. But we can study this case strictly considering the quadratic problem:

$$\text{find } P^* = \min[x_2^2 + a_3 x_1 x_2 + a_2 x_1^2 + a_1 x_1] \tag{9.13}$$

subject to the constraint

$$x_1^2 - x_2^2 = 0 \tag{9.14}$$

The detailed analysis shows that for the problem (9.13), (9.14) the dual quadratic bound ψ^* is equal to P^* for any polynomial $P_4(x_1) = x_1^4 + a_3 x_1^3 + a_2 x_1^2 + a_1 x_1$.

Example 2

$$k = 3 \; P_6 = x_1^6 + a_5 x_1^5 + a_4 x_1^4 + a_3 x_1^3 + a_2 x_1^2 + a_1 x_1 + a_0. \tag{9.15}$$

Let

$$P^* = \min_{x_1} P_6(x_1). \tag{9.16}$$

Introducing quadratic substitutions we may reduce the problem of finding P^* to a quadratic-type problem: find

$$P^* = \min_{x=\{x_1,x_2,x_3\}} \{x_3^2 + a_5 x_2 x_3 + a_4 x_1 x_2 + a_2 x_1^2 + a_1 x_1 + a_0\} \tag{9.17}$$

subject to the constraints:

$$f_1(x) = x_2 - x_1^2 = 0, \tag{9.18}$$

$$f_2(x) = x_3 - x_1 x_2 = 0. \tag{9.19}$$

We can try to receive the lower bound for P^* using the Lagrange function

$$L_1(x, u_1, u_2) = K(x_1, x_2, x_3) + u_1(x_1^2 - x_2) + u_2(x_1 x_2 - x_3),$$

where $x = \{x_1, x_2, x_3\}$, $K(x_1, x_2, x_3)$ is the minimized quadratic function in (9.17), u_1, u_2 are Lagrange multipliers. Let

$$\psi(u_1, u_2) = \inf_x L_1(x, u_1, u_2)$$

domψ contains all points of $\{u_1, u_2\}$, for which quadratic part in x of L_1 is positive definite (and may be some points for which it is semidefinite). It is worth noting that if $a_4 < 0$, domψ is empty. So, in general, $\psi(u_1, u_2)$ may give us only the trivial lower bound $(-\infty)$. It seems that in our case the Lagrange technique is useless. But such a conclusion is premature (hasty). Let us introduce the new constraint, which is a consequence of others:

$$f_3(x) = x_2^2 - x_1x_3 = 0 \tag{9.20}$$

and u_3 is the corresponding Lagrange multiplier. Let $u = \{u_1, u_2, u_3\}$ and one may form "renewed" Lagrange function

$$\begin{aligned} L_2(x, u) &= K(x_1, x_2, x_3) + u_1(x_1^2 - x_2) \\ &+ u_2(x_1x_2 - x_3) + u_3(x_2^2 - x_1x_3) \\ &= K_u(x) + l_u(x) + a_0 \end{aligned}$$

where $K_u(x) = (K_u x, x)$ is quadratic form of $L_2(\cdot, u)$ on x, l_u is a linear part of $L_2(\cdot, u)$ on x:

$$K_u = \begin{pmatrix} a_2 + u_1 & a_3 + u_2 & -u_3 \\ a_3 + u_2 & a_4 + u_3 & a_5 \\ -u_3 & a_5 & 1 \end{pmatrix}$$

$$l_u(x) = a_1x_1 - u_1x_2 - u_2x_3 = (l_u, x); \quad l_u^t = \begin{pmatrix} a_1 \\ -u_1 \\ -u_2 \end{pmatrix}$$

Let

$$\begin{aligned} \Omega &= \{u : K_u \text{ is positive semidefinite}\}, \\ \Omega^+ &= \{u : K_u \text{ is positive definite}\}. \end{aligned}$$

Ω^+ is nonempty set because for fixed $\overline{u}_2$ we can find $\overline{u}_3 > -a_4 + a_5^2$, and then find $\overline{u}_1$ such that det $K_{\overline{u}} > 0$, $\overline{u} = (\overline{u}_1, \overline{u}_2, \overline{u}_3)$, and so we obtain positiv definite $K_u(\cdot)$.

Let $\psi(u) = \inf_x L_2(x, u)$. $-\psi(u)$ is a convex function, dom$(-\psi)$ contains Ω^+ and those points of Ω/Ω^+, for which the range of the matrix $(K_u | l_u^T)$ is equal to that of K_u. Below we shall prove that $\psi^* = \max_{u \in \text{dom}\psi}$ equals to P^*.

We shall try to explain why the introducing of unnecessary, superfluous constraint (9.20) may influence so favorably on the precision of a dual bound.

Let us consider the problem (9.17), (9.18), (9.19) as an usual problem of nonlinear programming with constraints in the form of equalities. It is clear that in the optimal point $x^* = \{x_1^*, x_2^*, x_3^*\}$, where $x_2^* = (x_1^*)^2$, $x_3^* = (x_1^*)^3$, gradients of functions $f_1(x) = x_1^2 - x_2$ and $f_2(x) = x_1x_2 - x_3$ are linearly independent. So, by the Lagrange theorem, we have unique Lagrange multipliers $\overline{u}_1$, $\overline{u}_2$ such that $g_{L_1}^x(x^*, \overline{u}_1, \overline{u}_2) = 0$. (Here $g_{L_1}^x(x^*, \overline{u}_1, \overline{u}_2)$ is the gradient of the function $L_1(x, \overline{u}_1^1, \overline{u}_2^1)$ on x at the point $x = x^*$. Similarly we can consider the problems (9.17), (9.18), (9.20) and (9.17), (9.19), (9.20) and pairs of optimal Lagrange multipliers $(\overline{u}_1^{(2)}, \overline{u}_3^{(2)})$ and $(\overline{u}_2^{(3)}, \overline{u}_3^{(3)})$ correspondingly.

We have three equations:

$$\begin{aligned} g_K(x^*) + \overline{u}_1^{(1)} g_{f_1}(x^*) + \overline{u}_2^{(1)} g_{f_2}(x^*) &= 0 \\ g_K(x^*) + \overline{u}_1^{(2)} g_{f_1}(x^*) + \overline{u}_3^{(2)} g_{f_3}(x^*) &= 0 \\ g_K(x^*) + \overline{u}_2^{(3)} g_{f_2}(x^*) + \overline{u}_3^{(3)} g_{f_3}(x^*) &= 0 \end{aligned} \qquad (9.21)$$

Let $\alpha_1, \alpha_2, \alpha_3$ be real numbers and $\alpha_1 + \alpha_2 + \alpha_3 = 1$. We obtain equations of the following form as a consequence of the system (9.21):

$$\begin{aligned} & g_K(x^*) + (\alpha_1 \overline{u}_1^{(1)} + \alpha_2 \overline{u}_1^{(2)}) g_{f_1}(x^*) + \\ & +(\alpha_1 \overline{u}_2^{(1)} + \alpha_2 \overline{u}_2^{(3)}) g_{f_2}(x^*) + \\ & +(\alpha_2 \overline{u}_3^{(2)} + \alpha_3 \overline{u}_3^{(3)}) g_{f_3}(x^*) = 0 \end{aligned} \qquad (9.22)$$

Note, that vectors $\{g_{f_1}(x^*), g_{f_2}(x^*), g_{f_3}(x^*)\}$ are linearly dependent, because the surface $f_3(x) = 0$ contains the intersection of the surfaces $f_1(x) = 0$ and $f_2(x) = 0$. ($f_3(x) = 0$ is the consequence of the $\{f_1(x) = 0\} \bigvee \{f_3(x) = 0\}$), so the linear envelope M_L of the vectors $\{\overline{u}_1^{(1)}, \overline{u}_2^{(1)}, 0\}$, $\{\overline{u}_1^{(2)}, 0, \overline{u}_3^{(2)}\}$, $\{0, \overline{u}_2^{(3)}, \overline{u}_3^{(3)}\}$ has dimension 1. Due to (9.22) we can consider this family of points as vectors of optimal Lagrange multipliers of the problem (9.17), (9.18), (9.19). Geometrically M_L is the straight line l, containing the points $\overline{u}^{(1)} = \{\overline{u}_1^{(1)}, \overline{u}_2^{(1)}, 0\}$, $\overline{u}^{(2)} = \{\overline{u}_1^{(2)}, 0, \overline{u}_3^{(2)}\}$, and $u_3 = \{0, \overline{u}_2^{(3)}, \overline{u}_3^{(3)}\}$.

The region $\Omega = \{u \in R^3 : K_u$ is positive semidefinite $\}$ may not include any of the points $\overline{u}^{(1)}, \overline{u}^{(2)}, \overline{u}^{(3)}$ but if $M_L \cap \Omega$ is nonempty, then we may have an optimal vector of Lagrange multipliers of the form

$$u^* = \{\alpha_1^* u_1^{(1)} + \alpha_2^* u_1^{(2)}, \alpha_1^* u_2^{(1)} + \alpha_3^* u_2^{(3)}, \alpha_2^* u_3^{(2)} + \alpha_3^* u_3^{(3)}\},$$

$$\alpha_1^* + \alpha_2^* + \alpha_3^* = 1,$$

such that the dual estimate

$$\psi^* = \sup_u [\inf_x L_2(x, u)]$$

is attained at $u = u^*$, and

$$\psi(u^*) = \min_x L(x, u^*) = P^*.$$

Thus, we can obtain the exact dual bound adding the "superfluous" constraint (9.20) even in the case where dual bound generated by the Lagrange function $L_1(x, u_1, u_2)$ is equal to $-\infty$.

Indeed, as it is shown below, $\psi(u^*) = P^*$ for an arbitrary $P_6(x_1)$.

This example is a bright demonstration of the fact that superfluous constraints may essentially improve dual bounds.

Let us consider the general case of even degree polynomials of one variable.

$$P_{2k}(x_1) = x_1^{2k} + \sum_{i=0}^{2k-1} a_i x_1^i. \tag{9.23}$$

Let $P^* = \min_{x_1} P_{2k}(x_1)$. Introduce new variables $x_2, \ldots, x_k$ by the equalities:

$$q_{i+1}(x) \equiv x_{i+1} - x_i x_1 = 0,\ i = 1, \ldots, k. \tag{9.24}$$

So $x_i = x_1^i$ and the polynomial $P_{2k}(x_1)$ can be (nonuniquelly) represented as quadratic function in new variables $x_1, \ldots, x_k$. For example:

$$P_{2k}(x_1) = x_k^2 + \sum_{i=k+1}^{2k-1} a_i x_{i-1} x_k + \sum_{i=0}^{k} a_i x_i \tag{9.25}$$

Let $x = \{x_1, \ldots, x_k\}$. The right-hand part of (9.25) will be denoted by $K_0(x) = K_0(x_1, \ldots, x_k)$. For each degree i, $i = 2, \ldots, 2k$, one can record trivial equalities which are consequences of the equalities (9.24): $x_i = x_t x_{i-t}$ for all integers t, i, satisfying

$$0 < i - t \leq t \leq k,\ x_0 = 1 \tag{9.26}$$

If for certain i and k there are more than one t satisfying (9.26), we can obtain the equalities of the form:

$$x_{t_1} \cdot x_{i-t_1} - x_{t_2} \cdot x_{i-t_2} = 0, \; t_1 \neq t_2 \tag{9.27}$$

For example, if $i = 6$, $k = 5$, we have the identities:

$$x_1^6 \equiv x_5 \cdot x_1 = x_4 \cdot x_2 = x_3^2.$$

For given k and fixed $i \in \{2, 3, \ldots, 2k-2\}$ consider all kinds of equalities of the form

$$q_{st}(x) \equiv x_s x_{i-s} - x_t x_{i-t} = 0, \tag{9.28}$$

where $s = i$ for $i \leq k$ and $s = k$ for $i \geq k$; $0 \leq t < s$. Denote by T_{ik} the set of possible values t for given i and k.

For example, if $i = 6$, $k = 5$ we obtain the following equalities:

$$\begin{cases} q_{6,4}(x) = x_5 x_1 - x_4 x_2 = 0; \\ q_{6,3}(x) = x_5 x_1 - x_3^2 = 0; \end{cases}$$

In our case $|T_{65}| = 2$.

One can see that all the equalities of the form (9.27) are linear combinations of the equalities (9.28), and the equalities (9.24) are particular cases of the equalities (9.28).

Let $P(x_1)$ be an arbitrary polynomial of degree not exceeding $2k$:

$$P(x_1) = \sum_{i=0}^{2k} a_i x_1^i \tag{9.29}$$

$P(x_1)$ can be represented in variables $x = \{x_1, \ldots, x_k\}$ as a quadratic function, by ununique way, in general.

Theorem 119 *Let $K_1(x)$ and $K_2(x)$ be two different quadratic representations of the same polynomial $P(x_1)$ and $\delta(x) = K_1(x) - K_2(x)$. Then $\delta(x)$ can be represented as a linear combination of the quadratic polynomials $q_{st}(x)$ (see (9.28)).*

Proof. Let $\delta(x) = \sum_{r=0}^{2k} \sum_{s=0}^{2k} b_{rs} x_r x_s$ Since $K_1(x)$ and $K_2(x)$ are quadratic representations of the same polynomial, the sum of the coefficients b_{ij}, which corresponds to the same value of $i + j$, must be equal to zero.

So $\delta(x) = \sum_{r=0}^{2k}(\sum_{\{i,j:i+j=r\}} b_{ij}x_ix_j) = \sum_{r=0}^{2k}(\sum_{\{i,j:i+j=r\}} b_{ij}(x_ix_j - x_{s(r)}x_{r-s(r)}),$
where

$$s(r) = \begin{cases} r, & \text{for } r \le k \\ k, & \text{for } r > k \end{cases}$$

Hence $\delta(x) = -\sum_{r=0}^{2k} \sum_{t\in T_{rk}} b_{s(r)t}q_{s(r)t}$.

Corollary 7 *Let $K_0(x)$ be a "canonical" quadratic representation of $P(x)$:*

$$K_0(x) = \sum_{r=0}^{k} a_ix_i + \sum_{i=k+1}^{2k} a_ix_kx_{i-k}.$$

Then any quadratic representation of $P(x)$ has the following form:

$$K(x) = K_0(x) + \sum_{r=0}^{2k} \sum_{t\in T_{rk}} b_{s(r)t}q_{s(r)t}, \tag{9.30}$$

where $\{b_{s(r)t}\}$ is a vector of arbitrary multipliers.

As we saw above the problem of finding global minimum of polynomial

$$P_{2k}(x_1) = \sum_{i=0}^{2k} a_ix_i, \quad a_{2k} > 0$$

can be reduced to the quadratic-type problem in k variables of the following form: minimize

$$K_0(x) = \sum_{i=0}^{k} a_ix_i + \sum_{i=k+1}^{2k} a_ix_kx_{i-k} \tag{9.31}$$

subject to the constraints:

$$q_{st}(x) = x_s - x_{s-t}x_t = 0; \quad s = 2,\ldots,k; \quad 1 \le t \le \lceil\frac{s+1}{2}\rceil \tag{9.32}$$

$$q_{st}(x) = x_kx_{s-k} - x_{s-t}x_t; \quad \{s,t : k < s \le 2k-2; \ 1 < t \le \lceil\frac{s+1}{2}\rceil \tag{9.33}$$

Let M_k be a possible set of indeces $\{st\}$ on (9.32), (9.33). Consider the Lagrange function $L(x,u)$, $u = \{u_{st}\}_{st\in M}$, of the problem (9.31), (9.32), (9.33):

$$L(x,u) = K_0(x) + \sum_{(s,t)\in M} u_{st}q_{st}(x).$$

Let $\psi(u) = \inf_x L(x,u)$. Denote by Ω the domain of concave functionψ. $L(x,u)$ can be represented as function of x with parameter vector u. $L(x,u) = (K_u x, x)+(l_u, x)+c(u)$, where K_u, $l(u)$ and $c(u)$ are respectevly the square matrix $k \times k$, the vector of dimension k, and a number, which values are dependent of u. Let $U^+ = \{u : K_u \succ 0\}$, $\overline{U}^+ = \{u : K(u) \succeq 0\}$ and $\psi(u) = \inf_x L(x,u)$. Denote by Ω the domain of the concave function ψ.

It is clear that $\overline{U}^+ \supset \Omega \supseteq U^+$.

Theorem 120 *Let $P^* = \min_{x_1} P_{2k}(x_1)$ The lower dual bound for the corresponding quadratic optimization problem (9.31), (9.32), (9.33) is exact, i.e.*

$$\psi^* = \sup_u \{inf_x L(x,u)\} = P^*.$$

Proof. The function $\overline{P}_{2k}(x_1) = P_{2k}(x) - P^*$ is nonnegative. So, it can be represented as sum of squares of polynomials of degree not exceeding k (see Theorem 112). Thus the polynomial $\overline{P}_{2k}(x_1)$ can be representend as sum of squares of linear functions in variables $x_i = x_1{}^i$, $i = 1,\ldots,k$, i.e. $\overline{P}_{2k}(x_1)$ has a quadratic representation $K_1(x)$ in variables $x_1, x_2, \ldots, x_k$ with positive semidefinite quadratic part:

$$K_1(x) = (Bx,x) + (l,x) + c \geq 0 \text{ for all } x.$$

Due to the Lagrange theorem, there exists $\overline{u} = \{\overline{u}_{st}\}$ such that $K_1(x) = K_0(x) + \sum_{(s,t)\in M} \overline{u}_{st} q_{st}(x)$. For $x_i = x_1^i$, $i = 1,\ldots,k$, all $q_{st}(x) = 0$, and $K_1(x_1, x_2, \ldots, x_k) = \overline{P}_{2k}(x_1)$. Let x_1^* be the point of global minimum of the polynomial $\overline{P}_{2k}(x_1)$. Then $K_1(x_1^*, (x_1^*)^2, \ldots, (x_1^*)^*) = \overline{P}_{2k}(x_1^*) = 0$. Since $K_1(x) \geq 0$ for all x,

$$\min_x K_1(x) = \min_x [L(x,\overline{u}) - P^*] = \psi(\overline{u}) - P^* = 0.$$

Hence, $\psi^* = \psi(\overline{u})$ is an exact bound for P^*. The proof of Theorem 120 is over.

Remark 1 One can see that the proof of Theorem 120 is based on two statements: (i) on the structure of all quadratic representations of $P_{2k}(x_1)$ (Theorem 119); (ii) on the possibility to express a nonnegative polynomial of one variable as sum of squares of polynomials. In multidemensional case we can generalize rather easy Theorem 119 by construction rich enough family of variables and quadratic equalities. Many of considered equalities are superfluous and are used for obtaining more exact dual bounds. But not all nonnegative multidimensional polynomials can be decomposed into a sum of squares of polynomials. So the theorem, similar to Theorem 120, in multidimensional case is true only for polynomials that can be expressed as sum of squares of polynomials. Of course, in one-dimensional case the calculation of dual quadratic bounds is not the best way for obtaining global minimum of polynomials. There are more efficient procedures. We considered this case in detail in order to show the role of superfluous constraints for improving dual estimates and to prepare the reader to more hard technique for multidimensional polynomials.

Remark 2 Let us calculate how many constraints are sufficient to get exact dual quadratic bounds when we minimize the polynomial $P_{2k}(x_1)$. If s_{2k} means the number of such constraints (see (9.28)), one can obtain the reccurent formula:

$$s_{2k+1} = s_{2k} + k,\ s_2 = 0,$$

so, $s_{2k} = \frac{k(k-1)}{2}$.

In particular, $s_4 = 1$; $s_6 = 3$; $s_8 = 6$,, etc. The number of "superflous" constraints

$$\bar{s}_{2k} = s_{2k} - (k-1) = \frac{(k-1)(k-2)}{2}$$

So, $\bar{s}_4 = 0$; $\bar{s}_6 = 1$; $\bar{s}_8 = 3$, etc.

Remark 3 As a rule, the set U^* of optimal vectors of Lagrange multipliers u^* has the dimension equal to the number of superflous constraints. If the global minimum x_1^* is unique then there exists $u^* \in U^*$ such that $L(x, u^*)$ is positive definite quadratic function in x and $\min_x L(x, u^*)$ is attained at the point $x^* = (x_1^*, (x_1^*)^2, \ldots, (x_1^*)^k)$.

If the global minimum point is nonunique, then U^* contains only points u^* for which $L(x, u^*)$ is a degenerated positive semidefinite quadratic function, so in the process of maximization of $\psi(u)$ we cannot obtain the stable approximations for certain global minimum point of $P_{2k}(x_1)$. In this case we may recommend to use some $\varepsilon-$ perturbation for $P_{2k}(x_1)$ with the aim to obtain a polynomial close to $P_{2k}(x_1)$ with unique global minimum point. One may also use the information about the function $K_{u^*}(x) = L(x, u^*)$ for determination the affine manifold in R^k which corresponds to all global minimum points. By this information one can determine the number of global minimum points.

Remark 4 Let us try to find one of the real roots of polynomial equation of odd degree

$$P_{2k-1}(x_1) = \frac{x_1^{2k-1}}{2k} + \sum_{r=1}^{2k-1} \frac{a_{2k-r-1}}{2k-r} x_1^{2k-r-1} = 0$$

Using the integration we can deduce this problem to finding global minimum of the polynomial

$$P_{2k}(x_1) = x_1^{2k} + \sum_{r=1}^{2k-1} a_{2k-r} x_1^{2k-r}$$

and then use proposed above technique for obtaining the global minimum. When we have an even degree equation $\overline{P}_{2k}(x_1) = 0$, we may find the global minimum value P^* and if $P^* > 0$, this equation has not real roots. Otherwise, if $P^* < 0$, one can find two points $x_l < x^*$ and $x_r > x^*$ such that $P_{2k}(x_l)$ and $P_{2k}(x_r)$ are positive, and obtain at least two real roots by using dichotomy or more sofisticated methods in intervals (x_l, x^*) and (x^*, x_r) .

Global minimum of fractional polynomial function. Consider a rational function of one variable

$$f(x_1) = \frac{P_{2k_1}(x_1)}{P_{2k_2}(x_1)},$$

where the numerator $P_{2k_1}(x_1)$ is polynomial of even degree $2k_1$, the denominator $P_{2k_2}(x_1)$ is a polynomial of even degree $2k_2$, moreover, $P_{2k_2}(x_1) > 0$ for all x_1. Denote $\inf f(x_1)$ by f^*.

Let $P_f(x_1) = P_{2k_1}(x_1) - fP_{2k_2}(x_1)$. $\varphi(f) = \inf_{x_1} P_f(x_1)$ is a decreasing function of f. So, if $\varphi(f) > 0$, then $f^* > f$ and if $\varphi(f) < 0$, then $f^* < f$. Thus, for finding f^* one can use the combination of algorithm for obtaining dual quadratic bounds for polynomials and certain algorithm for solving the equation $\varphi(f) = 0$.

Let $T_k(\sin x, \cos x)$ be a trigonometrical polynomial in $\sin x$ and $\cos x$ of total degree k. using formulas:

$$\sin x = \frac{2\tan\frac{x}{2}}{1+\tan^2\frac{x}{2}}; \quad \cos x = \frac{1-\tan^2\frac{x}{2}}{1+\tan^2\frac{x}{2}}$$

and the substitution $y = \tan\frac{x}{2}$ one can reduce the problem of finding

$$T^* = \min_{x\in[0,2\pi]} T_k(\sin x, \cos x)$$

to that of finding global minimum of a fractional polynomial function $f(y)$ with the denominator $(1+y^2)^k$, which is positive for all $y \in (-\infty, +\infty)$.

9.2.1 The structure of the cone of convex polynomials in one variable

Consider polynomials in one variable of degree $2k$:

$$P_a(x) = \sum_{r=0}^{2k} a_r x^r, \tag{9.34}$$

where $a = \{a_r\}_{r=0}^{2k}$, $a_{2k} \neq 0$.

Let C_{2k}^{+} be a cone of nonnegative convex polynomials in the $(2k+1)$-dimensional space L_{2k+1} of coefficients of polynomials of degree $2k$ recorded in the form (9.34). We call a polynomial $P_a \in C_{2k}^{+}$ **extremal** if it cannot be decomposed into a sum of two noncolinear polynomials belonging to C_{2k}^{+}.

Let us study the set S_{2k} of extremal polynomials of degree $2k$ in the cone C_{2k}^{+}.

Consider polynomial $P_a(x)$ of the form (9.34) with $a_{2k} > 0$. The second derivative

$$P_a''(x) = \sum_{r=0}^{2(k-1)} (r+2)(r+1)a_{r+2}x^r.$$

If $P_a(x)$ is convex, $P_a''(x) \geq 0$ for all $x \in R$. Due to Theorem 112 P_a'' can be represented as sum of two squares of polynomials of degree $(k-1)$:

$$P_a''(x) = P_1^2(x) + P_2^2(x). \tag{9.35}$$

Moreover, $P_2(x) \equiv 0$ if and only if all roots of the polynomial $P_a''(x)$ are real numbers with even multiplicity.

In the case where $P_2 \not\equiv 0$, $P_a(x)$ can be represented as sum of two noncolinear convex functions which are result of double integration of nonnegative functions P_1^2 and P_2^2. So, in this case $P_a(x)$ cannot be extremal polynomial of the cone C_{2k}^{+}. Thus, if $P_a(x) \in S_{2k}$, $P_a''(x) = P_1^2(x)$, and all roots $\xi_1 \geq \xi_2 \geq \dots \geq \xi_{k-1}$ of polynomial $P_1(x)$ are real, i.e. P_a'' has the following form:

$$P_a''(x) = \frac{2k(2k-1)}{2} a_{2k} \prod_{i=1}^{k-1} (x - \xi_i)^2 \tag{9.36}$$

Besides, if $\overline{P}(x) \in S_{2k}$, $\min_x \overline{P}(x) = \overline{P}(x^*) = 0$, where x^* is minimum point of $\overline{P}(x)$.

Let a polynomial $\overline{P}(x)$ of degree $2k$ belongs to S_{2k} and

$$\overline{P}''(x) = b_0 \prod_{i=1}^{k-1} (x - \xi_i)^2 = b_0 x^{2(k-1)} + \sum_{r=0}^{2k-3} b_r x^r, \ b_0 > 0. \qquad (9.37)$$

Then

$$\overline{P}(x) = b_0 \frac{x^{2k}}{2k(2k-1)} + \sum_{i=2}^{2k-1} \frac{b_i}{(i-1)i} x^i + cx + d, \qquad (9.38)$$

where undetermined coefficients c and d are obtained by the conditions:

$$\overline{P}'(x^*) = 0; \ \overline{P}(x^*) = 0.$$

Now one can formulate the following proposition about extremal polynomials of the cone C_{2k}^+:

Theorem 121 *Due to formulas (9.37), (9.38) all extremal polynomials $\overline{P}(x)$ of the cone C_{2k}^+ of degree $(2k)$ with given elder coefficient $a_{2k} > 0$ are uniquely determined by points $\xi_1 \geq \xi_2 \geq \ldots \geq \xi_{k-1}$, in which the second derivative is equal to zero, and by the point of minimum x^*, in which $\overline{P}(x^*) = 0$ and $\overline{P}'(x^*) = 0$, besides $b_0 = 2k(2k-1)a_{2k}$.*

9.3 NECESSARY AND SUFFICIENT CONDITIONS FOR DUAL QUADRATIC BOUND TO BE EXACT. MULTIDIMENSIONAL CASE

Let R^n be n-dimensional linear space with real vectors $z = \{z_1, \ldots, z_n\}$, $P(z) = P(z_1, \ldots, z_n)$ be a polynomial real function determined on R^n. Consider the porblem of finding

$$f^* = \inf_{z \in R^n} P(z_1, \ldots, z_n)$$

We will be interested in nontrivial case where $f^* > -\infty$ i.e. $P(z)$ is bounded from below. Such polynomials will be called BB - polynomials. It is clear that if $P(z)$ belongs to BB - class, then for any $i, 1 \leq i \leq n$ the highest degrees S_i of variables z_i must be even and the corresponding coefficients must be nonnegative for any values of other variables. Moreover, specializations of BB - polynomials preserve the BB - property (we build polynomial specializations by substituting some of z_i by certain constants or the polynomial functions of the remaining free vafiables). Besides, the problem of the BB – property for a given polynomial is in general, similar by its computational complexity to the problem of finding f^*.

Let the highest degrees S_i of variables z_i in $P(z)$ be even: $S_i = 2l_i, i = 1, \ldots, n$, and let $P(z)$ be recorded in the standard form as a sum of monomials with certain real nonzero coefficients. For compact record of monomials we shall use vectors of degrees: $\alpha = (\alpha_1, \ldots, \alpha_n)$ with nonnegative integer entries and symbols $R[\alpha]$ of the corresponding monomials:

$$R[\alpha] = z_1^{\alpha_1} \ldots z_n^{\alpha_n}, \ \alpha_i \leq S_i, \ i = 1, \ldots, n \tag{9.39}$$

So

$$P(z) = \sum_{\alpha} c_\alpha R[\alpha] \text{ in new variables, } 0 \leq \alpha_i \leq 2l_i, i = 1, \ldots, n. \tag{9.40}$$

For any monomials $R[\alpha]$ one can write a system of identity relations:

$$R[\alpha^{(1)}]R[\alpha^{(2)}] - R[\alpha^{(3)}]R[\alpha^{(4)}] = 0 \tag{9.41}$$

for all combinations of $\alpha^{(i)}, i = 1, 2, 3, 4$ such that $\alpha^{(i)}$ are integer nonnegative vectors, and

$$\alpha^{(1)} + \alpha^{(2)} = \alpha^{(3)} + \alpha^{(4)} = \alpha.$$

Consider the set $M(l)$ of "feasible monomials" for a given positive integer vector $l = (l_1, \ldots, l_n)$:

$$M(l) = \{R[\alpha](z) : \ \alpha_i \leq l_i, \ i = 1, \ldots, n\} \tag{9.42}$$

and the corresponding set $V[l]$ of feasible variables $R[\alpha_]$

Every polynomial $P(z) = P(z_1, \ldots, z_n)$ with the highest degrees $S_i = 2l_i$ for the corresponding variables $i, 1 \leq i \leq n$, can be represented (in general, nonuniquely) as a quadratic function of feasible variables $R[\alpha]$ in the following way

$$P(z) = \sum_{t,j} c_{tj} R[\alpha^{(t)}]R[\alpha^{(j)}] + \sum_{k,l,m,p} u_{k,l,m,p}(R[\alpha^{(k)}]R[\alpha^{(l)}] - R[\alpha^{(m)}]R[\alpha^{(p)}]),$$

where $R[\alpha^{(t)}]R[\alpha^{(j)}]$ is one of the possible representations of monomial $R[\alpha]$ as a product of two feasible monomials: $\alpha = \alpha^{(t)} + \alpha^{(j)}$, $\alpha_i^{(t)}, \alpha_i^{(j)} \leq l_i$, for all

$i, 1 \leq i \leq n$, c_{ij} is the corresponding coefficient; $u = \{u_{k,l,m,p}\}$ is the vector of arbitrary multipliers corresponding to the system of identity equalities $[S]$ of the type (9.41) with feasible variables:

$$[S] \qquad R[\alpha^{(k)}]R[\alpha^{(l)}] - R[\alpha^{(m)}]R[\alpha^{(p)}] = 0,\ \alpha^{(k)} + \alpha^{(l)} = \alpha^{(m)} + \alpha^{(p)}.$$

Note, that we can exclude from the system $[S]$ the equations that are the linear combinations of remaining equations. For example, if we have the equations with two variables

$$R[2;1]R[1,2] - R[3;3]R[0,0] = 0$$

$$R[3;0]R[0;3] - R[3,3]R[0,0] = 0$$

then we may exclude the equation:

$$R[2;1]R[1;2] - R[3,0]R[0,3] = 0.$$

So we can consider instead of $[S]$ a shortened system of equations $[\overline{S}]$ after excluding some equations, linear dependent of remaining equations.

For example, as shortened system $[\overline{S}]$ we may use the following system of equations:

$$P[\alpha;\alpha^{(1)}](R) \equiv R[\beta(\alpha)] \cdot R[\alpha - \beta(\alpha)] - R[\alpha^{(1)}]R[\alpha - \alpha^{(1)}] = 0, \qquad (9.43)$$

where $\alpha, \alpha^{(1)} \leq \alpha$ are nonnegative integer n-dimensional vectors, $\alpha = (\alpha_1, \ldots, \alpha_n)$ with $\alpha_i \leq 2l_i, i = 1, \ldots, n,$

$$(\beta(\alpha))_i = \begin{cases} \alpha_i, & \text{if } \alpha_i \leq l_i, \\ l_i, & \text{if } \alpha_i > l_i, \end{cases}$$

$\alpha^{(1)} \geq \alpha - \alpha^{(1)}$ in lexicographical sense. Note, that the equations (9.43) are formulated in feasible variables (see (9.42)).

Due to the system (9.43) we have the "canonical" quadratic representation of $P(z)$:

$$P(z) = \sum_{\alpha} c_\alpha \prod_{i=1}^{n} z_i^{\alpha_i} \equiv \sum_{\alpha} c_\alpha R[\beta(\alpha)]R[\alpha - \beta(\alpha)] = K_0(R) \qquad (9.44)$$

and representationsof the Lagrange function:

$$L(R;u) = K_0(R) + \sum_{\{\alpha;\alpha_1\}} u(\alpha,\alpha^{(1)})P[\alpha;\alpha^{(1)}](R) \qquad (9.45)$$

where $u = \{u(\alpha;\alpha^{(1)})\}$ is a vector of Lagrange multipliers. We can consider $L(R,u)$ as Lagrange function for quadratic type (QT) optimization problem:

$$\text{find } f^* = \inf K_0(R) \tag{9.46}$$

subject to the system $\overline{S}$ of the equations (9.43). Decompose the Lagrange function $L(R;u)$ into quadratic and linear parts in variables R:

$$L(R;u) = (K(u)R,R) + (l(u),R) + c(u) \tag{9.47}$$

Let $\Omega^+ = \{u : K(u) \text{ is positive definite}\}$, $\overline{\Omega} = \{u : K(u) \text{ is positive semidefinite}\}$, $\psi(u) = \inf_R L(R,u)$. $\psi(u)$ is a concave function. If $\text{dom}\psi \neq \emptyset$, then it contains Ω^+ and (or) some subset of boundary points $\overline{u} \in \overline{\Omega}\backslash\Omega^+$ for which the singular system of linear equations

$$2K(\overline{u})R + l(\overline{u}) = 0$$

has solutions. $\text{dom}\psi$ is a convex or empty set. If $\text{dom}\psi = \emptyset$, we have a trivial lower bound for $f^* : \psi^* = -\infty$. Otherwise, finding $\psi^* = \sup_{u \in \text{dom}\psi} \psi(u)$ is a convex programming problem. Solving it we obtain a nontrivial lower bound ψ^* for f^* which we call **the dual lower quadratic bound** for f^*. The main problem is to find conditions when $f^* = \psi^*$.

In multidimensional case the situation is much more complex in comparison with one-dimensional case, because not all nonnegative polynomials even of two variables may be represented as sums of squares of other polynomials. But we can formulate some interesting statements for multidimensional case too.

Let $\pi(n, 2l, 2d)$ be the class of real polynomials $P(z)$ in n variables $z = \{z_0, \dots, z_n\}$ with maximum degree vector of monomials not exceeding $2l = \{2l_i\}_{i=1}^n$ and total degree of monomials not exceeding $2d$.

Consider $P(z) \in \pi(n, 2l, 2d)$ and let $K_1^P(R)$ and $K_2^P(R)$ be two different feasible quadratic representations of this polynomial written in feasible variables $V[l]$ (see (9.42)).

Theorem 122 *For two arbitrary feasible quadratic representations $K_1^P(R)$ and $K_2^P(R)$ of the same polynomial $P \in \pi(n, 2l, 2d)$ their difference $\delta(R)$ can be expressed as a linear combination of the quadratic polynomials $P[\alpha;\alpha^{(1)}](R)$ given by (9.43).*

Proof. Since $\delta(R) = K_1^{(P)} - K_2^{(P)}(R)$, where $K_1^{(P)}$ and $K_2^{(P)}$ are the feasible quadratic representations of the same polynomial $P(z)$, $\delta(R) = 0$ for $R \in M[\overline{S}]$, where $M[\overline{S}]$ is the set of solutions of the system $[\overline{S}]$ (see (9.43). So the sum of coefficients of quadratic function $\delta(R)$ that are related to the members $R[\alpha^{(1)}]\cdot R[\alpha^{(2)}]$ with the same $\alpha^{(1)}+\alpha^{(2)}$ must be equal zero.

Let

$$\delta(R) = \sum_{\{\alpha^{(1)},\alpha^{(2)}\}} u\left(\alpha^{(1)},\alpha^{(2)}\right) R\left(\alpha^{(1)}\right) R\left(\alpha^{(2)}\right)$$

Then

$$\begin{aligned}&\delta(R) = \\ &= \textstyle\sum_\alpha \sum_{\{\alpha^{(1)},\alpha^{(2)}:\alpha^{(1)}+\alpha^{(2)}=\alpha\}} \left[u\left(\alpha^{(1)},\alpha^{(2)}\right)\cdot \left(R[\alpha^{(1)}]R[\alpha^{(2)}] - R[\beta(\alpha)]R[\alpha-\beta(\alpha)]\right)\right] \\ &= -\textstyle\sum u\left(\alpha^{(1)},\alpha^{(2)}\right) P[\alpha^{(1)}+\alpha^{(2)},\alpha^{(1)}](R).\end{aligned}$$

The proof is over.

Theorem 123 *Let a real polynomial $P(z)$ in n variables, $z = \{z_1,\ldots,z_n\}$, with maximum degrees $2l_i$ of variables $z_i, i = 1,\ldots,n$, and maximal total degree of monomials $2d$ can be represented as sum of squares of m real polynomials $P_j(z), j \in \{1,\ldots,m\}$. Then for every representation of $P(z)$ as quadratic function $\overline{K}(R)$ in feasible variables $R[\alpha]$ there exists a vector of Lagrange multipliers $\overline{u} = \{\overline{u}(\alpha,\alpha^1)\}$ such that $\overline{K}(R) = L(R,\overline{u})$, where $L(R,u)$ is the Lagrange function for the quadratic-type optimization problem (9.46), (9.43).*

Proof. Let the polynomials $P(z)$ have the following form:

$$\begin{aligned} P_j(z) &= \sum_\alpha c(\alpha,j)R[\alpha](z), j = \{1,\ldots,m\}, \\ R[\alpha](z) &= \prod_{i=1}^n x_i^{\alpha_i}, i = 1,\ldots,n, \end{aligned}$$

for all feasible α:

$$0 \le \alpha_i \le l_i, i = 1,\ldots,n; \sum_{i=1}^n \alpha_i \le d.$$

Then

$$\begin{aligned} P(z) &= \sum_{j=1}^m \left[\sum_\alpha c(\alpha,j)R[\alpha](z)\right]^2 = \\ &= \sum_{j=1}^m \left[\sum_{\{\alpha^{(1)},\alpha^{(2)}\}} c(\alpha^{(1)},j)\cdot c(\alpha^{(2)},j)\cdot R[\alpha^{(1)}](z)\cdot R[\alpha^{(2)}](z)\right] \end{aligned} \quad (9.48)$$

Formula (9.48) gives us a reprezentation of $P(z)$ as quadratic function in feasible variables $R[\alpha]$. Note that this representation depends on certain decomposition $P(z)$ into a sum of squares of polynomials.

Let $K_0(R)$ be a canonical representation of $P(z)$ as quadratic function in feasible variables $\{R(\alpha)\}$.

Let $\overline{K}[R]$ be a quadratic function which corresponds to the right-hand side of (9.48).

Consider $F(R) = \overline{K}(R) - K_0(R)$. $\overline{K}(R) \equiv K_0(R)$ on the subset of feasible vectors R satisfying the equations (9.43). So, the sum of coefficients of quadratic function $F(R)$ on memebers, that correspond to the same monomial in $P(z)$, equal zero. Thus, due to the Lagrange theorem, there exists a vector $\overline{u} = \{u_{\alpha,\alpha^1}\}$ such that

$$\overline{K}(R) = K_0(R) + \sum \overline{u}\{\alpha, \alpha^1\}P[\alpha, \alpha^1](R),$$

so $\overline{K}(R) = L(R, \overline{u})$. **The theorem is proven.**

Now we can prove the main

Theorem 124 *Let $P(z) = P(z_1, \dots, z_n)$ be a BB-polynomial and $\inf_z P(z_1, \dots, z_n) = P^*$ and ψ^* be the dual quadratic bound for the problem (9.46), (9.43), i.e.*

$$\psi^* = \sup_u \inf_R L(R, u).$$

The $\psi^* = P^*$ if and only if $\overline{P}(z) = P(z) - P^*$ can be decomposed into a sum of squares of real polynomials.

Proof.

(A) Let polynomial $P(z) \in \pi(n, 2l, 2d)$ then $\overline{P}(z) \in \pi(n, 2l, 2d)$ also. Note that $\inf_z \overline{P}(z) = 0$.

Let

$$\overline{P}(z) = \sum_{j=1}^{m} [P_j(z)]^2,$$

where $P_j(z) \in \pi(n, l, d), j \in \{1, \dots, n\}$ and $\overline{K}(R)$ be the corresponding feasible quadratic representation of $\overline{P}(z)$. Due the Lagrange theorem there exists a vector of Lagrange multipliers $\overline{u}(\alpha, \alpha^{(1)})$ such that

$$\overline{K}(R) = \sum_{j=1}^{m} (K_j(R))^2 = L(R, \overline{u}) - P^*$$

$$\psi(\overline{u}) = \inf_R L(R, \overline{u}) - P^* \geq 0$$

But $\psi(\overline{u}) \leq \inf_z \overline{P}(z) - P(z) = 0$, so $\psi(\overline{u}) = P^*$. The first part (A) of the theorem is true.

(B) Let $\psi^* = \max_{u \in dom\psi(u)} L(R, u) = L(R, \overline{u}) = P^*$. Then $L(R, \overline{u}) - P^* \geq 0$, so $L(R, \overline{u})$ is a positive semidefinite quadratic function in feasible variables R. Hence, $L(R, \overline{u})$ can be represented as sum of squares of linear functions in variables R. Let us substitute in the corresponding expression $L(R, \overline{u}) = \sum_{j=1}^{m} l_j^2(R)$, where $l_j(R)$ are linear functions, instead of $R[\alpha]$ the corresponding monomials in z. By this way one obtains the decomposition of $\overline{P}(z)$ into a sum of squares of polynomials.

The proof is over.

As it is already known, every nonnegative homogeneous polynomial of degree $2k$ in two variables can be decomposed into a sum of squares of polynomials of degree k. For $2k = 4$ this fact can be made more precise in the case of convex polynomials.

Let $CH(2;4)$ be the cone of convex homogeneous polynomials in two variables of degree 4.

Lemma 12 *If $P(x) \neq 0$ belongs to the boundary of the cone $CH(2;4)$, then it can be represented as a sum of two polynomials that are fourth degree of linear forms or in degenerate case, as a fourth degree of linear form.*

Proof. Without loss of generality one can express $P(x)$ in the form:

$$P(x) = x_1^4 + a_1 x_1^3 x_2 + a_2 x_1^2 x_2^2 + a_3 x_1 x_2^3 + a_4 x_2^4$$

Since $P(x)$ is a boundary point of the cone $CH(2;4)$ there exist a vector $\overline{x} = (\overline{x}_1, \overline{x}_2) \neq 0$ and direction $\overline{\eta} \in R^2, \|\overline{\eta}\| = 1$ such that $(H(\overline{x})\overline{\eta}, \overline{\eta}) = 0$. Without loss of generality one can take $\overline{\eta} = (1, 0)$, i.e. $\frac{\partial^2 P}{\partial x_1^2}(\overline{x}) = 0$. $\frac{\partial^2 P}{\partial x_1^2}(x) = 12x_1^2 + 6a_1 x_1 x_2 + 2a_2 x_2^2 = 12(x_1 + \frac{1}{4}a_1 x_2)^2 + (2a_2 - \frac{3}{4}a_1^2)x_2^2 \geq 0$ for all $x \in R^2$. So

$(2a_2 - \frac{3}{4}a_1^2) \geq 0$. For $x = \overline{x}$, $\frac{\partial^2 P}{\partial x_1^2}(\overline{x}) = 0$. Thus $\overline{x}_1 + \frac{1}{4}a_1\overline{x}_2 = 0$; $2a_2 - \frac{3}{4}a_1^2 = 0$. So,

$$a_2 = \frac{3}{8}a_1^2 \tag{9.49}$$

Let us record the polynomial $P(x)$ in the form:

$$P(x) = (x_1 + \frac{1}{4}a_1x_2)^4 + P_0(x) = x_1^4 + 4a_1x_1^3x_2 + \frac{3}{8}a_1^2x_1^2x_2^2 + P_1(x),$$

where $P_1(x) = P_0(x) - \frac{1}{16}a_1^3x_1x_2^3 - \frac{1}{64}a_1^4x_2^4$. Since $\frac{3}{8}a_1^2 = a_2$,

$$\begin{aligned} P(x) &= x_1^4 + a_1x_1x_2^3 + a_2x_1^2x_2^2 + P_1(x), \\ P_1(x) &= a_3x_1x_2^3 + a_4x_2^4, \\ P_0(x) &= (a_3 + \frac{1}{16}a_1^3)x_1x_2^3 + (a_4 + \frac{1}{64}a_1^4)x_2^4 \end{aligned}$$

So $P(x)$ has the form:

$$P(x) = (x_1 + \frac{1}{4}a_1x_2)^4 + \beta_3x_1x_2^3 + \beta_4x_2^4.$$

Let $x_1 = x_1' - \frac{1}{4}a_1x_2$. Then $P(x) \equiv (x_1')^4 + \beta_3x'x_2^3 + \overline{\beta}_4x_2^4 = \overline{P}(x', x_2)$, $\frac{\partial^2 \overline{P}}{\partial x_2}(x', x_2) = 12\overline{\beta}_4x_2^2 + \beta_3x'x_2 \geq 0$ for all $(x', x_2) \in R^2$. Thus, $\beta_3 = 0, \overline{\beta}_4 \geq 0$. So $P(x) = (x_1 + \frac{1}{4}a_1x_2)^4 + \beta_4x_2^4$. Lemma is proved.□

Using the Caratheodory theorem one can deduce from the previous lemma the following result:

Theorem 125 *Every convex homogeneous polynomial in two variables of the fourth degree can be represented as sum of several polynomials of the form* $[l_\alpha(x)]^4$, *where* $l_\alpha(x)$ *are linear forms.*

Remark. In [CLR 95] some interesting results about representation of nonnegative polynomial forms by sums of squares of forms are described.

Let a real polynomial form f in n variables $x = \{x_1, \ldots, x_n\}$ of total degree $m = 2t$ can be represented as sum of squares of real polynomials $h_i, i \in I$ of total degree t:

$$f(x_1, \ldots, x_n) = \sum_{i \in I} [h_i(x_1, \ldots, x_n)]^2 .$$

The smallest $|I|$ for which such a representation is possible is called **the length** of f.

Let $F_{n,m}$ denote the polynomial subring $R[x_1, \ldots, x_n]$, formed by monomials of total degree m. In [CLR 95] the so-called pythagoras numbers $P(n, m)$ were studied:

$$P(n,m) = \sup_{f \in F_{n,m}} \text{length } f.$$

There are three well-known examples:

(1) $P(n, 2) = n$;

(2) $P(2, m) = 2; (m > 0)$

(3) $P(3, 4) = 3$ (the Hilbert theorem for ternary quartics [Hi 88], see also [Sw 95]).

R.M. Robinson [Ro 73] proved that

$$P(n,m) \leq \binom{n+m-1}{n-1}.$$

If a form f is written as a sum of squares:

$$f = \sum h_i^2,$$

then we can associate with this expression a "Gram matrix" obtained from dot products of the coefficients of the h_i's.

By considering all possible expressions of f as a sum of squares we get a family of associated Gram matrices. The minimum of the ranks of these Gram matrices turns out to be precisely the length of f (see [CLR 95], [Rez 92]).

This result provides the key for obtaining the lower and upper bounds for pythagoras numbers $P(n, m)$. The technique used in [CLR 95] is based on relations between dimensions of some algebraic constructions in the ring of polynomials of n variables. In particular, the following results are proved in [CLR 95]:

(i) if we fix m and let n vary, the $P(n, m)$ has the order of magnitude $n^{m/2}$. For quartic forms ($m = 4$) the more exact result is obtained:

$$0.092n^2 \leq P(n,4) \leq 0.289n^2,$$

when n is large.

For biquadratic forms

$$f(x,y) = \sum a_{ijkl} x_i x_j y_k y_l \ (1 \le i,j \le n_1; 1 \le k; l \le n_2)$$

the upper bound for pythagoras number is

$$\pi(n_1,n_2) = \left(\sqrt{1 + 2n_1 n_2 (n_1+1)(n_2+1)} - 1\right)/2.$$

In conclusion we can note that determination of the shortest representation of a nonnegative polynomial form as a sum of squares is a very hard problem even for $m = 4, n > 3$.

Remark. The generalized version of Theorem 124 is given in [Ferr 97]. In this article the extremal polynomial problems are considered in the following form: find

$$(\mathcal{P}) \begin{cases} \inf P_0(x) \\ x \in R^n \text{ st} \begin{cases} P_1(x) = 0 \\ \dots\dots\dots \\ P_m(x) = 0 \end{cases} \end{cases}$$

where $P_0, \dots, P_m$ are m polynomial forms.

Introducing new variables one may reduce (by ununique way) the problem $(\mathcal{P})$ to a quadratic-type problem $(\mathcal{Q})$ similarly as we performed for unconstrained case:

$$(\mathcal{Q}) \begin{cases} \inf K_0(R) \\ R \in R^p \text{ st} \begin{cases} M_1(R) = 0 \\ \dots\dots\dots \\ M_s(R) = 0 \end{cases} \end{cases}$$

where $K_0, M_1, \dots, M_s$ are real quadratic functions on R^p, p is the number of new variables. Let Υ be a quadratic manifold corresponding to the constraints $M_i(R) = 0, i = 1, \dots, m$. The manifold Υ may be described by different systems of quadratic equations. One can introduce nonlinear consequences of original system and obtain modified Lagrange functions and the corresponding dual bounds.

In [Ferr 97] the problem of finding the best dual bound is discussed (for the case where the nontrivial dual quadratic bound exists). The technique developed in this book for unconstrained case is used.

9.4 INSTANCES OF FINDING GLOBAL MINIMUM OF POLYNOMIAL FUNCTIONS USING R-ALGORITHM

I. Consider nonnegative homogeneous polynomial functions of degree 4 in 3 variables $\overline{P}(x) = \overline{P}(x_1, x_2, x_3) \in H(3,4)$ and function

$$f(x) = \frac{\overline{P}(x)}{x_3^4}.$$

Let $y_1 = \frac{x_1}{x_3}$; $y_2 = \frac{x_2}{x_3}$; $y = \{y_1, y_2\}$. Then $f(x)$ can be represented as a polynomial $P(y) = P(y_1, y_2)$ of the fourth degree in two variables. Let $P^* = \inf P(y_1, y_2)$. Due to the Hilbert theorem [Hi 88], $\overline{P}(x_1, x_2, x_3)$ can be represented as sum of squares of quadratic polynomials. Hence the polynomial $P(y_1, y_2) - P^*$, (nonnegative in general) also can be decomposed into a sum of squares of polynomials of two variables.

For testing we considered two families of polynomials:

$$P_\gamma(y_1, y_2) = (y_1 + 1)^4 - \gamma y_1^2 y_2^2 + (y_2 + 2)^4;$$

$$P_{\alpha_1,\alpha_2}(y_1, y_2) = \alpha_1 y_1^4 - 2y_1^3 + y_1^2 + \alpha_2 y_2^4 - 2y_2^3 + y_2^2$$

with the corresponding parameters: $\gamma(\gamma \in [0,2))$ for the first instance; $\alpha = (\alpha_1, \alpha_2), (\alpha_1, \alpha_2 \in [1-\varepsilon, 1+\varepsilon])$, where $\varepsilon > 0$, for the second instance.

The problem of finding global minimum for given polynomials was reduced to the problem of minimization of quadratic-type function in five variables:

$$x = \{x_1, x_2, x_3, x_4, x_5\},$$

$$x_1 = y_1; x_2 = y_2; x_3 = y_1^2; x_4 = y_2^2; x_5 = y_1 y_2$$

We consider quadratic representations

$$K_1(x) = 6x_1^2 + 4x_1x_3 + 24x_2^2 + 8x_2x_4 + x_3^2 - \gamma x_3x_4 + x_4^2 + 4x_1 + 32x_2 + 17$$

for family $P_\gamma(y_1, y_2)$ and

$$K_2(x) = x_1^2 - 2x_1x_3 + a_1x_3^2 + x_2^2 - 2x_2x_4 + a_2x_4^2$$

for family $P_{a_1,a_2}(y_1, y_2)$ and solve the problems: find

$$\min_x K_i(x), i = 1, 2,$$

subject to the following constraints:

$$x_1^2 - x_3 = 0; x_2^2 - x_4 = 0; x_1x_2 - x_5 = 0$$

$$x_5^2 - x_3x_4 = 0; x_3x_2 - x_1x_5 = 0; x_4x_1 - x_2x_5 = 0.$$

The Lagrange function has the form:

$$L(x,u) = K_i(x) + u_1(x_1^2 - x_3) + u_2(x_2^2 - x_4) + u_3(x_1x_2 - x_5) + \\ +u_4(x_5^2 - x_3x_4) + u_5(x_3x_2 - x_1x_5) + u_6(x_4x_1 - x_2x_5), i = 1,2$$

For obtaining dual bounds we used modification of r-algorithm described at the p.107 with the following parameters; $\alpha = 2$, $h_0 = 1.0$, $q_1 = 1.0$, $n_h = 3$, $q_2 = 1.1$. As a starting point we chose $u^{(0)} = (1,1,1,1,1,1)$. The stopping criterion was: $\varepsilon_x \leq 10^{-5}$ and $\varepsilon_g \leq 10^{-5}$ (see Chapter 3).

The results of calculations for $P_\gamma(y_1, y_2)$ with $\gamma \in [0, 1.9]$ are given in the table 9.1. Since $\gamma = 2$ is a critical value for family $P_\gamma(y_1, y_2)$ the results of calculations on segment $\gamma \in [0, 1.9]$ are given in table 9.2. In tables 9.1 and 9.2 we use the folowing notations: γ is the value of paramaters; *iter* is the number of iterations; $n_{\bar{\psi}}$ is the total number of calculations of function; n_λ – the number of calculations of eigenvalues; ψ_r^* is the record value of $\psi(u)$; y_1^*, y_2^* are approximate values of coordinates of global minimum; ε_1^* is the value of $||g_\psi||$ in approximate values of coordinates of the global minimum; $\varepsilon_2^* = \varepsilon_2^*/(\max(|y_1^*|^3, |y_2^*|))^3$.

In tables 9.3 the results of tests for polynomial $P_{a_1,a_2}(y_1, y_2)$ are given.

One can see that r-algorithm gives results with good accuracy even in a small neighborhood of critical points of parameters γ and $a = \{a_1, a_2\}$:

Table 9.1 Tests for polynomial $P_\gamma(y_1, y_2)$, $\gamma \in [0, 1.9]$

γ	$iter(n_{\tilde{\psi}}, n_\lambda)$	ψ_r^*	ε_1^*	y_1^*	y_2^*
.0	82(119,34)	$-.39888 * 10^{-11}$	$9.41 * 10^{-6}$	-1.0001	-1.9995
.2	69(93,5)	-5.9166	$4.56 * 10^{-6}$	-2.3399	-3.2063
.3	67(96,4)	-13.471	$4.32 * 10^{-6}$	-2.7468	-3.5967
.4	74(100,4)	-26.140	$7.81 * 10^{-6}$	-3.1634	-4.0006
.5	68(94,5)	-46.605	$9.12 * 10^{-6}$	-3.6080	-4.4347
.6	71(91,5)	-79.096	$7.16 * 10^{-6}$	-4.0953	-4.9131
.7	70(94,5)	-130.42	$3.93 * 10^{-6}$	-4.6406	-5.4505
.8	67(97,6)	-211.80	$5.80 * 10^{-6}$	-5.2623	-6.0652
.9	80(120,6)	-342.36	$5.52 * 10^{-6}$	-5.9842	-6.7808
1.0	82(118,12)	-555.83	$3.70 * 10^{-6}$	-6.8386	-7.6294
1.1	82(123,14)	-914.53	$8.93 * 10^{-6}$	-7.8716	-8.6572
1.2	90(130,16)	-1540.1	$5.24 * 10^{-6}$	-9.1520	-9.9326
1.3	92(137,23)	-2686.8	$4.54 * 10^{-6}$	-10.787	-11.563
1.4	91(148,26)	-4935.4	$8.13 * 10^{-6}$	-12.956	-13.728
1.5	103(156,30)	-9777.9	$3.59 * 10^{-6}$	-15.981	-16.748
1.6	103(166,38)	$-21724.$	$1.71 * 10^{-6}$	-20.504	-21.267
1.7	108(176,44)	$-58173.$	$1.08 * 10^{-6}$	-28.025	-28.785
1.8	118(198,57)	$-.22042 * 10^6$	$4.28 * 10^{-6}$	-43.046	-43.802
1.9	176(308,74)	$-.19687 * 10^7$	$8.23 * 10^{-6}$	-88.065	-88.818

Table 9.2 Test for polynomial $P_\gamma(y_1, y_2)$, $\gamma \to 2$

γ	$iter(n_{\tilde{\psi}}, n_\lambda)$	ψ_r^*	ε_2^*	y_1^* ($y_2^* - y_1^*$)
1.95000000	189(345,89)	$-1.66 * 10^7$	$4.2 * 10^{-12}$	-178.07(-.75158)
1.97500000	204(378,107)	$-1.36 * 10^8$	$2.2 * 10^{-12}$	-358.08(-.75079)
1.98750000	205(384,132)	$-1.11 * 10^9$	$3.9 * 10^{-8}$	-718.08(-.75040)
1.99375000	196(364,147)	$-8.90 * 10^9$	$2.6 * 10^{-9}$	-1438.1(-.75019)
1.99687500	218(416,177)	$-7.14 * 10^{10}$	$1.5 * 10^{-9}$	-2878.1(-.75010)
1.99843750	212(421,177)	$-5.72 * 10^{11}$	$1.5 * 10^{-11}$	-5758.1(-.75005)
1.99921875	228(484,207)	$-4.58 * 10^{12}$	$1.9 * 10^{-8}$	-11518.(-.74997)
2.00000000	153(523,523)	–	–	–

Table 9.3 Tests for polynomial $P_{a_1,a_2}(y_1,y_2)$

a_1	a_2	$iter(n_{\tilde{\psi}}, n_\lambda)$	ψ_r^*	y_1^*	y_2^*
1.000	1.000	152(244,58)	$-4.1 * 10^{-8}$	.59548	.44913
1.100	1.100	63(84, 8)	$-3.6 * 10^{-12}$	$5.5 * 10^{-6}$	$5.3 * 10^{-6}$
.9000	.9000	73(100,11)	−.29964	1.2060	1.2060
1.100	.9000	67(93, 7)	−.14982	$1.5 * 10^{-6}$	1.2060
.9000	1.100	71(104,12)	−.14982	1.2060	$-6.1 * 10^{-7}$
1.010	1.010	88(120,13)	$-5.3 * 10^{-13}$	$-5.0 * 10^{-6}$	$5.0 * 10^{-6}$
.9900	.9900	89(127,18)	−.020816	1.0200	1.0200
1.010	.9900	89(139,13)	−.010408	$4.9 * 10^{-6}$	1.0200
.9900	1.010	83(110,15)	−.010408	1.0200	$1.8 * 10^{-6}$
1.001	1.001	100(141,26)	$-2.7 * 10^{-14}$	$-6.0 * 10^{-6}$	$-1.4 * 10^{-6}$
.9990	.9990	110(162,21)	−.00201	1.0020	1.0020
1.001	.9990	109(159,22)	−.00100	$-2.7 * 10^{-6}$	1.0020
.9990	1.001	107(160,19)	−.00100	1.0020	$3.2 * 10^{-6}$

REFERENCES

[A 54] D. AGMON, *The relaxation method for linear inequalities.* Canad. J. Math. 6:382-392, 1954.

[AC 78] D. AVIS, V. CHVATAL, *Notes on Bland's pivoting Rule*, MP Study 8:24-38, 1978.

[AG 64] M.A. AIZERMAN AND F.R. GANTMACHER, *Absolute stability of regulator systems.* Information Systems. Holden-Day, San Francisco, 1964.

[AH 77a] K. APPEL AND W. HAKEN, *Every planar map is four colorable, part 1:discharging*, Ill. J. Math. 21:429-490, 1977.

[AH 77b] K. APPEL AND W. HAKEN , *Every planar map is four colorable, part 2:reducibility*, Ibid. pp.429-490, 1977.

[AHKS 87] E. ALLEN, R. HELGASON, R. KENNINGTON, AND B. SHETTY, *A generalization of Polyak's convergence result for subgradient optimization*, Mathematical Programming, 37:309-317, 1987.

[AHU 76] A.V. AHO, J.K. HOPCROFT, AND J.D. ULLMAN, *The design and analysis of computer algorithms*, Addison-Wesley, Reading, MA, 1976.

[Ak 59] H. AKAIKE, *On a successive transformation of probability distribution and its application to the optimum gradient method*, Ann. Inst. Statist. Math., pp.1-16, Tokyo, 1959.

[Akg 83] M. AKGUL, *Solving convex-concave games via ellipsoidal algorithm*, University of Waterloo Research Report CORR 81-16, 1983.

[Akg 84] M. AKGUL, *Topics in Relaxation and Ellipsoidal Methods*, volume 97 of Research Notes in Mathematics. Pitman, 1984.

[Al 91] F. ALIZADEH, *Combinatorial optimization with interior point methods and semi-definite matrices*, Ph.D. thesis, University of Minnesota, Minneapolis, MN, 1991.

[Al 92a] F. ALIZADEH, *Combinatorial optimization with semidefinite matrices*, In Proceedings of second annual Integer Programming and Combinatorial Optimization conference, Carnegie-Mellon University, 1992.

[Al 92b] F. ALIZADEH, *Optimization over the positive-definite cone: interior point methods and combinatorial applications*, In Panos Pardalos, editor, Advances in Optimization and Parallel Computing, North-Holland, 1992.

[Al 95] F. ALIZADEH, *Interior point methods in Semi-Definite Programming with applications to Combinatorial Optimization*, SIAM Journal on Optimization 5:13-51, 1995.

[All 89] J. ALLWRIGHT, *On maximizing the minimum eigenvalue of a linear combination of symmetric matrices*, SIAM J. on Matrix Analysis and Applications, 10:347-382, 1989.

[AM 90] B. ANDERSON AND J.B. MOORE, *Optimal Control: Linear Quadratic Methods*, Prentice-Hall, 1990.

[AMP 80] I. ADLER, R.P. MCLEAN, AND J.S. PROVAN, *An application of the Khachian-Shor algorithm to a class of linear complementary problems*, Cowles Foundation discussion paper no. 549, 1980.

[An 67] B.D. ANDERSON, *A system theory criterion for positive real matrices*, SIAM J. Control, 5:171-182, 1967.

[Ans 83] R.E. ANSTEE, *A polynomial algorithm for b-matchings: an alternative approach*: Research Report CORR 83-22, University of Waterloo, 1983.

[Anst 89] K.M. ANSTREICHER, *A combined Phase I-Phase 2 projective algorithm for linear programming*, Math. Programming, 43:209-223, 1989.

[Ar 74] L. ARNOLD, *Stochastic differential equations: theory and applications*, Wiley, New York, 1974.

[Art 27] E. ARTIN, *Ueber die Zerlegung definite Funktionen in Quadrate*, Hamburg Abh. 5:100-15, 1927.

[ASha 90] I. ADLER AND R. SHAMIR, *A Randomized Scheme for Speeding Up Algorithms for Linear and Convex Quadratic Programming Problems with a High Constraints-to-Variables Ratio*, Math. Programming 61:39-52, 1993.

[AShi 82] JA. I. AL'BER, S.V. SHILMAN, *A unified approach to the problem of minimizing smooth and nonsmooth functions*, Engineering Cybernetics 20, no. 1:21-27, 1982.

[AShil 80] B. ASPVALL, G. SHILOACH, *A polynomial algorithm for solving system of linear inequalities with two variables per inequality*, SIAM J. Computing 9:827- 845, 1980.

[Au 79] J.P. AUBIN, *Mathematical methods of game and economic theory*, North-Holland, Amsterdam 1979.

[Aus 77] A. AUSLENDER, *Programmation convexe avec erreurs: methodes d'epsilon-sous-gradients*, Compt. Rend. Acad. Sci. Ser. A 284: 109-112, 1977.

[Aus 78] A. AUSLENDER, *Minimisation de fonctions localement lipschitziennes: applications a la programmation mi- convexe, midifferentiable*, In: Nonlinear Programming 3 (O.L. Mangasarian, R.R. Meyer, S.M. Robinson, eds.), pp. 429-460. Academic Press: New York 1978.

[Aus 79] A. AUSLENDER, *Differential stability in nonconvex and nondifferentiable programming*, In: Point-to-Set Maps in Mathematical Programming (P. Huard, ed.) Math. Programming Study 10:29-41, North Holland: Amsterdam 1979.

[Aus 87] A. AUSLENDER, *Numerical methods for non-differentiable convex optimization*, Mathematical Programming Study 30:102-126, 1987.

[AV 73] B.D.O. ANDERSON AND S. VONGRANITLERD, *Network analysis and synthesis: a modern systems theory approach*,Prentice-Hall, 1973.

[AW 63] J.ABADIE, A.WILLIAMS, *Dual and parametric methods in decomposition*. In: Recent Advances in Mathematical Programming (R.L.Graves,P.Wolfe, eds.) pp.149-158. McGraw-Hill: New York 1963.

[Ba 91] L. BABEL, *Finding maximum clique in arbitrary and in special graphs*, Computing 46:321-341, 1991.

[Bak 80] G.M. BAKAN, *Filtration of a multivalued process with bounded noise*, Kibernetika i Vych. Teknika 49:7-18, 1980.

[Bal 85] M.L. BALINSKI, *Signature methods for the assignment problem*, Operations Research 33:527-536, 1985.

[Bar 89] N.E. BARABANOV, *Absolute stability of sample-data control systems*, Automation and Remote Control, pp.981-988, 1989.

[Barah 79] F. BARAHONA, *Sur la complexite du probleme du verre de spins*, Rapport de recherche n 177, Mathematiques Appliques et Informatiques, Universite Scientifique et Medicale de Grenoble, France, Octorber 1979.

[Barah 80] F. BARAHONA, *On the complexity of max cut*, Rapport de recherche n 186, Mathematiques Appliques et Informatiques, Universite Scientifique et Medicale de Grenoble, France, February 1980.

[Barah 81] F. BARAHONA, *Balancing signed graphs of fixed genus in polynomial time*, Depto. de Matematicas, Universidad de Chile, Santiago, Chile, June 1981.

[Barah 83] F. BARAHONA, *The max-cut problem in graphs not contractible to K_5*, Operations Research Letters, 2:107-111, 1983.

[Barah 93] F. BARAHONA, *On cuts and mateling in planar graphs*, Math. Progr., 60(1):53-68, 1993.

[Barm 83] B.R. BARMISH, *Stabilization of uncertain systems via linear control*, IEEE Trans. Aut. Control, AC-28(8):848-850, 1983.

[Barm 85] B.R. BARMISH, *Necessary and sufficient conditions for quadratic stabilizability of an uncertain system*, Journal of Optimization Theory and Applications, 46(4):399-408, 1985.

[Barm 93] B.R. BARMISH, *New Tools for Robustness of Linear Systems*, MacMillan, 1993.

[Barn 82] E.R. BARNES, *An algorithm for partitioning the nodes of a graph*, SIAM J.Alg. and Disc.Math., 3, 1982.

[Barn 86] E.R. BARNES, *A variation on Karmarkar's algorithm for solving linear programming problems*, Math. Programming, 36:174-182, 1986.

[Bart 81] R.R. BARTON, *Nondifferentiable optimization methods for a traffic assignment decomposition model*, Report 81-EM-018, RCA Labs., Princeton N.J., 1981.

[Baz 72] L.G.BAZHENOV, *On the conditions for convergence of methods for minimizing almost differentiable functions.* Kibernetika (Kiev), no.4:71-72, 1972.

[BB 90] S. BOYD AND V. BALAKRISHNAN, *A regularity result for the singular values of a transfer matrix and a quadratically converging algorithm for computing its L-norm*, Systems Control Lett., 15:1-7, 1990.

[BB 91] S. BOYD AND C. BARRATT, *Linear Controller Design: Limits of Performance*, Prentice-Hall, 1991.

[BBB 91] V. BALAKRISHNAN, S. BOYD, AND S.BALEMI, *Branch and bound algorithm for computing the minimum stability degree of parameter-dependent linear systems*, Int. J. of Robast and Nonlinear Control, 1(4):295-317, October-December 1991.

[BBFG 93] S. BOYD, V. BALAKRISHNAN, E. FERON, AND L.EL GHAOUI, *Control system analysis and synthesis via linear matrix inequalities*, In Proc. American Control Conf., 2:2147-2154, San Francisco, California, June 1993.

[BBFG 94] S. BOYD, V. BALAKRISHNAN, E. FERON, AND L.EL GHAOUI, *Linear matrix inequalities in system and control theory*, SIAM Studies in Applied Mathematics, Philadelphia, 1994.

[BBK 89] S. BOYD, V. BALAKRISHNAN, AND P.KABAMBA, *A bisection method for computing the H norm of a transfer matrix and related problems*, Mathematics of Control, Signals, and Systems, 2(3):207-219, 1989.

[BBP 78] G.P. BARKER, A. BERMAN, AND R.J. PLEMMONS, *Positive definite solutions to the Lyapunov equatios*, Linear and Multilinear Algebra, 5:249-256, 1978.

[BC 74] J.W. BANDLER, C. CHARALAMBOUS, *Nonlinear programming using minimax techniques.* J. Optim. Theory Appl. 13:607-619, 1974.

[BCHP 97] I. BOMZE, T. CSENDES, R. HORST AND P.M. PARDALOS, *Developments in Global Optimization*, Kluwer Academic Publishers, 1997.

[BD 94] J.M. Bomze and G. Danninger, *A finite algorithm for solving general quadratic problems*, J. Global Optimiz., 4:1-16, 1994.

[BDi 86] V. G Borisov and S.N. Diligenskii, *Numerical method of stability analysis of nonlinear systems*, Automation and Remote Control, pp. 1373-1380, 1986.

[BDDW 90] P. Boggs, P.D. Domich, J.R. Donaldson and C. Wrzgall, *Algorithmic enhancements to the method of centers for linear programming problems*, ORSA J. Comput., 1:159-171, 1990.

[Be 93] J.E.Beasley, *"Lagrangean relaxation"*, in Modern Heuristic Techniques for Combinatorial Problems, C.R. Reeves, ed., Blackwell Scientific Publications, Oxford, pp. 243-303, 1993.

[Bec 93] G. Becker, *Quadratic Stability and Performance of Linear Parameter Dependent Systems*, PhD thesis, University of California, Berkeley, 1993.

[Beh 37] F. Behrend, *Uber einige Affininvarianten konvexer Bereiche*, Math. Ann., 115:713-747, 1937.

[Bel 70] R. Bellman, *Introduction to Matrix Analysis* (McGraw-Hill, New York, 1970, 2nd ed.).

[Belo 77] E.G. Belousov, *Introduction to the convex analysis and integer programming*, Moscow University, 1977.

[Ben 62] J.F. Benders, *Partitioning procedures for solving mixed-variables programming problems*, Numerische mathematik, 4:238-252, 1962.

[Bens 92] A. Bensoussan, *Stochastic Control of Partially Observable Systems*, Cambridge University Press, Cambridge, Great Britain, 1992.

[Ber 90] O.A. Berezovski, *Dual algorithm for constructing optimal circumscribed ellipsoids*, In:

IK AN USSR, 1990.

[Berg 61] C. Berge, *Farbung von Graphen, deren samtliche bzw. deren ungerade Kreise starr sind (Zusammenfassung)*, Wissenschaftliche Zeitschrift, Martin Luther Universitat Halle-Wittenberg, Mathematisch-Naturwissenschaftliche Reihe, pp. 114-115, 1961.

[Berg 62] C. BERGE, *Sur une conjecture relative au probleme des codes optimaux*, Communication, 13eme assemblee generale de l'URSI, Tokyo, 1962.

[Bert 75a] D. BERTSEKAS, *Necessary and sufficient conditions for a penalty method to be exact*, Math. Programming 9:87-99, 1975.

[Bert 75b] D. BERTSEKAS, *Nondifferentiable optimization via approximation*. In: Math. Programming Stud. 3 (M.L. Balinski, P. Wolfe, eds.), pp. 1-25. North. Holland: Amsterdam 1975.

[Bert 76] D.P. BERTSEKAS, *On the Goldstein-Levitin-Polyak gradient projection method*, IEEE Transactions on Automatic Control, AC-21(2):174-184, 1976.

[Bert 77] D.P. BERTSEKAS, *Approximation procedures based on the method of multipliers*, J. Optim. Theory Appl. 23:487-510, 1977.

[Bert 82] D.P. BERTSEKAS, *Constrained Optimization and Lagrange Multiplier Methods*, Academic Press, New York, NY, 1982.

[Bes 75] M. BEST, *A method to accelerate the rate of convergence of a class of optimization algorithms*, Math. Programming 9:139-160, 1975.

[BF 63] R. BELLMAN AND K. FAN, *On systems of linear inequalities in matrix variables*, in: V.L. Klee, ed., Convexity, pp.1-11, American Mathematical Society, Providence, RI, 1963.

[BG 83] J.R. BIRGE, AND A.GANA, *Computational complexity of van der Heyden's variable dimension algorithm and Dantzig-Cottle's principal pivoting method for solving LCP's*, MP 26:316-325, 1983.

[BGFB 94] S. BOYD, L.EL GHAOUI, E. FERON, AND V.BALAKRISHNAN, *Linear matrix inequalities in system and control theory*, SIAM, Philadelphia, 1994.

[BGh 93] S. BOYD AND L.EL GHAOUI, *Method of centres for minimizing generalized eigenvalues*, Linear Algebra and Applications, special issue on Numerical Linear Algebra Methods in Control, Signals and Systems, 188:63-111, 1993.

[BGe 93] V. BLONDEL AND M. GEVERS, *Simultaneous stabilizability of three linear systems is rationally undecidable*, Mathematics of Control, Signals, and Systems, 6:135-145, 1993.

[BGK 82] A. BACHEM, M. GROTCHEL AND B. KORTE (EDS.), *Mathematical Programming: The State of Art*, Bonn 1982, (Springer, N.Y., 1983).

[BGLS 95] J.F. BONNANS, J.CH. GILBERT, C. LEMARECHAL, C.A. SAGASTIZABAL, *A family of variable metric proximal methods*, Math. Progr. 68:15-47, 1995.

[BGo 77] M.S.BAZARRA AND J.J. GOODE, *The traveling Salesman problem: a duality approach*, Mathematical programming 13(2):221-237, 1977.

[BGT 81] R.G. BLAND, D. GOLDFARB, AND M.J. TODD, *The ellipsoid method: A survey*, Operations Research, 29(6):1039-1091, 1981.

[BH 89a] D. BERNSTEIN AND W. HADDAD, *LQG control with an H performance bound: A Riccati equation approach*, IEEE Trans. Aut. Control, AC-34(3):293-305, 1989.

[BH 89b] D.B. BERNSTEIN AND W.D. HADDAD, *Robust stability and performance analysis for linear dynamic systems*, IEEE Trans. Aut. Control, 34(7):751-758, 1989.

[Bha 87] R. BHATIA, *Perturbation Bounds for Matrix Eigenvalues*, Pitman Research Notes in Mathematics series 162, 1987.

[BHL 89] A.C. BARLETT, C.V. HOLLOT, AND H.LIN, *Root locations of an entire polytope of polynomials: it suffices to check the edges*, Mathematics of Control, Signals, and Systems, 1(1):61-71, 1989.

[BHo 84] A.E. BARNES AND A.J. HOFFMAN, *Partitioning, spectral and linear programming*, Progress in combinatorial optimization, Academic Press: New York, 1984.

[Bi 88] G. BITSORIS, *Positively invariant polyhedral sets of discrete-time linear systems*, Int. J. Control, 47:1713-1726, 1988.

[BIG 74] A. BEN-ISRAEL, AND T.N.E. GREVILLE, *Generalized Inverses and its Applications*, (Wiley, N.Y., 1974).

[BK 93] B.R. BARMISH AND H.I. KANG, *A survey of extreme point results for robustness of control systems*, Automatica, 29(1):13-35, 1993.

[BKKM] H.J. BROESMA, T. KLOKS, D. KRATSCH AND H. MULLER, *Independent sets in asteroidal triple-free graphs*,

[BL 89] D.A. BAYER AND J.C. LAGARIAS, *The nonlinear geometry of linear programming*, I: Affine and projective scaling trajectories; II: Legendre transform coordinates and central trajectories, AT&T Bell Laboratories, Murray Hill, NJ, 1986; Trans. Amer. Math. Soc., 314:499-581, 1989.

[BL 91] D.A. BAYER AND J.C. LAGARIAS, *Karmarkar's algorithm and Newton's method*, Math. Programming, 50:291-330, 1991.

[Bl 77] R.B. BLAND, *New finite pivoting rules for the simplex method*, MOR 2:103-107, 1977.

[BM 73] D.BERTSEKAS, S. MITTER, *A descent numerical method for optimization problems with nondifferentiable cost functionals*. SIAM J. Control 11:637-652, 1973.

[BMW 89] A. BALAKRISHNAN, T.L. MAGNANTI, AND R.T. WONG, *A dual-ascent procedure for large-scale uncapacitated network design*, Operations Research, 37:716-740, 1989.

[BN 81] G.M. BAKAN, AND E.A. NIZHNICHENKO, *A finite-convergent algorithm for the solution of a countable system of inequalities*, Kibernetika Vych. Teknika 53:78-83, 1981. (Russian)

[BNo 89] R.H. BYRD AND J. NOCEDAL, *A tool for the analysis of quasi-Newton methods with application to unconstrained minimization*, SIAM Journal of Numerical Analysis 26:727-739, 1989.

[BNS 87] A. BIHAIN, V.H. NGUYEN, AND J.-J. STRODIOT, *A reduced subgradient algorithm*, Mathematical programming Study, 30:127-149, 1987.

[BNS 89] A. BERMAN, M. NEUMANN, AND R. STERN, *Nonnegative Matrices in Dynamic Systems*, Wiley-Interscience, 1989.

[BO 93] J.V. BURKE AND M.L. OVERTON, *On the subdifferentiability of functions of a matrix spectrum: I: Mathematical foundations; II: Subdifferential formulas*, in: F. Giannessi, ed., Nonsmooth Optimization, Proceedings of a Conference Held at Erice, Italy, June 1991, to appear(1993).

[Bom 92] JM.M. BOMZE, *Copositivity conditions for global optimality in general quadratic problems*, Czechoslovac J. of OR, 1:7-19, 1992.

[Bom 97] JM.M. BOMZE, *Evolution towards the maximum clique*, Journal of Global Optimization 10:143-164, 1997.

[Bop 87] R.B. BOPPANA, *Eigenvalues and graph bisection: an average case analysis*, in Proc. 28th JEEE Annual Symposium on Foundations of Computer Science, 1987.

[Bor 82] K.H. BORGWARDT, *The average number of pivot steps required by the simplex method is polynomial*, Zeitschrift fur Operations Research, A 26:157-178, 1982.

[BP 88] A.V. BOGATYREV AND E.S. PYATNITSKII, *Construction of piecewise quadratic Lyapunov functions for nonlinear control systems*, Automation and Remote Control, pp. 1292-1299, 1988.

[BPG 89] J. BERNUSSOU, P.L.D. PERES, AND J.C. GEROMEL, *A linear programming oriented procedure for quadratic stabilization of uncertain systems*, Syst. Control Letters, 13:65-72, 1989.

[BPPB 93] G. BECKER, A.PACKARD, D. PHILBRICK, AND G. BALAS, *Control of parameter dependent systems: applications to H gain scheduling*, In 1993 IEEE Regional Conference on Aerospace Control Systems, pp. 329-333, May 1993.

[Br 95] U. BRANNLUND, *A generalized subgradient method with relaxation step*, Mathematical Programming, 71:207-219, 1995.

[Bre 65] L.M. BREGMAN, *The method of successive projection for finding a common point of convex sets*, SMD 6:688-692, 1965.

[Bre 67] L.M. BREGMAN, *The relaxation method of finding the common point of convex sets and its application to the solution of problems in convex programming*, USSR Computational Mathematics and Mathematical Physics, 7:200-217, 1967.

[Brel 79] D. BRELAZ, *New methods to color the vertices of a graph*, communications of ACM 4:251-256, 1979.

[BS 79] M.S. BAZARAA, C.M. SHETTY, *Nonlinear programming, Theory and applications.* Wiley: New York 1979.

[BS 81] M.S. BAZAARA, H.D. SHERALI, *On the choice of the step size in subgradient optimization*, European J. O.R. 7:380-388, 1981.

[BSS 93] M.S. BAZAARA, H.D. SHERALI, AND C.M. SHETTY, *Nonlinear Programming. Theory and Algorithms*, Wiley, New York, NY, 1993.

[BT 79] R.K. BRAYTON AND C. H. TONG, *Stability of dynamical systems: A constructive approach*, IEEE Trans. Circuits Syst., CAS-26(4):224-234, 1979.

[BT 80] R.K. BRAYTON AND C.H. TONG, *Constructive stability and asymptotic stability of dynamical systems* IEEE Trans. Circuits Syst., CAS-27(11):1121-1130, 1980.

[BTr 76] J. BOROSH, L.B. TREYBIG, *Bounds on positive integral solutions of linear diophantine equations* Proc. Amer. Math. Soc., 55, N2, 1976.

[BTN 92] A. BEN-TAL AND A. NEMIROVSKII, *Interior point polynomial time methods for truss topology design*, Research Report # 3/92, Faculty of Industrial Engineering and Management, The Israel Institute of Technology, Haifa 32000, Israel, June 1992.

[Bu 79] V.A. BULAVSKII, *Relaxation in problems with inequalities*, Optimizacija 23 (40):32-40, 128-129, 1979. (Russian) MR 81h, 90122.

[Bus 58] H. BUSEMANN, *Convex surfaces*. Interscience: New York 1958.

[BV 95] S. BOYD AND L. VANDENBERGHE, *A primal-dual potential reduction method for problems involving matrix inequalities*, Mathematical Programming, Series B 69:205-236, 1995.

[BW 75] M.L. BALINSKI, P. WOLFE,(EDS.), *Nondifferentiable optimization*. Math. Programming Stud. 3. North Holland, Amsterdam 1975.

[BW 92] M. BAKONYI AND H.J. WOERDEMAN, *The central method for positive semi-definite, contractive and strong Parrott type completion problems*, Operator Theory: Advances and Applications, 59:78-95, 1992.

[BY 86] E. BALAS AND C. YU, *Finding a maximum clique in an arbitrary graph*, SIAM J. of computing 15, N4:1054-1068, 1986.

[BZS 76] L.V. BELAEVA, N.G. ZHURBENKO, N.Z. SHOR, *On a certain nonlinear transpotation problem*. In:Teor. Optimal. Resenii, Trudy Sem. Nauc.Sov. Akad. Nauk Ukrain. SSSR po Kibernet.,pp. 83-89. 1976.

[BZS 78] L.V. BELAEVA, N.G. ZHURBENKO, N.Z. SHOR, *On a method for solving a class of dynamic distribution problems*, Econom. i Mat. Metody 14:137-146, 1978.

[CD 68] R.W. COTTLE AND G.B. DANTZIG, *Complementary pivot Theory of mathematical programming*, Linear Algebra and its Applications 1:103-125, 1968.

[CD 92] G.E. COXSON AND C.L. DEMARCO, *Computing the real structured singular value is NP-hard*, Technical Report ECE-92-4, Dept. of Elec. and Comp. Eng., Univ. of Wisconsin-Madison, June 1992.

[CDW 75] J.CULLUM, W.E. DONATH AND P.WOLFE, *The minimization of certain nondifferentiable sums of eigenvalues of symmetric matrices*, Mathematical Programming Study 3:35-55, 1975.

[CDZ 86] R.W. COTTLE, S.G. DUVALL, AND K. ZIKAN, *A Lagrangean relaxation algorithm for the constrained matrix problem*, Naval Research Logistics Quarterly, 33:55-76, 1986.

[CF 82] W.H. CUNNINGHAM, AND A. FRANK, *A primal-dual algorithm for submodular flows*, Report no.82224-OR, (1982), Inst. Okon. Oper. Res.., Universitat Bonn.

[CFM 75] P. CAMERINI, L. FRATTA, F. MAFFIOLI, *On improving relaxation methods by modified gradient techniques*, In: Math. Programming Study, 3 (M.L.Balinski, P Wolfe eds.), pp. 26-34. North Holland: Amsterdam 1975.

[CG 59] E.W. CHENEY, A.A. GOLDSTEIN, *Newton's method for convex programming and Chebyshev approximation*, Numer. Math. 1:253-268, 1959.

[CGG 77] A. CHARNES, D. GRANOT AND F. GRANOT, *A primal algorithm for interval programming*, Linear Algebra and its Applications 17:65-78, 1977.

[CH 93] E.B. CASTELAN AND J.C. HENNET, *On invariant polyhedra of continuous-time linear systems*, IEEE Trans. Aut. Control, 38(11):1680-1685, December 1993.

[Ch 88] F. CHERNOUSKO, *Evaluation of phase conditions of dynamic systems (Ellipsoid method)*, Nauka, Moscow, 1988, 320p. (in Russian).

[Ch 91] F. CHERNOUSKO, *Optimal estimation theory for dynamic systems with set membership uncertainty*: An overview. Automatica, 27:997-1009, November 1991.

[Cha 77] C. CHARALAMBOUS, *Nonlinear least p-th optimization and nonlinear programming*, Math. Programming 12:195-225, 1977.

[Che 68] S.N. CHERNICOV, *Linear Inequalities*, Russian, Nauka, Moscow, 1968.

[CHL 78] J. CHATELON, D. HEARN, T.J. LOWE, *A subgradient algorithm for certain minimax and minimum problems*. Math. Programming 14:130-145, 1978.

[Chu 90] K.E. CHU, *On multiple eigenvalues of matrices depending on several parameters*, SIAM J. on Numerical Analysis 5:1368-1385, 1990.

[Chv 75] V. CHVATAL, *On certain polytopes associated with graphs*, J.C.T. B 18:138-154, 1975.

[Cl 73] F.H. CLARKE, *Necessary conditions for nonsmooth problems in optimal control and the calculus of variations*, Diss. Doctor Philos. Washington 1973.

[Cl 75] F.H. CLARKE, *Generalized gradients and applications*. Trans. Amer. Math. Soc. 205:47-262, 1975.

[Cl 83] F.M. CLARKE, *Optimization and nonsmooth analysis*, Wiley, New York 1983.

[Clar 95] K.L. CLARKSON, *Las Vegas Algorithms for Linear and Integer Programming When the Dimension Is Small*, J. of ACM 42:488-499, 1995.

[CLem 93] R. CORREA, AND C. LEMARECHAL, *Convergence of some algorithms for convex minimization*, Mathematical Programming, 62:261-275, 1993.

[CLen 82] Y. CENCOR, AND A. LENT, *Cyclic subgradient projections*, MP 24:232-235, 1982.

[CLR 95] M.D. CHOI, T.Y. LAM AND B. REZNICK, *Sums of squares of real polynomials*, Symp. in Pure Math, 58:103-126, 1995.

[CMa 93] B. CHAZELLE AND J. MATOUSEK, *On Linear-Time Deterministic Algorithms for Optimization Problems in Fixed Dimension*, Proc. of the 4th ACM-SIAM Symp. on Discrete Algorithms, pp. 281-290, 1993.

[CMo 81] B.D. CRAVEN AND B. MOND, *Linear programming with matrix variables*, Linear Algebra and its Applications 38:73-80, 1981.

[CMo 87] P.H. CALAMAI, AND J.J.MORE, *Projected gradient methods for linearly constrained problems*, Mathematical Programming, 39:93-116, 1987.

[CMu 81] S.J. CHUNG AND K.G. MURTY, *A polynomially bounded ellipsoid algorithms for convex quadratic programming*, Nonlinear Programming 4:439-485 Academic Press, N.Y., 1981.

[CMS 90] C. CHOI, C. L. MONMA, AND D.F. SHANNO, *Further development of primal-dual interior point methods*, Report 60-88, Rutgers Center for Operation Research, Rutgers University, New Brunswick, NJ, 1990.

[Co 73] A.R. CONN, *Constrained optimization uzing a nondifferentiable penalty function*, SIAM J. Numer. Anal. 10:760-784, 1973.

[Co 76] A.R. CONN, *Linear programming via a nondifferentiable penalty function*, SIAM Jr. of Numerical Analysis 13:145-154, 1976.

[Coo 71] S.A. COOK, *The complexity of theorem-proving procedures*, Proc.ACM Symp. Theory of Computing 3:151-158, 1971.

[CPa 90] R. CARRAGHAN, AND P.M. PARDALOS, *An exact algorithm for the maximum clique problem*, Oper. Res. Lett., 9:375-382, 1990.

[CPM 75] W.H. CUNNINGHAM, W. PULLEYBLANK, A.B. MARSH, *A scaling algorithm for optimum matching*, (1975).

[CPS 92] R. COTTLE, J.S. PANG, AND R.E. STONE, *The Linear Complementarity Problem*, Academic Press, 1992.

[Cr 76] H. CROWDER, *Computational improvements for subgragient optimization*, Symposia Mathematica, XIX:357-372, 1976.

[CT 89] H.-D. CHIANG AND J.S. THORP, *Stability regions of nonlinear dynamical systems: A constructive methodology*, IEEE Trans. Aut. Control, 34(12):1229-1241, 1989.

[CW 86] D. COPPERSMITH AND S. WINOGRAD, *Matrix multiplication via arithmetic progression*, Dept. of Math. Sci., IBM Thomas J.Watson Res. Center, Nov. 1986, preprint.

[CYP 93] M. CHEUNG, S. YURKOVICH, AND K.M. PASSINO, *An optimal volume ellipsoid algorithm for parameter estimation*, IEEE Trans. Aut. Control, AC-38(8):1292-1296, 1993.

[D 92] J. DANCIS, *Positive semidefinite completions of partial hermitian matrices*, Linear Algebra and Appl. 175:97-114, 1992.

[Da 67] J. M. DANSKIN, *The theory of max-min*, Springer, New York 1967.

[Dan 63] G.B. DANTZIG, *Linear programming and extensions*, Princeton University Press. Princeton 1963.

[De 77] V.F. DEMYANOV, *Extremal basis method in minimax problems*, Vycisl. Mat. i Mat. Fiz. 17:512-517, 1977.

[De 81a] V.F. DEMYANOV, *Subgradient method and saddle points*, Vest. Leningr. Univer. 13:17-23, 1981.

[De 81b] V.F. DEMYANOV, *On a relation the Clarke subdifferential and the quasidifferential*, Vestnik Leningrad University, 13:183-189, 1981.

[DG 80] L.C.W. DIXON, M. GAVIANO, *Reflections on nondifferentiable optimization*, part II, convergence. J. Optim. Theory Appl. 32:259-276, 1980.

[DGP 97] DINGZHU DU, JUN GU AND P.M. PARDALOS, *Satisfiability Problems*, DIMACS Series, American Mathematical Society, (in press, 1997).

[DH 72] W.E.DONATH, AND A.J. HOFFMAN, *Algorithm for partitioning graphs and computer logic based on eigenvectors of connection matrices*, Math. Prog. Study 3:35-55, 1972.

[DH 73] W.E.DONATH, AND A.J. HOFFMAN, *Lower bounds for the partitioning of graphs*, IBM J.Res.Dev., 17, 1973.

[Di 67] I. DIKIN, *Iterative solution of problems of linear and quadratic programming*, Soviet Math. Dokl., 8(3):674-675, 1967.

[Di 74] I. DIKIN, *On the speed of an iterative process*, Upravlyaemye sistemi, 12:54-60, 1974.

[Dix 80] L.C.W. DIXON, *Reflections on nondifferentiable optimization*, part 1: the ball-gradient. J. Optim. Theory Appl. 32:123-134, 1980.

[DKW 82] C. Davis, W.M. Kahan, and H.F. Weinberger, *Norm-preserving dilations and their applications to optimal error bounds* SIAM J. Numerical Anal., 19(3):445-469, 1982.

[DLL 57] L. Danzer, D. Laugwitz, and H. Lenz, *Uber das Lownersche Ellipsoid und sein Analogen unter den einem Eikorper einbeschriebner Ellipsoiden*, Archiv der Mathematik, 8:214-219, 1957.

[DM 72] V.F. Demyanov, V.N. Malozemov, *Introduction to minimax*, Nauka: Moscow 1972.

[DMi 95] O. Damberg and A. Migdalas, *A data parallel space dilation algorithm for the concentrator location problem*, DIMACS Series in Discr. Math., 22:57-80, DMS 1995.

[Do 83] J.L. Doob, *Stochastic processes.* Wiley, New York 1983

[DP 93] DingZhu Du and P.M. Pardalos, *Network Optimization Problems: Algorithms, Complexity and Applications*, World Scientific, 1993.

[DP 95] DingZhu Du and P.M. Pardalos, *Minimax and Applications*, Kluwer Academic Publishers, 1995.

[DP 97] DingZhu Du and P.M. Pardalos, *Network Design: Connectivity and Facilities Location*, DIMACS Series, American Mathematical Society, (in press, 1997).

[DPo 93] C. Delorme and S. Poljak, *Laplacian eigenvalues and the maximum cut problem*, Mathematical Programming, G2:557-574, 1993.

[DPop 95] I.I. Dikin and O.M. Popova, *Determination of interior feasible points of systems of linear constraints*, Siberian Energy Institute, Irkutsk, Russia, 1995.

[Dr 83] Z. Drezner, *The nested ball principle for the relaxation method*, O.R. 31:587-590, 1983.

[DR 80] V.F. Demyanov, A.M. Rubinov *On quasidifferentiable functionals*, Soviet Mathematics Doklady 21:14-17, 1980.

[DS 78] V.F. Demyanov, and V.K.Shomesova, *Conditional subdifferentials of convex functions*, Soviet Mathematics Doklady, 19:1181-1185, 1978.

[DS 80] V.F. DEMYANOV, AND V.K.SHOMESOVA, *Subdifferentials of functions on sets*, Cybernetics, 16(1):24-31, 1980.

[DV 85] V.F. DEMYANOV, L.V. VASILIEV, *Nondifferentiable optimization.* Optimization Software Inc./Springer, New York, 1985, Russian edition: Nauka, Moscow, 1981.

[DW 60] G. DANTZIG, P. WOLFE, *Decomposition principles for linear programs.* Oper. Res. 8:101-111, 1960.

[DW 61] G.B. DANTZIG, AND P. WOLFE, *The decomposition algorithm for linear programming*, Econometrica 29:767-778, 1961.

[DZ 80] I.I. DIKIN AND ZORKALTSEV, *Iterative solving mathematical programming problem (interior point method:algorithms)*, Nauka, 1980.

[EA 76] I.I. EREMIN, N.N. ASTAF'EV, *Introduction to the theory of linear and convex programming*, Nauka: Moscow 1976.

[Ed 65a] J. EDMONDS, *Paths, trees and flowers*, Canad. J. Math., 17:449-467, 1965.

[Ed 65b] J. EDMONDS, *Maximum matching and a polyhedron with 0-1 vertices*, J. Res. Nat. Bur. Standards 69B:125-130, 1965.

[Ed 67] J. EDMONDS, *Optimum branchings*, J. Res. Bur Standards, 71B:233-240, 1967.

[EFS 56] P. ELIAS, A. FEINSTEIN AND C.E. SHANNON, *A note on the maximum flow through a network*, IRE Transactions on Information Theory IT 2:117-119, 1956.

[EHM 92] L.ELSNER, C. HE, AND V. MEHRMANN, *Minimization of the norm, the norm of the inverse and the condition number of a matrix by completion*, Technical report 92-028, Universitat Bielefeld, POB 8640, D-4800 Bielefeld 1, Germany, 1992.

[EHM 93] L.ELSNER, C. HE, AND V. MEHRMANN, *Minimization of the condition number of a positive-definite matrix by completion. Technical report*, TU-Chemnitz-Zwikau, Chemnitz, Germany, September 1993.

[EHW 84] CH. EBENEGGER, P.L. HAMMER, AND D. DE WERRA, *Pseudo-Boolean functions and stability of graphs*, Ann. Discrete Math. 19:83-98, 1984.

[EK 72] J. EDMONDS, AND R.M. KARP, *Theoretical improvements in algorithmic efficiency for network flow problems*, J. ACM 19:248-264, 1972.

[EK 83] J.G. ECKER AND M. KUPFERSCHMID, *An implementation and generalization of Shor's ellipsoid algorithm for convex programming*, MP 27:83-106, 1983.

[EM 75] J. ELZINGA, T. MOORE, *A central cutting plane algorithm for the convex programming problem.* Math. Programming 8:134-145, 1975.

[Em 77] EMELIN, *On the speed of convergence of the speedest descent method*, Uspekhi Mathem. Nauk, v32, N1:163-164, 1977. (in Russian)

[Er 62] I.I. EREMIN, *Iterative method for Chebyshev approximations of inconsistent systems of linear inequalities*, Dokl. Akad. Nauk SSSR 143:1254-1256, 1962.

[Er 65] I.I. EREMIN, *The relaxation method of solving systems of inequalities with convex function on the left side*, SMD 6:219-221, 1965.

[Er 66] I.I. EREMIN, *The application of the method of Fejer approximation to the solution of problems of convex programming with non-smooth constraints*, CMMP 9:225-235, 1966.

[Er 67] I.I. EREMIN, *On a penalty method in convex programming*, Kibernetika (Kiev), no. 4:63-67, 1967.

[Er 68a] I.I. EREMIN, *Methods of Fejer approximations in convex programming*, (Ph.D.thesis) Mathematical Notes 3:139-149, 1968.

[Er 68b] I.I. EREMIN, *On the speed of convergence in the method of Fejer approximation*, Mathematical Notes 4:522-527, 1968.

[Er 70] I.I. EREMIN, *On systems of inequalities with convex functions in the left sides*, American Mathematical Society Translations, 88:67-83, 1970.

[Erm 66] YU.M. ERMOL'EV, *Methods for solving nonlinear extremal problems*, Kibernetika (Kiev), no. 4:1-17, 1966.

[Erm 76] YU.M. ERMOL'EV, *Methods of stochastic programming.* Nauka: Moscow 1976.

[ES 67] YU.M. ERMOL'EV, N.Z. SHOR, *On minimization of nondifferentiable functions*, Kibernetika (Kiev), no. 1:101-102, 1967.

[ES 68] YU.M. ERMOL'EV, N.Z. SHOR, *A random search method for two-stage problems of stochastic programming and its generalization*, Kibernetika (Kiev), no. 1:90-92, 1968.

[ET 73] J. EVANS, J. TOLLE, *Exact penalty functions in nonlinear programming*, Math. Programming 4:72-97, 1973.

[ET 74] I.EKELAND, R.TEMAM, *Analyse convexe et problemes variationnels*, Dunod: Paris 1974.

[EW 87] YU. ERMOL'EV, AND R.J.-B. WETS (EDS), *Numerical procedures for stochastic optimization*, Springer-Verlag, Berlin, 1987.

[EZ 71] B.C. EAVES, W.I. ZANGWILL, *Generalized cutting plane algorithms*, SIAM J. Control 9:529-542, 1971.

[EZKN 90] S.V. EMELYANOV, P.V. ZHIVOGLADOV, S.K. KOROVIN AND S.V. NIKITIN, *Group of isometries and some stabilizability aspects of uncertain systems* Problems of Control and Information Theory, 19(5-6):451-477, 1990.

[F 81] T.L. FAM, *On multistep methods of linear programming*, Dok.Akad.Nauk BSSR 25, no.9:784-787,859, 1981.

[Fan 49] K.FAN, *On a theorem of Weyl concerning the eigenvalues of linear transformations*, Proceedings of the National Academy of the Sciences of U.S.A. 35:652-655, 1949.

[Fan 93] M.K.H.FAN, *A quadratically convergent algorithm on minimizing the largest eigenvalue of a symmetric matrix*, Linear Algebra and Appl., 188-189:231-253, 1993.

[FC 68] A. FIACCO, G. MC CORMICK, *Nonlinear programming*, Sequential unconstrained minimization techniques, Wiley:New York 1968.

[Fed 71] F.F. FEDOROV, *The Theory of Optimal Design*, Nauka, Moscow, 1971.

[Fer 93] E. FERON, *Linear Matrix Inequalities for the Problem of Absolute Stability of Control Systems*, PhD thesis, Stanford University, Stanford, CA 94305, October 1993.

[Ferr 97] CH. FERRIER, *Hilbert's 17th problem and best dual bounds in quadratic minimization*, reprint of Laboratory of approximation and optimization of Toulouse Paul Sabatier University, March 1997.

[FF 60] D.K. FADDEEV, V.N. FADDEEVA, *Computational methods of linear algebra*, Fizmatgiz: Moscow 1960.

[FFul 56] L.R. FORD,JR. AND D.R. FULKERSON, *Maximal flow through a network*, Canadian Journal of Mathematics, 8:399-404, 1956.

[FG 88] M.E. FISHER AND J.E. GAYEK, *Estimating reachable sets for two-dimensional linear discrete systems*, Journal of Optimization Theory and Applications, 56(1):67-88, 1988.

[FGLS 91] U. FEIGE, S. GOLDWASSER, L. LOVASZ, AND S. SAFRA, *Approximating the maximum clique is almost NP-complete*, Proceedings 32nd IEEE Symposium of foundations of computer science, IEEE, pp. 2-12, 1991.

[Fi 81] M.L. FISHER, *The Lagrangian relaxation method for solving integer programming problems*, Management Sci. 27:1-18, 1981.

[Fi 85] M.L. FISHER, *An applications oriented guide to Lagrangean relaxation*, Interfaces 15:10-21, 1985.

[Fi 94] M.L. FISHER, *Optimal solution of vehicle routing problems using minimum k-trees*, Operations Research, 42:626-642, 1994.

[Fis 73] N. FISHER, *Optimal solution of scheduling problems using Lagrange multipliers*, Part 1. Oper. Res. 21:1114-1127, 1973.

[FJ 95] A. FRIEZE AND M. JERRUM, *Improved approximation algorithms for MAX k-CUT and MAX BISECTION*, Preprint, Department of Mathematics, Carnegie Mellon University, 1995.

[FK 97] S.D. T. FUJIE AND M. KOJIMA, *Semidefinite Programming Relaxation for Nonconvex Quadratic Programs*, Journal of Global Optimization 10:367-380, 1997.

[Fl 92] S.D. FLAM, *On finite convergence and constraints identification of subgradient projection methods*, Mathematical Programming, 57:427-437, 1992.

[Fle 81] R. FLETCHER, *A nonlinear programming problem in statistics (educational testing)*, SIAM J.Stat. comput., 2:257-267, 1981.

[Fle 85] R. FLETCHER, *Semidefinite matrix constraints in optimization*, SIAM Journal on Control and Optimization 23:493-513, 1985.

[Fle 87] R. FLETCHER, *Practical Methods of Optimization* (Wiley, Chichester, 1987, 2nd ed.).

[FM 68] A. FIACCO AND G.MCCORMICK, *Nonlinear programming: sequential unconstrainted minimization techniques*, Wiley, 1968. Reprinted 1990 in the SIAM Classics in Applied Mathematics series.

[FMi 78] I.S. FILOTTI AND G.L. MILLER, *On determining the genus of a graph in $O(V^{O(g)})$ steps*, Proc. 10th Ann. ACM Symp. on Theory of Computing, Association of Computing Machinery, New York, 1978.

[FN 92] M.K.FAN AND B. NEKOOIE, *On minimizing the largest eigenvalue of a symmetric matrix*, In Proc. IEEE Conf. on Decision and Control, pp. 134-139, December 1992.

[FNO 87] S. FRIEDLAND, J.NOCEDAL AND M.L.OVERTON, *The formulation and analysis of numerical methods for inverse eigenvalue problems*, SIAM Journal on Numerical Analysis 24:634-667, 1987.

[FNS 75] N. FISHER, W. NORTHUP, J. SHAPIRO, *Using duality to solve discrete optimization problems*, In: Math. Programming Stud. 3 (M. L. Balinski, P. Wolfe, eds.), pp. 56-94, North Holland, Amsterdam 1975.

[FNT 72] D.R FULKERSON, G.L. NEMHAUSER AND L.G. TROTTER, *Two computationally difficult set covering problems that arise in computing the 1-width of incidence matrices of Steiner triple systems*, MP Study 2:137-163, 1972.

[FP 90] C.A. FLOUDAS AND P.M. PARDALOS, *Collection of Test Problems for Constrained Global Optimization Algorithms*, Springer-Verlag, Lecture Notes in Computer Science 455, 1990.

[FP 92] C.A. FLOUDAS AND P.M. PARDALOS, *Recent Advances in Global Optimization*, Princeton University Press, 1992.

[FP 96a] C.A. FLOUDAS AND P.M. PARDALOS, *State of the Art in Global Optimization: Computational Methods and Applications*, Kliwer Academic Publishers, 1996.

[FP 96b] C.A. FLOUDAS AND P.M. PARDALOS, *Solving Combinatorial Optimization Problems in Parallel: Methods and Techniques*, Springer-Verlag, Lectire notes in computer science, Vol.1054, 1996.

[Fra 81] A. FRANK, *How to make a digraph strongly connected*, Combinatorica 1:145-143, 1981.

[Fre 94] R.M. FREUND, *Complexity of an algorithm for finding an approximate solution of a semidefinite program with no regularity assumption*, Technical report OR 302-94, Operations Research Center, MIT, 1994.

[Fri 56] R. FRISCH, *La resolution des problemes de programme lineaire par la methode du potential logarithmique*, Cahiers du Seminaire D'Econometrie, 4:7-20, 1956.

[Frie 81] S. FRIEDLAND, *Convex spectral functions*, Linear and Multilinear Algebra 9:299-316, 1981.

[FSM 89] R. FLETCHER AND E. SAINZ DE LA MAZA, *Nonlinear programming and nonsmooth optimization by successive linear programming*, Math. Progr. 43:235-256, 1989.

[FTa 87] M.L. FREDMAN AND R.E. TARJAN, *Fibonacci heaps and their uses in improved network optimization algorithms*, Journal of the Association for Computing Machinery 34, 1987.

[FTi 86] M.K. FAN AND A. TITS, *Characterization and efficient computation of the structured singular value*, IEEE Trans. Automat. Control, AC-31:734-743, 1986.

[FTD 91] M.K.H. FAN, A.L. TITS, AND J.C.DOYLE, *Robustness in the presence of mixed parametric uncertainty and unmodeled dynamics*, IEEE Trans. Aut. Control, 36(1):25-38, 1991.

[Fuk 84] M. FUKUSHIMA, *A descent algorithm for non-smooth convex programming*, Math. Progr. 30:163-175, 1984.

[Fu 74] D.R FULKERSON, *Packing rooted directed cuts in a weighted directed graph*, MP 6:1-13, 1974.

[FW 71] P.A. FILLMORE AND J.P. WILLIAMS, *Some convexity theorems for matrices*, Glasgow Mathematical Journal 12:110-117, 1971.

[FWo 56] M. FRANK, AND P. WOLFE, *An algorithm for quadratic programming*, Naval research Logistics Quarterly, 3:95-110, 1956.

[FY 73] A.L. FRADKOV AND V.A. YAKUBOVICH, *The S-procedure and duality relations in nonconvex problems of quadratic programming*, Vestnik Leningrad Univ. Math., 6(1):101-109, 1979. In Russian, 1973.

[G 92] P. GAHINET, *A convex parametrization of H suboptimal controllers*, In Proc. IEEE Conf. on Decision and Control, pp. 937-942, 1992.

[Ga 53] F.R. GANTMAKHER, *Theory of matrices*, Gostekhizdat, Moscow 1953.

[Gau 80] J. GAUVIN, *Shadow prices in nonconvex mathematical programming*, Math. Programming 19:300-312, 1980.

[GCG 93] F. GAROFALO, G. CELENTANO, AND L. GLIELMO, *Stability robustness of interval matrices via Lyapunov quadratic forms*, IEEE Trans. Aut. Control, 38(2):281-284, 1993.

[Ge 70] A.M. GEOFFRION, *Primal resource-directive approaches for optimizing nonlinear decomposable systems*, Oper. Res. 18:375-403, 1970.

[Ge 77] V.I. GERSHOVICH, *On the experience in solving the problem of coal distribution for coking*, In: Teor. Optimal. Resenii, Trudy Sem. Nauc, Sov. Akad. Nauk Ukrain SSSR po Kibernet. pp. 22-35. 1977.

[Ge 79a] V.I. GERSHOVICH, *About a cutting method with linear space transformations*, In Theory of Optimal Solutions, (Kiev, Inst. of Cybernetics, Ac.Sci.Ukr.SSR., pp. 15-23, 1979.

[Ge 79b] V.I. GERSHOVICH, *One way of representing space transformation operators in high-speed versions of generalized gradient methods*, In the book: Numerical methods of nonlinear programming, Proceedings of the III all-union Seminar, Kharkov, pp.64-66, 1979.

[Ge 80] V.I. GERSHOVICH, *One optimization method using linear space transformations*, In Theory of Optimal Solutions (Kiev, Inst. of Cybernetics, Ac.Sci.Ukr. SSR., pp. 38-45, 1980.(Russian).

[Ge 81a] V.I. GERSHOVICH, *Quadratic smoothing in iterative decomposition algorithms of solution of large-scale linear programming problems*, In II. Republican Symp. on Methods of Solution of Nonlinear Equations and Optimization Problems. pp. 188-190, Tallin, 1981 (Russian).

[Ge 81b] V.I. GERSHOVICH, *About an ellipsoid algorithm*, In the book: Some algorithms of nonsmooth optimization and discrete programming, Preprint 81-6, Inst. of Cybernetics, pp. 8-13, Kiev, 1981.

[Ger 60] M. GERSTENHABER, *Solution of large scale transportation problems*, In: Combinatorical analysis (R. Bellman, M. Hall, eds.), pp. 251-260. Amer. Math. Soc.: Providence 1960.

[GH 83] J.L. GOFFIN, AND A.J. HOFFMAN, *On the relationship between the Hausdorff distance and matrix distances of ellipsoids*, Linear Algebra and its Applications, 52/53:301-313, 1983.

[GHP 96] L.E. GIBBONS, D.W. HEARN, AND P.M.PARDALOS, *A continuous-based heuristic for the maximum clique problem*, In: D.S. Johnson and M. Trick (eds), Second DIMACS Implementation Challenge, DIMACS Series in discrete mathematics and theoretical computer science, 1996.

[GHPR 96] L.E. GIBBONS, D.W. HEARN, P.M.PARDALOS, AND M.V. RAMANA, *Continuous characterization of the maximum clique*, to appear in Math. Oper. Res., 1996.

[GHV 92] J.L. GOFFIN, A. HAURIE, AND J.-P. VIAL, *Decomposition and nondifferentiable optimization with the projective algorithm*, Management Sciennce, 38:284-302, 1992.

[GJ 79] M. R. GAREY AND D.S. JOHNSON, *Computers and Intractability: A Guide to the Theory of NP-Completeness*, W. H. Freeman, San Francisko, 1979.

[GJSW 84] R. GRONE, C.R. JOHNSON, E.M. SA, AND H.WOLKOWICZ, *Positive definite completions of partial Hermitian matrices*, Linear Algebra and Appl., 58:109-124, 1984.

[GKM 83] R. GABASOV, F.M. KIRILOVA, B.S. MORDUKHOVICH, *The - maximum principle for suboptimal controls*, SMD 27, 1983.

[GLe 89] CH. GILBERT AND C. LEMARECHAL, *Some nimerical experiments with variable-storage quasi-Newton algorithms*, Math. Progr. 45:407-435, 1989.

[GLi 88] D. GOLDFARB AND S. LIU, *An O(n L) primal interior point algorithm for convex quadratic programming*, Tech. Report, Dept. of IEOR, Columbia University, New York, 1988.

[GLo 93] K.GU AND N.K. LOH, *Direct computation of stability bounds for systems with polytopic uncertainties*, IEEE Trans. Aut. Control, 38(2):363-366, 1993.

[GLov 81] P. GACS AND L. LOVASZ, *Khachiyan's algorithm for linear programming*, Math. Program. Studies, 14:61-68, 1981.

[GLS 81] M. GROTSCHEL, L. LOVASZ, AND A. SCHRIJVER, *The ellipsoid method and its consequences in combinatorial optimization*, Combinatorica, 1:125-156, 1981.

[GLS 86] M. GROTSCHEL, L.LOVASZ, AND A. SCHRIJVER, *Relaxations of vertex packing*, Journal of combinatorial Theory, Seies B 40:330-343, 1986.

[GLS 88] M. GROTSCHEL, L.LOVASZ, AND A. SCHRIJVER, *Geometric Algorithms and Combinatorial Optimization*, volume 2 of Algorithms and Combinatorics. Springer-Verlag, 1988.

[GMSTW 86] P.E. GILL, W. MURRAY, M.A. SAUNDERS, J.A. TOMLIN, AND M.H. WRIGHT, *On projected Newton barrier methods for linear programming and an equivalence to Karmarkar's projective method*, Math. Programming, 36:183-209, 1986.

[GNS 73] G. GORRY, W.NORTHUP, J. SHAPIRO, *Computational experience with a group integer programming algorithm*, Math. Programming 4:171-192, 1973.

[Go 77] J.L. GOFFIN, *On convergence rates of subgradient optimization methods*. Math. Programming 13:329-347, 1977.

[Go 78] J.L. GOFFIN, *Nondifferensiable optimization and the relaxation method*, in: Nonsmooth Optimization, C. Lemareshal, R. Mifflin, eds., Proceedings of a IIASA Workshop, pp. 31-49, Pergamon Press, Oxford, 1978.

[Go 80] J.L. GOFFIN, *The relaxation method for solving systems of linear inequalities*, Mathematics of Operations Research, 5:388-414, 1980.

[Go 81] J.L. GOFFIN, *Convergence results in a class of variable metric subgradient methods*, in Nonlinear Programming 4, Editors: O.L. Mangasarian, R.R. Meyer and S.M. Robinson, pp. 283-326, Academic Press 1981.

[Go 82a] J.L. GOFFIN, *On the nonpolynomiallity of the relaxation method for systems of linear inequalities*, MP 22:93-103, 1982.

[Go 82b] J.L. GOFFIN, *Convergence of a cyclic ellipsoid algorithm for systems of linear equalities*, MP 22:239-260, 1982.

[Go 83] J.L. GOFFIN, *Convergence rates of the ellipsoid method on general convex functions*. Math. Oper. Res. 8:135-150, 1983.

[Gol 75] A. GOLDSTEIN, *Optimization with corners*, In: Nonlinear Programming 3 (O. L. Mangasarian, R.R. Meyer, S.M. Robinson, eds.), pp. 215-230. Academic Press: New York 1975.

[Gol 77] A.A. GOLDSTEIN, *Optimization of Lipschitz continuous functions*, Math. Programming 13:14-22, 1977.

[Goll 87] B. GOLLAN, *Eigenvalue perturbation and nonlinear parametric optimization*, Math. programming study, 30:67-81, 1987.

[Gols 72] E.G. GOL'SHTEIN, *Generalized gradient method for saddle point seeking*. Ekonom. i Mat. Metody 8:569-579, 1972.

[Gon 88] C.C. GONZAGA, *Polynomial affine algorithm for linear programming*, Report ES-139/88, Universidade Federal do Rio de Janeiro, Brazil, 1988.

[Gon 89] C.C. GONZAGA, *Conical projection algorithm for linear programming*, Math. Programming, 43:151-173, 1989.

[Grz 85] S.M. GRZEGOSKI, *Orthogonal projections on convex sets for Newton-like methods*, SIAM Journal on Numerical Analysis 22:1208-1219, 1985.

[GPB 91] J.C. GEROMEL, P.L.D. PERS, AND J.BERNUSSOU, *On a convex parameter space method for linear control design of uncertain systems*, SIAM J. Control and Optimization, 29(2):381-402, 1991.

[GPR 67] L. Gubin, B. Poljak, E. Raik, *The method of projections for finding the common point of convex sets*, CMMP 7, 1967.

[Gr 60] B. Grunbaum, *Partitions of mass-distributions and of convex bodies by hyperplanes*, Pacific J. Math., 10:1257-1261, 1960.

[Gr 67] B. Grunbaum, *Convex Polytopes*, (Wiley, N.Y., 1967).

[GS 79] D. Goldfarb and W.Y. Sit, *Worst case behavior of the steepest edge simplex method*, Discrete Applied Math. 1:277-285, 1979.

[GT 88] C.Goh and D.Teo, *On minimax eigenvalue problems via constrained optimization*, Journal of Optimization Theory and Applications, 57(1):59-68, 1988.

[GTo 82] D. Goldfarb and M.J. Todd, *Modifications and implementation of the Shor-Khachiyan algorithm for linear programming*, MP.23: 1-19, MR 83d: 90116, zbl 477.90038, 1982.

[GTo 89] C.C. Gonzaga and M.J. Todd, *An O(n L)-iteration large-step primal-dual affine algorithm for linear programming*, Tech. Report No. 862, School of Operations Research and Industrial Engineering, Cornell University,Ithaca, NY, 1989.

[GTr 89] E.G. Golshtein and N.V. Tretyakov, *Modified Lagrageans*, Moscow, Nauka, 1989. (In Russian).

[GTs 61] I.N. Gel'fand, M.L. Tsetlin, *The principle of nonlocal search in the problems of automatic optimization*, Dokl. Akad. Nauk SSSR 137:295-298, 1961.

[GTT 90] A.V. Goldberg, E. Tardos, and R.E. Tarjan, *Network flow algorithms*, in: Paths, Flows and VLSI - Layout, Shpringer-Verlag, pp. 101-164, 1990.

[Gu 92] K.Gu, *Designing stabilizing control of uncertain systems by quasiconvex optimization*, IEEE Trans. Aut.Control, 38(9), 1992.

[Gu 94a] K.Gu, *Designing stabilizing control of uncertain systems by quasiconvex optimization*, IEEE Trans. Aut.Control, 39(1):127-131, 1994.

[Gu 94b] K.Gu, *Global optimization for satisfiability (SAT) problem*, IEEE Transactions on Knowledge and Data Engineering, 6(3):361-381, 1994.

[Gup 79] A.M. GUPAL, *Stochastic methods for solving nonsmooth extremum problems*, Kiev: Naukova Dumka 1979.

[GV 74] G.GOLUB, J.VARAH, *On characterization of the best L2 scaling of a matrix*, SIAM J. on Numerical Analysis, 11(3):472-479, 1974.

[GVi 86] G. DE GHELLINK AND J.-P. VIAL, *A polynomial Newton method for linear programming*, Algorithmica, 1:425-454, 1986.

[GVi 87] G. DE GHELLINK AND J.-P. VIAL, *An extension of Karmarkar's algorithm for problems in standart form*, Math. Programming, 37:81-90, 1987.

[GW 94] M.X. GOEMANS AND D.P. WILLIAMSON, *0.878-Approximation algorithms for MAX-CUT and MAX 2SAT*, Proceedings of the 26th Annual ACM Symposium on Theory of Computing, pp. 422-431, 1994.

[GW 95] M.X. GOEMANS AND D.P. WILLIAMSON, *Improved approximation algorithms for maximum cut and satisfiability problems using semidefinite programming*, Journal of the ACM, 42(6):1115-1145, 1995.

[Gw 81] J. GWINNER, *Bibliography on nondifferentiable optimization and non-smooth analysis*, J. Comput. Appl. Math. 7:277-285, 1981.

[GY 66] E.G. GOL'SHTEIN, D.B.YUDIN, *New directions in linear programming*, Sovetskoe Radio, Moscow 1966.

[Had 75] F.O. HADLOCK, *Finding a maximum cut of a planar graph in polynomial time*, SIAM J. Comput. 4:221-225, 1975.

[Ham 82] S.J.HAMMARLING, *Numerical solution of the stable, non-negative definite Lyapunov equation*, IMA J. of Numerical Analysis, 2:303-323, 1982.

[Han 81] S.-P. HAN, *Variable metric methods for minimizing a class of minimax optimization*, Math. Programming 20:1-13, 1981.

[HC 71] M. HELD, R. CARP, *The travelling salesman problem and minimum spanning trees*. Part 2. Math. Programming 1:6-25, 1971.

[HHP 94] W. HAGER, D. HEARN AND P.M. PARDALOS, *Large Scale Optimization: The State of the Art*, Kluwer Academic Publishers, 1994.

[Hi 35] D. HILBERT, Gessamelte Abhandlungen, v.3, 290-329, 1935.

[Hi 88] D. HILBERT, *Uber die Darstellung definiter Formen als Summe von Formen-quadraten*, Mathematical Annals 32:342-350, 1888.

[HJ 91] R.HORN, C.JOHNSON, *Topics in Matrix Analysis*, Cambridge University Press, Cambridge, 1991.

[HK 71] M. HELD, AND R.M. KARP, *The treveling-salesman problem and minimum spanning trees: Part II*, Mathematical Programming, 1:6-25, 1971.

[HKL 80] R. HELGASON, J. KENNIGTON, AND H. LALL, *A polynomially bounded algorithm for a singly constrained quadratic program*, MP. 18:338-343, 1980.

[HKr 56] A.J. HOFFMAN, AND J.B. KRUSKAL, *Integral boundary points of convex polyhedra*, in Linear Inequalities and Related Systems, H.W. Kuhn and A.W. Tucker (eds.), Princeton University Press, Princeton, N.J., 1956.

[HLP 62] G.H. HARDY, J.E. LITTLEWOOD, AND G. POLYA, *Inequalities*, Second edition, Cambridge University Press, 1962.

[HM 81] J. HALD, K. MADSEN, *Combined LP and quasi-Newton methods for minimax optimization*, Math. Programming 20:49-62, 1981.

[Ho 79] A.J. HOFFMAN, *On the use of unimodularity in applying linear inequalities to combinatorial problems*, Annals of Discrete Math. 4:73-84, 1979.

[Hog 73] W.W. HOGAN, *Directional derivatives of convex functions with applications to the completely convex case*, Oper. Res. 21: 188-209, 1973.

[HP 95] R. HORST AND P.M. PARDALOS, *Handbook of Global Optimization*, Kluwer Academic Publishers, 1995.

[HPRW 95] C. HELMBERG, S. POLJAK, F. RENDL, AND H. WOLKOWICZ, *Combining semidefinite and polyhedral relaxation for integer progtams*, Lecture Notes in Computer Science 920:124-134, 1996.

[HPT 95] R. HORST, P.M. PARDALOS AND N.V. THOAI, *Introduction to Global Optimization*, Kluwer Academic Publishers, 1995.

[HPV 93] J. HASSELBERG, P.M. PARDALOS, AND G. VAIRAKTARAKIS, *Test case generators and computational results for the maximum clique problem*, Journal of Global Optimization 3:463-482, 1993.

[HRo 80] M.S. AND W.O. ROM, *Solving the assignment problem by relaxation*, Operations Research 28:969-982, 1980.

[HRVW 96] H. HELMBERG, F. RENDL, R.J. VANDERBEY, AND H. WOLKOWICZ, *An interior-point method for semidefinite programming*, SIAM Journal on optimization, 6(2):342-361, 1996.

[HS 96] J.L. HIGLE, AND S.SEN, *Stochastic decomposition: A statistical method for large scale stochastic linear programming*, Nonconvex Optimization and its Applications, 8, Kluwer Academic Publishers, Dordrecht, 1996.

[HSi 88] J. HOFBAUER, AND K. SIGMUND, *The theory of evolution and dynamical systems*, Cambridge University Press, 1988.

[HU 78] J.B. HIRIART-URRUTY, *Generalized gradients of marginal value functions*, SIAM J. Control Optim. 16:301-316, 1978.

[Hu 87] H. HU, *An algorithm for rescaling a matrix positive definite*, Linear Algebra and its Applications, 96:131-147, 1987.

[Hua 70] H.Y. HUANG, *Unified approach to quadratically convergent algorithms for function minimization*, J. Optim. Theory Appl. 5: 402-423, 1970.

[Huar 67] P HUARD, *Resolution of mathematical programming with nonlinear constraints by the method of centers*, North Holland, Amsterdam, 1967.

[HUL 93] J.-B. HIRIART-URRUTY AND C. LEMARECHAL, *Convex Analysis and Minimization Algorithms* I, volume 305 of Grundlehren der mathematischen Wissenschaften. Springer-Verlag, New York, 1993.

[HY 91] J.B.HIRRIART-URRUTY AND D.YE, *Sensitivity analysis of the sum of the largest eigenvalues of a symmetric matrix*, Preliminary Research Report, Laboratoire d'Analyse Numerique, Universite Paul Sabatier, Toulouse, 1991.

[HY 92] J.-B. HIRIART-URRUTY AND D. YE, *Sensitivity analysis of all eigenvalue of a symmetric matrix*, Technical report, Univ. Paul Sabatier, Toulouse, 1992.

[HWC 74] M. HELD, P. WOLFE AND H. CROWDER, *Validation of subgradient optimization*, MP. 6:62-88, 1974.

[I 81] A.D. IOFFE, *Nonsmooth analysis: differential calculus of nondifferentiable mappings*, Trans. Amer. Math. Soc. 266: 1-56, 1981.

[II 86] M.IRI AND H.IMAI, *Multiplicative barrier function method for linear programming*, Algorithmica, 1:455-482, 1986.

[IK 92] C.-M. IP AND J. KYPARISIS, *Local convergence of quasi-Newton methods for B-differentiable equations*, Math. Progr. 56:71-89, 1992.

[IN 82] Y. IKURA AND G.L. NEMHAUSER, *An efficient primal simplex algorithm for maximum weighted vertex packing on bipartite graphs*, Annals of Discrete Mathematics 16:149-168, 1982.

[IS 68] L.V. IVANOVA, N.Z. SHOR, *A certain iterative method of solution of linear programming problems and matrix games*, in Theory of Optimal Solutions 3:22-30, 1968,(Russian) MR 46. 4930.

[IT 79] A.D.IOFFE AND V.M.TIHOMIROV, *Theory of Extremal Problems* (North-Holland, Amsterdam,1979).

[J 87] F.JARRE, *On the convergence of the method of analytic centers when applied to convex quadratic programs*, Manuscript, Institut fur Angewandte Mathematik und Statistik, Universitat Werzburg, 1987.

[J 90] F.JARRE, *Interior point methods for convex programming*, Report SOL 90-16, Dept. of Operations Research, Stanford University, Stanford, CA, 1990; Appl. Math. Optim., to appear.

[J 93] F. JARRE, *An interior-point method for minimizing the maximum eigenvalue of a linear combination of matrices*, SIAM J. Control and Opt., 31:1360-1377, 1993.

[Ja 79] S. JAYARAYAN, *A nonlinear optimization problem in educational testing*, MSc thesis, Dept. Methematics, Univ. Dundee, Scotland, 1979.

[Je 73] R.G. JEROSLOW, *The simplex algorithm with the pivot rule of maximizing criterion improvement*, Discrete Math. 4:367-377, 1973.

[Je 76] R.G. JEROSLOW, *There cannot be any algorithm for integer programming with quadratic constraints*, Op. Res., 21(1), 1976.

[JN 76a] D.B. JUDIN AND A.S. NEMIROVSKII, *Estimation of the informational complexity of mathematical programming problems*, EMM 12(1), 1976.

[JN 76b] D.B. JUDIN AND A.S. NEMIROVSKII, *Informational complexity and effective methods for the solution of convex extremal problems*, EMM 12(2), 1976.

[JN 77] D.B. JUDIN AND A.S. NEMIROVSKII, *Informational complexity of strict convex programming*, EMM 13(3), 1977,(Russian).

[Jo 48] F. JOHN, *Extremum problems with inequalities as subsidary conditions*, Studies and essays, presented to R.Courant on his 60th birthday, N.Y., pp. 187-204, 1948.

[K 68] W. KAHAN, *Circumscribing an ellipsoid about the intersection of two ellipsoids*, Canad. Math. Bull., 11(3):437-441, 1968.

[KA 91] S. KIM, AND H. AHN, *Convergence of a generalized subgradient method for nondifferentiable convex optimization*, Mathematical Programming, 50:75-80, 1991.

[Ka 63] R.E. KALMAN, *Lyapunov functions for the problem of Lur'e in automatic control*, Proc. Nat. Acad. Sci., USA, 49:201-205, 1963.

[Kam 83] V.A. KAMENETSKII, *Absolute stability and absolute instability of control systems with several nonlinear nonstationary elements*, Automation and Remote Control, 44(12):1543-1552, 1983.

[Kan 47] L.V. KANTOROVICH, *On the method of steepest descent*, Dokl. Akad. Nauk SSSR 56:233-236, 1947.

[Kan 59] L.V. KANTOROVICH, *Economic calculation of the best utilization of resources*, Izdatel'stwo Akad. Nauk SSSR Moscow 1959.

[Kar 84] N. KARMARKAR, *A new polynomial-time algorithm for linear programming*, Combinatorica, 4(4):373-395, 1984.

[Karp 72] R.M.KARP, *Reducibility among combinatorial problems*, in Complexity of Computer Computations, R.E.Miller and J.W.Thatcher, eds., Plenum Press, New York, 1972.

[Karp 75] R.M.KARP, *On the computational complexity of combinatorial problems*, Networks, 5:45-68, 1975.

[Kat 82] T. KATO, *A Short Introduction to Perturbation Theory for Linear Operators*, Springer, New York, 1982.

[Ke 60] J. KELLEY, *The cutting plane method for solving convex programs*, J. Soc. Ind. Appl. Math. 8:703-712, 1960.

[Kh 79] L.G. KHACHIYAN, *A polynomial algorithm in linear programming*, SMD 20:191-194, 1979, MR 80g. 90071.

[Kh 80] L.G. KHACHIYAN, *Polynomial algorithms in linear programming*, CMMP 20:53-72, 1980.

[Kh 82] L.G. KHACHIYAN, *Convexity and algorithmic complexity of polynomial programming problems*, Izvestiya AS USSR, Technical cybernetics, N6:46-56, 1982.

[Kha 78] V.L. KHARITONOV, *Asymptotic stability of an equilibrium position of a family of systems of linear differential equations*, Differential'nye Uravneniya, 14(11):1483-1485, 1978.

[KI 85] D.YA. KHUSAINOV AND E.B. IVOHIN, *On estimating solutions of linear systems with application of Lyapunov's functions*, Kibernetika, 2:7-10, 1985.

[Ki 83a] K.C. KIWIEL, *A phase I-phase II method for inequality constrained minimax problems*, Control Cyb. 12:55-75, 1983.

[Ki 83b] K.C. KIWIEL, *An aggregate subgradient method for nonsmooth convex minimization*, Math. Programming 27:320-341, 1983.

[Ki 85] K.C. KIWIEL, *Methods of descent for nondifferentiable optimization*, Lecture Notes in Mathematics, Springer-Verlag, Berlin, 1133:362, 1985.

[Ki 90] K.C. KIWIEL, *Proximity control in bundle methods for convex nondifferentiable minimization*, Mathematical Programming, 46:105-122, 1990.

[Ki 95] K.C. KIWIEL, *Approximations in bundle methods and decomposition of convex programs*, Journal of Optimization Theory and Applications, 84:529-548, 1995.

[Ki 96a] K.C. KIWIEL, *The efficiency of subgradient projection methods for convex optimization, part I: General level methods*, SIAM Journal on Control and Optimization, 34:660-676, 1996.

[Ki 96b] K.C. KIWIEL, *The efficiency of subgradient projection methods for convex optimization, part II:Implementations and extensions*, SIAM Journal on Control and Optimization, 34:677-676, 1996.

[KK 92] L. KHACHIYAN AND B. KALANTARI *Diagonal matrix scaling and linear programming*, SIAM J. on Optimization, 2(4):668-672, 1992.

[KKr 66] H.P. KUNZI, W. KRELLE, *Nonlinear programming*. Blaisdell, Waltham 1966.

[KL 85] B. KOUVARITAKIS AND D. LATCHMAN, *Singular value and eigenvalue techniques in the analysis of systems with structured perturbations*, Int. J. Control, 41(6):1381-1412, 1985.

[KLa 89] C. KENNEY AND A. LAUB, *Condition estimates for matrix functions*, SIAM J. Matrix Anal. Appl., 10(2):191-209, 1989.

[KLSW 89] N.K. KARMARKAR, J.C. LAGARIAS, L. SLUTSMAN, AND P. WANG, *Power series variants of Karmarkar-type algorithms*, AT&T Technical Journal 68:20-36, 1989.

[KM 72] V. KLEE AND G.L. MINTY, *How good is the simplex algorithm?*, in O. Shisha (ed.) Inequalities III, pp. 159-175, Academic Press, New York, 1972.

[KMS] D. KARGER, R. MOTWANI, AND M. SUDAN, *Improved graph coloring by semidefinite programming.* (In preparation.)

[KMY 88] M.S.KOJIMA, S.MIZUNO, AND A.YOSHISE, *An O(n L) iteration potential reduction algorithm for linear complementarity problems*, Research Reports on Information Sciences B-217, Dept. of Information Sciences, Tokyo Institute of Technology, Tokyo, 1988.

[KMY 89] M.S.KOJIMA, S.MIZUNO, AND A.YOSHISE, *A polynomial time algorithm for linear complementarity problems*, Math. Programming, 44:1-26, 1989.

[Ko 82] B. KORTE, *Modern Applied Mathematics*, (ed.) (North-Holland, N.Y., 1982).

[KP 82] R.M. KARP, AND C.H. PAPADIMITRIOU, *On linear characterizations of combinatorial optimization problems*, SIAM Jr. Comp. 11:620-632, 1982. Mr 83k.90064.

[KPy 87] V.A. KAMENETSKII AND E.S. PYATNITSKII, *Gradient method of constructing Lyapunov functions in problems of absolute stability*, Automation and Remote Control, 48(1):1-9, 1987.

[KR 91] P.P.KHARGONEKHAR AND M.A.ROTEA, *Mixed H /H control: A convex optimization approach*, IEEE Trans. Automat. Control, 36:824-837, 1991.

[Kr 56] M.A. KRASNOSEL'SKII, *Topological methods in the theory of nonlinear integral equations*, Gostekhizdat, Moscow 1956.

[KRR 91] N. KARMARKAR, M.G.C. RESENDE, AND K.G RAMAKRISHNAN, *An interior point algorithm to solve computationally difficult set covering problem*, Math. Progr. 52:597-618, 1991.

[KSc 95] P. KLEINSCHMIDT, H. SCHANNATH, *A strongly polynomial algorithm for the transportation problem*, Math. Progr. 68:1-13, 1995.

[KSch 80] B. KORTE AND R. SCHRADER, *A note on convergence proofs of Shor-Khachiyan methods*, in Lecture notes in Control and Information Sciences 30:51-57, Springer, N.Y., 1980.

[KSh 77] J. KENNINGTON, AND M. SHALABY, *An effective subgradient procedure for minimal cost multicommodity flow problems*, Management Science, 23:994-1004, 1977.

[KT 90] L. KHACHIYAN AND M. TODD, *On the complexity of approximating the maximal inscribed ellipsoid for a polytope*, Cornell University, Ithaka, New York, Technical report 839, 1990.

[KTE 88] L.G.KHACHIYAN, S.P.TARASOV, AND A.I.ERLICH, *The inscribed ellipsoid method*, Dokl. Akad. Nauk SSSR, Vol.298, 1988.

[KTK 79] M.K. KOZLOV, S.P. TARASOV, L.G. KHACHIYAN, *Polynomial solvability of convex quadratic programming*, SMD 20:1108-1111, 1979.

[KU 89] S. KIM, AND B.-S. UM, *Polyak's subgradient method with simplified projection for nondifferentiable optimization with linear constraints*, Optimization, 20:451-456, 1989.

[Kum 91] B. KUMMER, *Newton's method based on a generalized derivatives for non-smooth functions: convergence: convergence analysis*, in: Lecture Notes in Economics and Mathematical Systems 382:171-194, Springer-Verlag, 1991.

[KW 59] J. KIEFER, AND J. WOLFOWITZ, *Optimum desighns in regression problems*, Ann. Math. Statist., 30:271-294, 1959.

[L 70] L.S. LASDON, *Optimization theory for large systems*, Macmillan: London 1970.

[La 72] P.J. LAURENT, *Approximation et optimization*, Herman: Paris 1972.

[Le 74] C. LEMARECHAL, *An algorithm for minimizing convex functions*, In: Proceedings of the IFIP Congress, pp. 552-556. North Holland:Amsterdam 1974.

[Le 75] C. LEMARECHAL, *An extention of Davidon methods to nondifferentiable problems*, In: Math. Programming Stud. 3:95-100 (M.L. Balinski, P. Wolfe, eds.), North Holland: Amsterdam 1975.

[Le 78a] C. LEMARECHAL, *Bundle methods in nonsmooth optimization*, In: Nonsmooth Optimization (C. Lemarechal. R. Mifflin, eds.), pp. 79-102. Pergamon Press: Oxford 1978.

[Le 78b] C. LEMARECHAL, *Nonsmooth optimization and descent methods*, RR-78-4 International Institute for Applied Systems Analysis, Laxenburg, Austria 1978.

[Le 81] C. LEMARECHAL, *A view of line-searches*, In: Optimization and Optimal Control(A. Auslender, W. Oettli, J. Stoer, eds.), pp. 59-78. Lecture Notes in Control and Information Sciences 30, Springer: Berlin 1981.

[Le 82] C. LEMARECHAL, *Numerical experiments in nonsmooth optimization*, In: Progress in Nondifferentiable Optimization (E.A. Nurminski, ed.), pp. 61-84. CP-82-S8, International Institute for Applied Systems Analysis, Laxenburg, Austria 1982.

[Le 89] C. LEMARECHAL, *Nondifferentiable optimization*, in: Optimization, G.L. Nemhauser, A.H.G. Rinnooy Kan, M.J. Todd, eds., Handbooks in Operations Research and Management Science. Volume 1:529-572, 1989.

[Len 81 65] H.W. LENSTRA JR, *Integer programming with a fixed number of variables*, Dept of Math., Univ. of Amsterdam. Rept. N 81-03, 1981.

[Lev 65] A. YU. LEVIN, *On an algorithm for minimizing convex functions*, Dokl. Akad. Nauk SSSR 160:1244-1247, 1965.

[LL 96] T. LARSSON, AND Z. LIU, *A Lagrangean relaxation scheme for structured linear programs with application to multicommodity network flows*, Optimization (to appear).

[LLP 92] T. LARSSON, Z. LIU, AND M. PATRIKSSON, *A dual scheme for traffic assignment problem*, Report LiTH-MAT-R-92-21, Department of Mathematics, Linkoping Institute of Technology, Linkoping, Sweden. (Revised for publication in Optimization.)

[LM 70] YU.I. LYUBICH, G. D. MAISTROVSKII, *General theory of relaxation processes for convex functions*, Uspehi Mat. Nauk 1:57-112, 1970.

[LMi 82] C. LEMARECHAL, R. MIFFLIN, *Global and superlinear convergence of an algorithm for one-dimensional minimization of convex functions*, Math. Programming 24:241-256, 1982.

[LMS 90] I.J.LUSTIG, R.MARSTEN, AND D.F.SHANNO, *On implementation Mehrotra's predictor-corrector interior point method for linear programming*, Report SOR-90-3, Department of Civil Engineering and Operations Research, Princeton University, Princeton, NJ, 1990.

[LNN 95] C. LEMARECHAL, A. NEMIROVSKII, AND YU. NESTEROV, *New variants of bundle methods*, Mathematical Programming, 69:111-147, 1995.

[Lo 72] L.LOVASZ, *Normal hypergraphs and the perfect graph conjecture*, Discrete Mathematics, 2:253-267, 1972.

[Lo 79] L.LOVASZ, *On the Shannon capacity of a graph*, IEEE Trans. Inform. Theory, 25:355-381, 1979.

[Lo 82] L. LOVASZ, *Bounding the independence number of a graph*, Annals of Discrete Mathematics 16, 1982.

[LS 81] L. LOVASZ, A. SCHRIJVER, *Some combinatorial applications of the new linear programming algorithm*, In Lecture Notes in Math., 885:33-41, Springer, Berlin, 1981, MR 83h.90071.

[LS 91] L. LOVASZ, A. SCHRIJVER, *Cones of matrices and set functions and 0-1 optimization*, SIAM Journal on Optimization 1:166-190, 1991.

[LSB 81] C. LEMARECHAL J.-J STRODIOT, A. BIHAIN, *On a bundle algorithm for nonsmooth minimization*, In: Nonlinear Programming 4:245-281 (O.L. Mangasarian, R.R. Meyer, S.M. Robinson, eds.), Academic Press: New York 1981.

[Lu 57] A.I. LUR'E, *Some Nonlinear Problems in the Theory of Automatic Control*, H.M. Stationery Off., London, 1957. In Russian, 1951.

[Ly 47] A.M. LYAPUNOV, *Probleme general de la stabilite du mouvement*, volume 17 of Annals of Mathematics Studies. Princeton University Press, Princeton, 1947.

[LZ 83] C. LEMARECHAL, J.ZOWE, *Some remarks on the construction of higher order algorithms in convex optimization*, Appl. Math. Opt., 10:51-68, 1983.

[MA 87] R.D.C.MONTEIRO AND I.ADLER, *An $O(n^3L)$ interior point algorithm for convex quadratic programming*, Manuscript, Department of Industrial Engineering and Operations Research University of California, Berkeley, CA, to appear in Mathematical Programming 44, 1989.

[MA 89a] R.D.C.MONTEIRO AND I.ADLER, *Interior path-following primal-dual algorithms*, Part 1: Linear programming, Math. Programming, 44:27-42, 1989.

[MA 89b] R.D.C.MONTEIRO AND I.ADLER, *Interior path-following primal-dual algorithms*, Part 2: Convex Quadratic programming, Math. Programming, 44:43-66, 1989.

[MA 90] R.D.C.MONTEIRO AND I.ADLER, *An extension of Karmarkar-type algorithm to a class of convex separable programming problems with global linear rate of convergence*, Math. Oper. Res., 15:408-422, 1990.

[Mad 75] K. MADSEN, *An algorithm for minimax solution of overdetermined systems of nonlinear equations*, J. Inst. Math. Appl. 16: 321-328, 1975.

[Mai 91] R.S. MAIER, *Computational results with nonsmooth methods for large-scale nonlinearly-constrained optimization*, University of Minnesota, AHPCRC, preprint 91-68, 1991.

[Mais 76] D. MAISTROVSKII, *Gradient methods for finding saddle points*, EMM 12:917-929, 1976.

[Man 83] O.L. MANGASARIAN, *Least norm linear programming solution as an unconstrained minimization problem*, Jr. Math. Analys. Applications 92:240-251, 1983.

[Man 81] O.L. MANGASARIAN, *A condition number for linear inequalities and linear programs*, Methods of O.R. 43:3-15, 1981.

[Mat 72] YU.V. MATIYASEVICH, *Diophantine sets*, Uspekhi Mathematicheskich nauk, V.XXII, N5, 1972.

[Meh 91] S. MEHROTRA, *Higher order methods and their performance*, Technical report 90-16R1, Dept. of Industrial Engineering and Management Science, Northwestern University, 1991.

[MESS 67] V.S. MIKHALEVICH, YU. M. ERMOL'EV, V.V. SHKURBA, N.Z. SHOR, *Complex systems and the solution of extremal problems*, Kibernetika (Kiev), no. 5:29-39, 1967.

[MHB 75] R.E. MARSTEN, W.W. HOGAN, J.W. BLANKENSHIP, *The boxstep method for large-scale optimization*, Oper. Res. 23:389-405, 1975.

[Mi 77a] R. MIFFLIN, *An algorithm for constrainted optimization with semismooth functions*, Math. Oper. Res. 2:191-207, 1977.

[Mi 77b] R. MIFFLIN, *Semismooth and semiconvex functions in constrained optimization*, SIAM J. Control Optim. 15:959-972, 1977.

[Mi 82] R. MIFFLIN, *A modification and extension of Lemarechal's algorithm for nonsmooth minimization*, Mathematical Programming Study, 17:77-90, 1982.

[Mi 83] R. MIFFLIN, *A superlinearly convergent algorithm for one-dimensional constrained minimization with convex functions*, Math. Oper. Res. 8:185-195, 1983.

[Mi 84] R. MIFFLIN, *Stationarity and superlinear convergence of an algorithm for univariate locally Lipschitz constrained minimization*, Math. Programming 28:50-71, 1984.

[Mi 93] R. MIFFLIN, *A quasi-second-order proximal bundle algorithm*, Technical Report 93-3, University of Washington, 1993.

[Mig 82] M. MIGNOTTE, *Some Useful Bounds*, in B. Buchberger, G.E. Collins and R. Loos (eds.) in cooperation with R. Albrecht, Computer Algebra, Symbolic and Algebraic Computation (Second Edition), Springer, Wien, 1992.

[Min 86] M. MINOUX, *Mathematical Programming: Theory and Algorithms*, Wiley, Chichester, 1986.

[Mit 68] B.S. MITIAGIN, *Two inequalities for volumes of convex bodies*, Mathematical Notes, 5:61-65, 1968.

[MM 64] M. MARCUS AND H. MINC, *A Survey of Matrix Theory and Matrix Inequalities*, Allyn and Bacon, Boston, MA, 1964.

[MMS 89] K.A.MCSHANE, C.L.MONMA, AND D.F.SHANNO, *An implementation of a primal-dual interior point method for linear programming*, ORSA J.Comput., 1:70-83, 1989.

[Mo 71] N.N. MOISEEV, *Numerical methods in the theory of optimal systems*, Nauka, Moscow 1971.

[Mol 87] A.P. MOLCHANOV, *Lyapunov functions for nonlinear discrete-time control systems*, Automation and Remote Control, pp. 728-736, 1987.

[MP 89] A.P. MOLCHANOV AND E.S. PYATNITSKII, *Criteria of asymptotic stability of differential and difference inclusions encountered in control theory*, Syst. Control Letters, 13:59-64, 1989.

[MPS 97] A. MIGDALAS, P.M. PARDALOS AND S. STOROY, *Parallel Computing in Optimization*, Kluwer Academic Publishers, 1997.

[MPV 97] A. MIGDALAS, P.M. PARDALOS AND P. VARBRAND, *Multilevel Optimization: Algorithms and Applications*, Kluwer Academic Publishers, 1997.

[MSa 83] B. MURTAUGH AND M. SAUNDERS, *MINOS 5.0 user's guide*, Report SOL 83-20, Stanford University, Palo Alto, CA, 1983.

[MSGZ 77] V.S. MIKHALEVICH, N.Z. SHOR, L.A. GALUSTOVA N.G. ZHURBENKO AND OTH, *Computational methods for choosing optimal design decisions*, Naukova Dumka, Kiev 1977.

[MSh 54] T. MOTZKIN, I. SHOENBERG, *The relaxation method for linear inequalities*, Canad. J. Math. 6:393-404, 1954.

[MSJ 78] K. MADSEN, H. SCHJAER-JACOBSEN, *Linearly constrained minimax optimization*, Math. Programming 14:208-223, 1978.

[MSo 83] J.J. MORE AND D.C. SORENSEN, *Computing a trust region step*, SIAM, J. Sci. Statist. Comput. 4: 553-572, 1983.

[MSS 81] V.S. MIKHALEVICH, I.V. SERGIENKO AND N.Z. SHOR, *Investigation of optimization methods and their applications*, Cybernetics 17:522-548, 1981.

[MSt 65] T.S. MOTZKIN AND E.G. STRAUS, *Maxima for graphs and a new proof of a theorem of Turan*, Canadian J. Math., 17:533-540, 1965.

[MSu 87] S.MEHROTRA AND J.SUN, *An algorithm for convex quadratic programming that requires $O(n^{3.5}L)$ arithmetic operations*, Manuscript, Department of Industrial Engineering and Management Science, Northwestern University, Evanston, IL, 1987.

[MSu 88a] S.MEHROTRA AND J.SUN, *A method of analytic centers for quadratically constrained convex quadratic programs*, Manuscript, Department of Industrial Engineering and Management Science, Northwestern University, Evanston, IL, 1988.

[MSu 88b] S.MEHROTRA AND J.SUN, *An interior point algorithm for solving smooth convex programs based on Newton's method*, Manuscript, Department of Industrial Engineering and Management Science, Northwestern University, Evanston, IL, 1988.

[MSu 91] S.MEHROTRA AND J.SUN, *A method of analytic centers for quadratically constrained convex quadratic programs*, Tech. Report 88-01, Dept. Siam J.Numer.Anal., 28:529-544, 1991.

[MTS 86] V.S. MIKHALEVICH, V.A. TRUBIN, N.Z. SHOR, *Optimization problems of production-transportation planning. Models, methods, algorithm*, Moscow, Nauka, 1986 (Russian).

[MW 97] H. VAN MAAREN, J. WARNERS, *Bounds and fast approximation algorithms for binary quadratic optimization problems with application to MAX2SAT and MAXCUT*, 5th Twent Workshop on graphs and combinatorial optimization, May 1997 (extended abstract).

[N 86] J.L.NAZARETH, *Homotopy techniques in linear programming*, Algorithmica, 1:529-535, 1986.

[Ne 93] A. NEMIROVSKII, *Several NP-Hard problems arising in robust stability analysis*, Mathematics of Control, Signals, and Systems, 6:99-105, 1993.

[Nes 83] YU.E.NESTEROV, *An O(1/k)-rate of convergence method for smooth convex functions minimization*, Dokl. Acad. Nauk SSSR, 269: 543-547, 1983.

[Nes 89] YU.E.NESTEROV, *Dual polynomial time algorithms for linear programming*, Kibernetika, 1:34-54, 1989.

[Nes 97] YU.E.NESTEROV, *Quality of semidefinite relaxation for nonconvex quadratic optimization*, CORE Discussion Paper # 9719, CORE, March 1997.

[New 65] D.J. NEWMAN, *Location of the maximum on unimodal surfaces*, JACM 12: 395-398, 1965.

[NF 93] B. NEKOOIE AND M. FAN, *A quadratically convergent local algorithm on minimazing sums of the largest eigenvalues of a symmetric matrix*, Computational Optimization and applications, 2(2):107-128, 1993.

[NM 90] G. NARASIMHAN, AND R.MANBER, *A generalization of Lovasz's sandwich theorem*, Polyhedral combinatorics: Proced. of a DIMACS Workshop AMS, 1990.

[NN 88] YU.E.NESTEROV AND A.S.NEMIROVSKY, *Polynomial-time barrier methods in convex programming*, Ekonomika i Mathem. Metody, 24(7):1084-1091, 1988.

[NN 89] YU.E.NESTEROV AND A.S.NEMIROVSKY, *Self-concordants function and polynomial time methods in convex programming*, USSR Acad. Sci. Central Economic & Methematical Institute, Moscow, 1989.

[NN 94] YU. NESTEROV AND A. NEMIROVSKY, *Interior-point polynomial methods in convex programming*, volume 13 of Studies in Applied Mathematics. SIAM, Philadelphia, PA, 1994.

[NY 83] A. NEMIROVSKY AND D. YUDIN, *Problem Complexity and Method Efficiency in Optimization*, John Wiley & Sons, 1983.

[Nu 73] E.A. NURMINSKI, *A quasi-gradient method for solving the nonlinear programming problem*, Kibernetika (Kiev) no. 1:122-125, 1973.

[Nu 77] E.A. NURMINSKI, *On the continuity of -subgradient mappings*, Kibernetika (Kiev), no. 5:148-149, 1977.

[Nu 79] E.A. NURMINSKI, *Numerical methods for solving deterministic and stochastic minimax problems*, Naukova Dumka: Kiev 1979.

[Nu 82] E.A. NURMINSKI, ED. *Progress in nondifferentiable optimization*, CP-82-S8, International Institute for Applied Systems Analysis, Laxenburg, Austria 1982.

[NY 78] A.S.NEMIROVSKY AND D.B.YUDIN, *Informational Complexity and Efficient Methods for Solution of Convex Extremal Problems*, Nauka, 1978.

[NY 79] A.S. NEMIROVSKI, D.B. YUDIN, *Complexity and efficiency of optimization methods*, Nauka: Moscow 1979.

[OC 87] A.I. OVSEEVICH AND F.L. CHERNOUSKO, *On optimal ellipsoids approximating reachable sets*, Prob. of Control and Info. Th., 16(2):125-134, 1987.

[OD 72] G.I. ORLOVA AND Y.G. DORFMAN, *Finding the maximum cut in a planar graph*, Engrg. Cybernetics 10:502-506, 1972.

[Oh 94] Y. OHTA, *Polytope Lyapunov functions and dynamic convex hull algorithm*, Systems, Control and Information, 38(3), 1994. Special issue on Numerical Approaches in Control Theory.

[Or 88] J.B. ORLIN, *A faster strongly polynomial minimum cost flow algorithm*, in: Proceedings 20th ACM Symposium on the theory of Computation, pp. 377-387, 1988.

[OS 76] I.G. OVRUTSKII, N.Z. SHOR, *Application of methods for minimizing nonsmooth functions to the solution of the problem of gravimetric data interpretation*, Kibernetika (Kiev), no. 2:57-64, 1976.

[OS 94] A. OHARA AND T. SUGIE, *Control systems synthesis via convex optimization*, Systems, Cotrol and Information, 38(3), 1994. Special issue on Numerical Approach in Control Theory.

[Ov 88] M.L. OVERTON, *On minimizing the maximum eigenvalue of a symmetric matrix*, SIAM Journal on Matrix Analysis and Applications 9: 256-268, 1988.

[Ov 92] M.L. OVERTON, *Large-scale optimization of eigenvalues*, SIAM Journal on Optimization 2:88-120, 1992.

[OW 88] M.L.OVERTON AND R.S. WOMERSLEY, *On minimizing the spectral radius of a nonsymmetric matrix function: Optimality conditions and duality theory*, SIAM Journal on Matrix Analysis and Applications 9:473-498, 1988.

[OW 92] M. OVERTON AND R. WOMERSLEY, *On the sum of the largest eigenvalues of a symmetric matrix*, SIAM J. on Matrix Analysis and Applications, pp. 41-45, 1992.

[OW 93] M.L. OVERTON AND R.S. WOMERSLEY, *Optimality conditions and duality theory for minimizing sum of the largest eigenvalues of symmetric matrices*, MP, 62:321-358, 1993.

[OZ 95] J. OUTRATA, J. ZOWE, *A numerical approach to optimization problems with variational inequality constrains*, Math. Progr. 68:105-130, 1995.

[Pan 90] J.-S. PANG, *Newton's method for B-differentiable equations*, Mathematics of Operations Research 15:311-341, 1990.

[Pap 79] C.H. PAPADIMITRIOU, *Efficient search for rationals*, Inf.Proc.Letters 8(1):1-4, 1979.

[Pap 81] G. PAPAVASSILOPOULOS, *Algorithms for a class of nondifferentiable problems*, J. Optim. Theory Appl. 34:41-82, 1981.

[Par 92] P.M. PARDALOS, *Advances in Optimization and Parallel Computing*, North-Holland, 1992.

[Par 93] P.M. PARDALOS, *Complexity in Numerical Optimization*,World Scientific, 1993.

[PD 75] B.N. PSHENICHNY, YU. M. DANILIN, *Numerical methods in extremal problems*, Nauka, Moscow 1975.

[PDe 91] P.M. PARDALOS AND N. DESAI, *An algorithm for finding a maximum weighed independent set in an arbitrary graph*, Intern. J. Computer Math. 38:163-175, 1991.

[Pe 87] I.R. PETERSEN, *A stabilization algorithm for a class of uncertain systems*, Syst. Control Letters, 8(4):351-357, 1987.

[PHH 97] P.M. PARDALOS, D. HEARN AND W. HAGER, *Network Optimization*, Lecture Notes in Economics and Mathematical Systems, Vol. 450, Springer-Verlag, 1997.

[PK 97] L. PORKOLAB AND L. KHACHIYAN, *On the complexity of semidefinite programs*, JOGO, vol. 10, No. 4:351-365, 1997.

[PM 78] V.M.POLTEROVICH AND I.I.MITJUSHIN, *A criterion of monotonicity of a demand function*, Ekonomika i Matem. Metody, 14:122-128, 1978.

[PMW 83] E.POLAK, D.Q.MAYNE, Y.WARDI, *On the extension of constrained optimization algorithm from differentiable to nondifferentiable problems*, SIAM J. Control Optim. 21:179-203, 1983.

[Po 64] B.T. POLJAK, *Gradient methods for solving equations and inequalities*, CMMP 4:17-32, 1964.

[Po 67] B.T. POLJAK, *A general method of solving extremum problems*, SMD 8: 593-597, 1967.

[Po 69] B.T. POLJAK, *Minimization of unsmooth functionals*, CMMP 9(3): 14-29, 1969.

[Po 78] B.T. POLJAK, *Subgradient methods: a survey of Soviet research*, in: Nonsmooth Optimization, Pergamon Press, 1978.

[Po 87] B.T. POLJAK, *Introduction to Optimization*, Optimization Software, New York, NY, 1987.

[Pol 73] V.M.POLTEROVICH, *Economical equilibrium and optimum*, Ekonomika i Matem. Metody, 9:835-845, 1973.

[Poly 28] G. POLYA, *Ueber positive Darstellung von Polynomen*, Vierteljahrsschrift d. naturforschenden Gesellsch. Zurich 73:141-145.

[Pom 82] J.C.POMEROL, *The Lagrange multiplier set and the generalized gradient set of the marginal function of a differentiable program in a Banach space*, J. Optim. Theory Appl. 38:307-317, 1982.

[Pop 73] V.M. POPOV, *Hyperstability of Control Systems*, Springer-Verlag, New-York, 1973.

[PP 90] P.M. PARDALOS AND A.T. PHILLIPS, *A global optimization approach for solving the maximum clique problem*, Inter. J. Computer Math. 33:209-216, 1990.

[PPR 92] P.M. PARDALOS, A.T. PHILLIPS, AND J.B. ROSEN, *Topics in Parallel Computing in Mathematical Programming*, Science Press, 1992.

[PQ 93] J.S. PANG AND L.Q. QI, *Non-smooth equations: motivation and algorithms*, SIAM Journal on Optimization 3:443-465, 1993.

[PR 82] M.W. PADBERG, M. RAO, *Odd minimum cut-sets and b-matchings*, M.O.R. 7:67-80, 1982.

[PRen 95] S. POLJAK AND F. RENDL, *Solving the max-cut problem using eigenvalues*, Discrete Applied Mathematics, 62:249-278, 1995.

[PRes 96] P.M. PARDALOS AND M.G.C. RESENDE, *Interior point methods for global optimization problems*, in Terlaky, T. (ed.), Interior Point Algorithms, Kluwer Academic Publishers, Dordrecht, The Netherlands, pp. 467-500, 1996.

[PRod 92] P.M. PARDALOS AND G.P. RODGERS, *A branch and bound algorithm for the maximum clique problem*, Comp. and Oper. Research 19:363-375, 1992.

[PRoh 93] S. POLJAK AND J. ROHN, *Checking robust nonsingularity is NP-hard*, Mathematics of Control, Signals, and Systems, 6:1-9, 1993.

[PRos 87] P.M. PARDALOS AND J.B. ROSEN, *Constrained Global Optimization: Algorithms and Applications*, Lecture Notes in Computer Science 268, Springer-Verlag, New York, 1987.

[PRos 90] P.M. PARDALOS AND J.B. ROSEN, *Computational Methods in Global Optimization*, Annals of Operations Research Vol. 25, 1990.

[Pr 77] M.E. PRIMAK, *On convergence of a modified method of Chebyshev's centers for solving the convex programming problem*, Kibernetika (Kiev), 5:100-102, 1977.

[PRR 95] P.M. PARDALOS, M.G.C. RESENDE AND K.G. RAMAKRISHNAN, *Parallel Processing of Discrete Optimization Problems*, DIMACS Series Vol. 22, American Mathematical Society, 1995.

[PRW 95] S. POLJAK, F. RENDL AND H. WOLKOWICZ, *A resipe for semidefinite relaxation for (0,1)-quadratic programming*, Journal of Global Optimization 7:51-73, 1995.

[PS] C.H. PAPADIMITRIOU AND K. STEIGLITZ, *Combinatorial Programming: Algorithm and Complexity*, (Wiley, N.Y., 198 ?).

[PS 82] E.S. Pyatnitskii and V.I. Skorodinskii, *Numerical methods of Lyapunov function construction and their application to the absolute stability problem*, Syst. Control Letters, 2(2):130-135, August 1982.

[PS 83] E.S. Pyatnitskii and V.I. Skorodinskii, *Numerical method of construction of Lyapunov functions and absolute stability criteria in the form of numerical procedures*, Automation and Remote Control, 44(11):1427-1437, 1983.

[Psh 69] B.N. Pshenichnyi, *Necessary conditions for an extremum*, Marcel Dekker, 1971.

[Psh 93] B.N. Pshenichnyi, *Methods of Linearization*, Springer-Verlag, 1993.

[PSX 996] P.M. Pardalos, D. Shalloway and G. Xue, *Global Minimization of Nonconvex Energy Functions: Molecular Conformation and Protein Folding*, DIMACS Seris Vol. 23, American Mathematical Society, 1996.

[PSZ 95] P.M. Pardalos, Y. Siskos and C. Zopounidis, *Advances in Multicriteria Analysis*, Kliwer Academic Publishers, 1995.

[Pu] W. Pulleyblank, *Polyhedral Combinatorics*, in [B3] 312-345.

[PW 82] E. Polak and Y. Wardi, *A nondifferentiable optimization algorithm for the design of control systems subject to singular value inequalities over a frequency range*, Automatica, 18(3):267-283, 1982.

[PWol 94] P.M. Pardalos and H. Wolkowicz, *Quadratic Assignment and Related Problems*, DIMACS Series Vol. 16, American Mathematical Society, 1994.

[PWol 97] P.M. Pardalos and H. Wolkowicz, *Topics in Semidefinite and Interior-Point Methods*, Fields Institute Communocations Series, American Mathematical Society (in press 1997).

[PWor 94] V. Powers and T. Wormann, *An algorithm for sums of squares of real polynomials*, 1994, preprint.

[PX 94] P.M. Pardalos and J. Xue, *The maximum clique problem*, Journal of Global Optimization 4: 301-328, 1994.

[PV 91] P.M. PARDALOS AND S. VAVASIS, *Quadratic programming with one negative eigenvalue is NP-hard*, Journal of Global Optimization 1: 15-23, 1991.

[Qi 93] L.Q. QI, *Convergence analysis of some algorithms for solving non-smooth equations*, Mathematics of Operations Research, 18:227-244, 1993.

[Qian 92] M. QIAN, *The variable metric proximal point algorithm: global and superlinear convergence*, Manuscript GN 50, University of Washington, Department of Mathematics, 1992.

[QS 93] L. QI AND J. SUN, *A non-smooth version of Newton's method*, Math. Progr. 58: 353-367, 1993.

[Raj 93] A.R. RAJWADE, *Squares*, London Math. Soc. Lecture Note Ser., vol. 171, Cambridge Univ. Press, London and New York, 1993.

[Ram 93] M. RAMANA, *An algorithmic analysis of multiquadratic and semidefinite programming problems*, PhD thesis, Johns Hopkins University, Baltimore, MD, 1993.

[Ram 95] M. RAMANA, *An Exact Duality Theory for Semidefinite Programming and its Complexity Implications*, DIMACS Technical Report 95-02.

[Rap 88] L.B. RAPOPORT, *Sign-definiteness of a quadratic form with quadratic constraints and absolute stability of nonlinear control systems*, Sov. Phys. Dokl., 33(2):96-98, 1988.

[Rap 93] L.B. RAPOPORT, *On existence of non-smooth Lyapunov functions in absolute stability problem*, Automation and Remote Control, 53(3), 1993.

[Re 87a] J.RENEGAR, *On the worst-case arithmetic complexity of approximating zeros of polynomials*, Journal of complexity, 3:90-113, 1987.

[Re 87b] J.RENEGAR, *A polynomial time algorithm, based on Newton's method for linear programming*, Math. Programming, 40:59-93, 1987.

[Re 92a] J. RENEGAR, *On the Computational Complexity of Approximating Solutions for Real Algebraic Formulae*, SIAM J. on Computing 21:1008-1025, 1992.

[Re 92b] J.RENEGAR, *On the computational Complexity and Geometry of the First Order Theory of the Reals. Part I: Introduction; Preliminaries; the Geometry of Semi-Algebraic Sets; the Decision Problem for the Existential Theory of the Reals*, J. of Symbolic Computation 13:255-299, 1992.

[Rez 92] B. REZNICK, *Sums of even powers of linear forms*, Mem. Amer. Math. Soc., vol 96, Amer. Math. Soc., Providence, RI, 1992.

[Ro 73] R.M. ROBINSON, *Some definite polynomials which are not sums of squares of real polynomials*, Selected Questions of Algebra and Logic, Acad. Sci. USSR, pp.264-282, 1973.

[Ro 76] S.M. ROBINSON, *A subgradient algorithm for solving K-convex inequalities*, In: Optimization and Operations Research (P. Wolfe, ed.), pp.237-245. Lecture Notes in Economics and Mathematical Systems 117, Springer: Berlin 1976.

[Ro 87] S.M. ROBINSON, *Local structure of feasible sets in non-linear programming, part III: Stability and sensitivity*, Mathem. Programming Study 30:45-66, 1987.

[Roc 70] R.T. ROCKAFELLAR, *Convex Analysis*, Princeton University Press, Princeton, NJ, 1970.

[Roc 81] R.T. ROCKAFELLAR, *The Theory of Subgradients and its Applications to Problems of Optimization: Convex and Nonconvex Functions*, Helderman Verlag, Berlin, 1981.

[Roc 82a] R.T. ROCKAFELLAR, *Convex Analysis and Optimization*, Pitman, Boston, 1982.

[Roc 82b] R.T. ROCKAFELLAR, *Lagrange multiplier and subderivatives of optimal value functions in mathematical programming*, In: Nondifferential and Variational Techniques in Optimization (D.C. Sorensen, R.J.-B. Wets, eds.), pp.28-66. Mathematical Programming Study 17, North-Holland: Amsterdam 1982.

[Roc 93] R.T. ROCKAFELLAR, *Lagrange multipliers and optimality*, SIAM Review, 35:183-283, 1993.

[Ros 60] J.B. ROSEN, *The gradient projection method for nonlinear programming, part I: linear constraints*, Journal of the Society for Industrial and Applied Mathematics 8:181-217, 1960.

[Ros 61] J.B. ROSEN, *The gradient projection method for nonlinear programming, part II:*, Journal of the Society for Industrial and Applied Mathematics 8:181-217, 1960.

[Ros 63] J.B. ROSEN, *Convex partition programming*, In: Recent advances in mathematical programming (R.L. Graves, P. Wolfe, eds.), pp. 159-176, McGraw Hill, New York 1963.

[RP 96] M. RAMANA AND P.M. PARDALOS, *Semidefinite programming*, in Terlaky, T. (ed.), Interior Point Algorithms, Kluwer Academic Publishers, Dordrecht, The Netherlands, 369-398, 1996.

[RW 87] R.T. ROCKAFELLAR AND R.J.-B. WETS, *The principle of scenario aggregation in optimization under uncertainty*, Manuscript, University of Washington, 1987.

[RWo 90] F. RENDY AND H. WOLKOWICZ, *A projection technique for partitioning the nodes of a graph*, Department of Combinatorics and Optimization Report 90-20, University of Waterloo (Waterloo, Ont., 1990).

[SA 92] H.D. SHERALI AND A.R. ALAMEDDINE, *A new reformulation-linearization technique for bilinear programming problems*, Journal of Global Optimization 2:379-410, 1992.

[Sanc 92] L. SANCHIS, *Test case construction for the vertex cover problem*, Working paper, DIMACS, March 1992.

[Sant 94] E. DE SANTIS, *On positively invariant sets for discrete-time linear systems with disturbance: An application of maximal disturbance sets*, IEEE Trans. Aut. Control, 39(1):245-249, January 1994.

[SB 69] N.Z. SHOR, V.I. BILETSKII, *A method of space dilation for accelerating convergence in gully-type problems*. In: Teor. Optimal. Resenii, Trudy Sem. Nauk. Sov. Akad. Nauk Ukrain. SSR po Kibernet. 2:3-18, Kiev 1969.

[SB 88] A.SHAPIRO AND J.D.BOTHA, *Dual algorithms for ortogonal procrustes rotations*, SIAM Journal on Matrix Analysis and Applications 9: 378-383, 1988.

[SB 89] N.Z. SHOR AND O.A. BEREZOVSKI, *Application of subgradient type method with space dilation for constructing maximal volume ellipsoid*, Kibernetika, 6:119-120, 1989.

[SB 91] N.Z. Shor and O.A. Berezovski, *Finding global extremum of a quadratic function on a quadratic surface*, In: preprint IK AN USSR, 1991.

[SB 92a] N.Z. Shor and O.A. Berezovski, *New algorithms for constructing optimal circumscribed and iscribed ellipsoids*, Optimization Methods and Software, 1:283-299, 1992.

[SB 92b] N.Z. Shor and O.A. Berezovski, *Constructing the maximal inscribed ellipsoid for a polytope using successive space dilation*, Kibernetika i Vych. Tehnika, 93:1-6, 1992.

[SB 93] N.Z. Shor and O.A. Berezovski, *An algorithm for solving the problem of minimization of quadratic functions on intersection of two quadratic surfaces*, Teor. i Vych. Prob. Optim. pp. 30-35, IK AN USSR, Kiev, 1993.

[SB 94] N.Z. Shor and O.A. Berezovski, *Using technique of dual quadratic bounds for solving the system of polynomial equations in complex numbers*, Kibernetika i sistemny analiz, 5:67-75, 1994.

[SB 95] N.Z. Shor and O.A. Berezovski, *New algorithms for solving weight max-cut problem*, Kibernetika i sistemny analiz, 2:100-106, 1995.

[Sch 86] A. Schrijver, *Theory of Linear and Integer Programming*, Wiley, New York, 1986.

[SD 85] N.Z.Shor and A.Davydov, *On a method for bounding in quadratical problems*, Kibernetika, 2:48-54, 1985.

[Se 90] A. Seeger, *Sensitivity analysis of nondifferentiable sums of singular values of rectangular matrices*, Technical Report, Department de Matematica Aplicada i Analisi, Universitat de Barcelona (Barcelona, 1990).

[Sey 80] P.D. Seymour, *Decomposition of regular matroids*, JCT B 28:305-359, 1980.

[SGe 79] N.Z. Shor, and V.J. Gershovich, *On one family of algorithms for solving of convex programming problems*, Kibernetika, 4:62-67, 1979.

[SGo 67] N.Z. Shor, G.I. Gorbach, *The solution of distribution type problems by the method of generalized gradient descent*, In: Teor. Optimal. Resenii, Trudy Sem. Nauk. Sov. Akad. Nauk Ukrain. SSR po Kibernet., 1:59-71, Kiev 1967.

[SGM 78] N.Z. SHOR, L.A. GALUSTOVA, A.I. MOMOT, *An application of mathematical methods to the optimal design of the central gas supply system, accounting for the dynamics of its development*, Kibernetika (Kiev), 1:69-74, 1978.

[Sh 85] A. SHAPIRO, *Extremal problems on the set of nonnegative definite matrices*, Linear Algebra and its Applications 67:7-18, 1985.

[Sha 73] L.P.SHABASHOVA, *Gradient methods for solving nonlinear minimax problems*, Diss. Doctor Philos. Dnepropetrowsk 1973.

[Shan 56] C.E. SHANNON, *The zero-error capacity of a noisy channel*, Composium on Information Theory, I.R.E. Transactions, 3, 1956.

[Shc 87] M.B. SHCHEPAKIN, *On the method of orthogonal descent*, Kibernetika, pp.58-62, 1987.

[She 74] M.A. SHEPILOV, *On Gradient and penalty methods in mathematical programming problems*, Diss Doctor Philos. Moscow 1974.

[She 76] M.A. SHEPILOV, *On a method of generalized gradient for finding the absolute minimum of a convex function*, Kibernetika (Kiev), 4: 52-57, 1976.

[Shi 80] Y. SHILOACH, *A polynomial solution of the undirected two paths problem*, Journal of the ACM, 27(3):445-457, 1980.

[Shor 62] N.Z. SHOR, *An application of the method of gradient descent to the solution of the network transportation problem*, In: Materialy Nauchnogo Seminara po Teoret. i Priklad. Voprosam Kibernet. i Issled. Operacii, Nauchnyi Sov. po Kibernet. Akad. Nauk Ukrain. SSSR, 1:9-17, 1962.

[Shor 64] N.Z. SHOR, *On the structure of algorithms for numerical solution of problems of optimal planning and design*, Diss. Doctor Philos. Kiev 1964.

[Shor 67a] N.Z. SHOR, *An application of the generalized gradient descent in block programming*, Kibernetika (Kiev), 3:53-55, 1967.

[Shor 67b] N.Z. SHOR, *Multistage convex stochastic programming*, In: Teor. Optimal. Resenii, Trudy Sem. Nauc. Sov. Akad. Nauk Ukrain. SSR po Kibernet., pp. 48-58., 1967.

[Shor 68] N.Z. SHOR, *On the speed of convergence of the generalized gradient descent.* Kibernetika (Kiev), 3:98-99, 1968.

[Shor 69] N.Z. SHOR, *The generalized gradient descent*, In: Trudy 1 Zimnei Skoly po Mat. Programmirovaniyu, 3:578-585, 1969.

[Shor 70a] N.Z. SHOR, *On the speed of convergence of the method of generalized gradient descent with space dilation*, Kibernetika (Kiev), 2:80-85, 1970.

[Shor 70b] N.Z. SHOR, *Methods for minimizing nondifferentiable functions and their applications*, dis. Doctor Sci. Kiev 1970.

[Shor 70c] N.Z. SHOR, *An application of the operation of space dilation to the problems of minimizing convex functions*, Kibernetika (Kiev), 1:6-12, 1970.

[Shor 72] N.Z.SHOR, *On a method for minimizing almost differentiable functions*, Kibernetika (Kiev), 4:65-70, 1972.

[Shor 75] N.Z. SHOR, *The analysis of convergence of a gradient type method with space dilation in the direction of the difference of two successive gradients*, Kibernetika (Kiev), 4:48-53, 1975.

[Shor 76] N.Z. SHOR, *Generalized gradient methods for minimizing nonsmooth functions and their application to mathematical programming problems*, Econ. i Mat. Metody, 12:337-356, 1976.

[Shor 77] N.Z. SHOR, *A method of section with space dilation for solving convex programming problems*, Kibernetika (Kiev), 1: 94-95, 1977.

[Shor 79] N.Z. SHOR, *Minimization Methods for Non-Differentiable Functions and their Applications*, Naukova Dumka, Kiev, 1979.

[Shor 85] N.Z. SHOR, *Minimization Methods for Non-Differentiable Functions*, Springer-Verlag, Berlin, 1985.

[Shor 87] N.Z.SHOR, *Quadratic optimization problems*, Izvestija AN SSSR, Tekhnitcheskaya kibernetika, 1, 1987.

[Shor 87b] N.Z.SHOR, *One Idea of Getting Global Extremum in Polynomial Problems of Mathematical Programming*, Kibernetica 26(5):102-106, 1987 (in Russian).

[Shor 87c] N.Z.SHOR, *One Class of Estimates for Global Minimum of polynomial Functions*, Kibernetica 26(6):9-11, 1987 (in Russian).

[Shor 90] N.Z.SHOR, *Dual quadratic estimates in polynomial and boolean programming*, Annals of Operations Research 25:163-168, 1990.

[Shor 95] N.Z.SHOR, *Minimization of matrix functions and nondifferentiable optimization*, Obozrenie prikladnoi matematiki, 2:113-138, 1995.

[Shr 86] A.SHRIJVER, *Theory of linear and integer programming*, Wiley-Interscience, John Wiley and Sons, Inc., 1986.

[SI 69] N.Z. SHOR, L.V. IVANOVA, *On a certain iterative method for solving linear programming problems and matrix games.* In: Teor. Optimal Resenii, Trudy Sem. Nauk.Sov. Akad.Nauk Ukrain. SSR, po Kibernet. 3:22-30, Kiev 1969.

[Sk 74] V.A. SKOKOV, *Note of minimization methods employing space stretching*, Cybernetics 10:689-692, 1974.

[Sko 90] V.I. SKORODINSKII, *Iterational method of construction of Lyapunov-Krasovskii functionals for linear systems with delay*, Automation and Remote Control, 51(9):1205-1212, 1990.

[Sl 66] S.G. SLOBODNIK, *The continuity and differential properties of functions*, Dis. D. Phil. Moscow 1966.

[SL 93] T.J. SU AND P.L. LIU, *Robust stability for linear uncertain time-delay systems with delay-dependence*, Int. J. Systems Sci., 24(6):1067-1080, 1993.

[Sm 82] S. SMALE, *The problem of the average speed of the simplex method*, in: Mathematical Programming: The State of Art, Bonn 1982.

[SM 85] D.F. SHANNO AND R.E. MARSTEN, *On implementing Karmarkar's algorithm*, Working Paper, Graduate School of Administraion, University of California, Davis, CA, 1985.

[SNH 83] J.-J. STRODIOT, V.H.NGUYEN, N.HEUKEMES, *e-Optimal solution in nondifferentiable convex programming and some related questions*, Math. Programming 25:307-328, 1983.

[So 74] G. SONNEVEND, *On optimization of algorithms for function minimization*, Colloq. Math. Soc. Janos Bolyai, 865-893, 1974.

[So 88] G. SONNEVEND, *New algorithms in convex programming based on a notion of 'centre' (for systems of analytic inequalities) and on rational extrapolation*, International Series of Numerical Mathematics, 84:311-326, 1988.

[Sor 82] D.C. SORENSEN, *Newton's method with a model trust region modification*, SIAM J. Numer. Anal. 19 (1982), 2:409-426, 1982.

[SR 67] N.Z. SHOR, N.I. ROSINA, *Scheme for partitioning linear and convex programming problems and its application to the solution of transportation planning problems*, In: Dokl. I. Vsesoyuz. Konf. po Optimizacii i Modelirovaniyu Transport. Setei, pp. 225-237, Kiev 1967.

[SS 72] N.Z.SHOR, L.I. SHABASHOVA, *On the solution of minimax problems by the method of generalized gradient descent with space dilation*. Kibernetika (Kiev), 1:82-88, 1972.

[SSh 68] N.Z. SHOR, M.B. SHCHEPAKIN, *An algorithm for solving the two-stage problem of stochastic programming*, Kibernetika (Kiev), 3: 56-58, 1968.

[SShu 93] M.B. SHCHEPAKIN AND I.A. SHUBENKOVA, *Investigation of the modified method of orthogonal descent for searching of the null of a convex function*, Kibernetika i sistemny analiz, Kiev, 4:63-72, 1993.

[SSt 89] N.Z. SHOR, S.I. STETSENKO, *Quadratic extremal problems and nondifferentiable optimization*, Naukova dumka, Kiev, 1989 (in Russian).

[SSt 90] N.Z. SHOR, S.I. STETSENKO, *An algorithm with successive space pressing for constructing minimal on volume circumscribed ellipsoid*, In: preprint IK AN USSR, Kiev, 1990.

[SSt 97] N.Z. SHOR, P.I. STETSYUK,, *Using of r-algorithm modifications for obtaining of global minimum of polynomial functions*, Kibernetika i sistemny analiz, 4, Kiev, 1997.

[SSZ 89] G.SONNEVEND, J.STOER, AND G.ZHAO, *On the complexity of following the central path of linear programs by linear extrapolation*, Math. Oper. Res., 63:19-31, 1989.

[St 96] P.I. STETSYUK,, *r-algorithms and ellipsoids*, Kibernetika i sistemny analiz, N1, Kiev, 1996.

[St 96a] P.I. STETSYUK,, *Subgradient methoda with space transformation for minimization of nonsmooth convex functions*, Diss. Doct. Philos., Kiev, 1996.

[ST 95] H.D. SHERALI AND C.H. TUNCBILEK, *A reformulation-convexification approach for solving nonconvex programming problems*, Journal of Global Optimization 7:1-31, 1995.

[Su 93] SUBRAMANI, *Sums of singular values*, M.Sc. Thesis, School of Mathematics, University of New South Wales, Kensington, Austalia, 1993.

[Sw 95] R.G. SWAN, *Hilbert's theorem on positive ternary quartics*, Procedings of Symposium in Pure Math., 58, 1995.

[SZ 71] N.Z. SHOR, N.G. ZHURBENKO, *A minimization method using the operation of space dilation in the direction of the difference of two successive gradients*, Kibernetika (Kiev), 3:51-59, 1971.

[T 78] R.E. TARJAN, *Complexity of combinatorial algorithms*, SIAM Review 20:457-491, 1978.

[Ta 51] A. TARSKI, *A decision method for elementary algebra and geometry*, Berkelcy, Univ. of califormia Press, 1951.

[TCL 92] K.L.TEO, A.CANTONI, AND X.G. LIN, *A nonsmooth optimization problem in envelope constrained filtering*, Appl. Math. Lett., 5(6):79-82, 1992.

[Te 82] J. TELGEN, *On relaxation methods for systems of linear inequalities*, European Jr. of O.R. 9:184-189, 1982.

[Ti 75] D.M. TITTERINGTON, *Optimal design: some geometric aspects of D-optimality*, Biometrica, 62:313-320, 1975.

[TK 80] S.P. TARASOV, AND L.G. KHACHIYAN, *Bounds of solutions and algebraic complexity of systems of convex diaphantine inequalities*, SMD 22:700-704, 1980.

[To 88] M.J.TODD, *Improved bounds and containing ellipsoids in Karmarkar's linear programming algorithm*, Math. Oper. Res., 13: 650-659, 1988.

[Top 70] D.M. TOPKIS, *Cutting-plane methods without nested constraint sets*, Oper. Res. 18:404-413, 1970.

[Top 82] D.M. TOPKIS, *A cutting-plane algorithm with linear and geometric rates of convergence*, J. Optim. Theory Appl. 36:1-22, 1982.

[Tru 87] K. TRUEMPER, *A decomposition theory for matroids V: testing of matrix total unimodularity*, Working Paper, University of Texas at Dallas, Richardson, Texas, January 1987.

[Tu 56] W.T. TUTTE, *A theorem on planar graphs*, Trans. Amer. Math. Soc. 82:99-116, 1956.

[TY 90] M.J.TODD AND Y.YE, *A centered projective algorithm for linear programming*, Tech. Report No.763, Math. Oper. Res., 15:508-529, 1990.

[V 90] P.M.VAIDYA, *An algorithm for linear programming which requires O((m+n)n +(m+n) n)L) arithmetic operations*, Math. Programming, 47:175-202, 1990.

[Vav 91] S.A. VAVASIS, *Nonlinear Optimization: Complexity Issues*, Oxford University Press, New York, 1991.

[Wag 64] K. WAGNER, *Beweis einer Abschwachung der Hadwiger-Vermutung*, Math. ANN. 153:139-141, 1964.

[Wag 70] K. WAGNER, *Graphentheorie*, Hochschultakshenbucher-Verlag, 1970.

[Wat 91] G.A. WATSON, *Minimizing a smooth function plus the sum of the k largest eigenvalues of a symmetric matrix*, Numerical analysis report NA/132A (revised), University of Dundee, Dundee, Scotland, 1991.

[Wel 93] D.J.A. WELSH, *Complexity: Knots, Colourings and Counting*, London Mathematical Society Lecture Notes 186, Cambridge University Press, 1993.

[Wie 55] H. WIELANDT, *An extremum property of sums of eigenvalues*, Proceedings of the American Mathematical Society 6:106-110, 1955.

[Wier 82] A.P.WIERZBICKI, *Lagrangian functions and nondifferentiable optimization*, In: Progress in Nondifferentiable Optimization (E.A. Nurminski, ed.), pp. 173-213. CP-82-S8, International Institute for Applied Systems Analysis, Laxenburg, Austria 1982.

[Wo 90] H.J. WOERDEMAN, *Strictly contractive and positive completions for block matrices*, Linear Algebra and Appl., 105:63-105, 1990.

[Wol 80] P. WOLFE, *A bibliography for the ellipsoid algorithm*, IBM Research Center, Yorktown Heights, N.Y., USA, 1980.

[Wol 75] P. WOLFE, *A method of conjugate subgradients for minimizing nondifferentiable functions*, In: Math. Programming Stud. 3 (M.L. Balinski, P. Wolfe, eds.), pp. 145-173. North Holland, Amsterdam 1975.

[Wol 76] P. WOLFE, *Finding a nearest point in a polytope*, Math. Progr., v.11(2): 128-149, 1976.

[Wols 80] L.A. WOLSEY, *Heuristic analysis, linear programming and branch and bound*, Mathematical Programming Study 13: Combinatorial Optimization II, pp. 121-134, North-Holland, 1980.

[Y 67] V.A. YAKUBOVICH, *The method of matrix inequalities in the stability theory of nonlinear control systems*, I,II,III. Automation and Remote Control, 25-26(4):905-917, 577-592, 753-763, April 1967.

[Y 71] V.A. YAKUBOVICH, *The S-procedure in non-linear control theory*, Vestnik Leningrad Univ. Math., 4:73-93, 1977. In Russian, 1971.

[Yan 78] M. YANNAKAKIS, *Node- and edge-deletion NP-complete problems*, Proc. 10th Ann. ACM Symp. on Theory of Computing, Association of Computing Machinery, pp. 253-264, New York, 1978.

[Ye 88] Y.YE, *An $O(n^3L)$ potential reduction algorithm for linear programming*, Manuscript, Department of Management Sciences, The University of Iowa, Iowa City, IA, 1988.

[Ye 90] Y.YE, *Complexity analysis on Karmarkar's algorithm*, Working Paper Series No.90-5, College of Business Administration, The University of Iowa, Iowa City, IA, March 1990.

[Ye 92] Y.YE, *A new complexity result on minimization of a quadratic function with a sphere constraint*, In "Recent Advances in Global Optimization" (Editors: Floudas C.A. and Pardalos P.M.), pp.19-31, Princeton University Press, 1992.

[YL 82] B. Yamnitsky, L.A. Levin, *A 1965 linear programming algorithm runs in polynomial time*, in Proc. IEEE/FOCS, Chicago, 1982, also, in Waterloo Silver Jubilee Conf. on Combinatorics 1982.

[YN 76] D.B. Yudin, A.S. Nemirovskii, *Informational complexity and efficient methods for solving convex extremal problems*, Ekonom. i Mat. Metody 12:357-369, 1976.

[YP 91] Y.Ye and P. Pardalos, *A class of linear complementarity problems solvable in polynomial time*, Linear Algebra Appl., 152: 1-10, 1991.

[YT 89] Y.Ye and E. Tse, *An extension of Karmarkar's projective algorithm for convex quadratic programming*, Math. Programming, 44: 157-179, 1989.

[Yu 74] D.B. Yudin, *Mathematical methods of control under incomplete information*, Sovetskoe Radio, Moscow 1974.

[Z 73] N. Zaden, *A bad network problem for the simplex method and other minimum cost flow algorithms*, MP 5:255-266, 1973.

[Z 80] N. Zaden, *What is the worst case behaviour of the simplex?*, Department of Operations Research, Stanford University TR 27, 1980.

[Zag 58] V.L. Zaguskin, *On circumscribed and inscribed ellipsoids of extremum volume*, UMN V. XIII, 6:89-93, 1958.

[ZF 83] G. Zames and B.A. Francis, *Feedback, minimax sensitivity and optimal robustness*, IEEE Trans. Aut. Control, AC-28(5):585-601, May 1983.

[Zh 95] G. Zhao, *On the choice of parameter for power-series interior point algorithms for linear programming*, Math. Progr. 68:49-71, 1995.

[Zhu 77] N.G. Zhurbenko, *Analysis of a class of algorithms for minimizing nonsmooth functions and their application to the solution of large scale problems*, Diss. Doctor Philos. Kiev 1977.

[ZK 88] K.Zhou and P.P. Khargonekar, *Robust stabilization of linear systems with norm-bounded time-varying uncertainty*, Syst. Control Letters, 10:17-20, 1988.

[Zo 85] J. ZOWE, IN K. SCHITTKOWSKI (ED.), *Computational Mathematical Programming*, pp.323-356, Springer-Verlag, Berlin, 1985.

[ZP 75a] S.I. ZUHOVITSII, AND M.E. PRIMAK, *On the convergence of the method of Chebyshev centers and the method of centered sections for solving a convex programming problem*, SMD 16:615-618. D.J.A. Welsh, Matroid Theory, 1975, (A.P., N.Y. 1976).

[ZPSY 76] N.G. ZHURBENKO, E.G. PINAEV, N.Z. SHOR, G.N. YUN, *The choice of the set of public service aircrafts and their distribution among airlines*, Kibernetika (Kiev), 4:138-141, 1976.

[ZS 93] G. ZHAO AND J. STOER, *Estimating the complexity of path-following methods for solving linear programs by curvature integrals*, Applied Mathematics and Optimization 27:85-103, 1993.

[Zou 60] G. ZOUTENDIJK, *Methods of feasible directions*, Elsevier, Amsterdam 1960.

Nonconvex Optimization and Its Applications

1. D.-Z. Du and J. Sun (eds.): *Advances in Optimization and Approximation.* 1994. ISBN 0-7923-2785-3
2. R. Horst and P.M. Pardalos (eds.): *Handbook of Global Optimization.* 1995 ISBN 0-7923-3120-6
3. R. Horst, P.M. Pardalos and N.V. Thoai: *Introduction to Global Optimization* 1995 ISBN 0-7923-3556-2; Pb 0-7923-3557-0
4. D.-Z. Du and P.M. Pardalos (eds.): *Minimax and Applications.* 1995 ISBN 0-7923-3615-1
5. P.M. Pardalos, Y. Siskos and C. Zopounidis (eds.): *Advances in Multicriteria Analysis.* 1995 ISBN 0-7923-3671-2
6. J.D. Pintér: *Global Optimization in Action.* Continuous and Lipschitz Optimization: Algorithms, Implementations and Applications. 1996 ISBN 0-7923-3757-3
7. C.A. Floudas and P.M. Pardalos (eds.): *State of the Art in Global Optimization.* Computational Methods and Applications. 1996 ISBN 0-7923-3838-3
8. J.L. Higle and S. Sen: *Stochastic Decomposition.* A Statistical Method for Large Scale Stochastic Linear Programming. 1996 ISBN 0-7923-3840-5
9. I.E. Grossmann (ed.): *Global Optimization in Engineering Design.* 1996 ISBN 0-7923-3881-2
10. V.F. Dem'yanov, G.E. Stavroulakis, L.N. Polyakova and P.D. Panagiotopoulos: *Quasidifferentiability and Nonsmooth Modelling in Mechanics, Engineering and Economics.* 1996 ISBN 0-7923-4093-0
11. B. Mirkin: *Mathematical Classification and Clustering.* 1996 ISBN 0-7923-4159-7
12. B. Roy: *Multicriteria Methodology for Decision Aiding.* 1996 ISBN 0-7923-4166-X
13. R.B. Kearfott: *Rigorous Global Search: Continuous Problems.* 1996 ISBN 0-7923-4238-0
14. P. Kouvelis and G. Yu: *Robust Discrete Optimization and Its Applications.* 1997 ISBN 0-7923-4291-7
15. H. Konno, P.T. Thach and H. Tuy: *Optimization on Low Rank Nonconvex Structures.* 1997 ISBN 0-7923-4308-5
16. M. Hajdu: *Network Scheduling Techniques for Construction Project Management.* 1997 ISBN 0-7923-4309-3
17. J. Mockus, W. Eddy, A. Mockus, L. Mockus and G. Reklaitis: *Bayesian Heuristic Approach to Discrete and Global Optimization.* Algorithms, Visualization, Software, and Applications. 1997 ISBN 0-7923-4327-1
18. I.M. Bomze, T. Csendes, R. Horst and P.M. Pardalos (eds.): *Developments in Global Optimization.* 1997 ISBN 0-7923-4351-4
19. T. Rapcsák: Smooth Nonlinear Optimization in R^n. 1997 ISBN 0-7923-4680-7
20. A. Migdalas, P.M. Pardalos and P. Värbrand (eds.): *Multilevel Optimization: Algorithms and Applications.* 1998 ISBN 0-7923-4693-9
21. E.S. Mistakidis and G.E. Stavroulakis: *Nonconvex Optimization in Mechanics.* Algorithms, Heuristics and Engineering Applications by the F.E.M. 1998 ISBN 0-7923-4812-5

Nonconvex Optimization and Its Applications

22. H. Tuy: *Convex Analysis and Global Optimization.* 1998 ISBN 0-7923-4818-4
23. D. Cieslik: *Steiner Minimal Trees.* 1998 ISBN 0-7923-4983-0
24. N.Z. Shor: *Nondifferentiable Optimization and Polynomial Problems.* 1998 ISBN 0-7923-4997-0

KLUWER ACADEMIC PUBLISHERS – DORDRECHT / BOSTON / LONDON

GPSR Compliance

The European Union's (EU) General Product Safety Regulation (GPSR) is a set of rules that requires consumer products to be safe and our obligations to ensure this.

If you have any concerns about our products, you can contact us on ProductSafety@springernature.com

In case Publisher is established outside the EU, the EU authorized representative is:

Springer Nature Customer Service Center GmbH
Europaplatz 3
69115 Heidelberg, Germany

Batch number: 10755284

Printed by Printforce, the Netherlands